Principles of Ecotoxicology

SCOPE 12

Executive Committee of SCOPE

President: Professor G. White, Director, Institute of Behavioral Science, University of Colorado, Boulder, Colorado 80302, U.S.A.

Past-President: Professor V. Kovda, Moscow State University, Department of Pedology, Moscow V 234, U.S.S.R.

Vice-President: Professor M. Kassas, Department of Botany, Faculty of Science, Cairo University, Cairo, Egypt.

Secretary General: Dr. F. Fournier, Inspecteur Général de Recherches, ORSTOM, 24 rue Bayard, 75008 Paris, France.

Treasurer: Dr. A. H. Meyl, Deutsche Forschungsgemeinschaft, Kennedyallee 40, 53 Bonn, West Germany.

Members

Dr. R. W. J. Keay, Executive Secretary, The Royal Society, 6 Carlton House Terrace, London SW1Y 5AG, U.K.

Professor V. Landa, Assistant Secretary General, Czechoslovak Academy of Sciences, Národni tř. 3, Prague 1, Czechoslovakia.

Dr. R. Munn, Institute for Environmental Studies, University of Toronto, Ontario M5S 1A4, Canada.

Professor J. W. M. La Rivière, International Institute for Hydraulic and Sanitary Engineering, onde Delft 95, Delft, Netherlands

Editor-in-Chief

Professor F. J. Fenner, Director, Centre for Resource and Environmental Studies, The Australian National University, Box 4, P.O., Canberra, A.C.T. 2600, Australia.

Principles of Ecotoxicology

SCOPE 12

Edited by
G. C. Butler
*Director, National Research Council,
Division of Biological Sciences, Ottawa, Canada*

*Published on behalf of the
Scientific Committee on Problems of the Environment (SCOPE)
of the
International Council of Scientific Unions (ICSU)
by*

JOHN WILEY & SONS
Chichester · New York · Brisbane · Toronto

Copyright © 1978 by the
Scientific Committee on Problems of the Environment (SCOPE)

All rights reserved

No part of this book may be reproduced by any means, nor transmitted, nor translated into a machine language without the written permission of the copyright holder.

Library of Congress Cataloging in Publication Data:
Main entry under title:

Principles of ecotoxicology.

    (SCOPE report; 12)
    'Published on behalf of the Scientific Committee on Problems of the Environment (SCOPE) of the International Council of Scientific Unions (ICSU)'
    Includes index.
    1. Pollution—Environmental aspects. 2. Pollution —Toxicology. I. Butler, Gordon Cecil, 1913–
II. Series: Scientific Committee on Problems of the Environment. SCOPE report; 12.
QH545.AlP74   574.4'222   78-4045
ISBN 0 471 99638 6

Typeset in Great Britain by Preface Ltd, Salisbury, Wilts and printed by Unwin Brothers Ltd, The Gresham Press, Old Woking, Surrey.

Foreword

The Scientific Committee on Problems of the Environment (SCOPE) was established by the International Council of Scientific Unions (ICSU) in 1969 to advance knowledge of the influence of the human race upon its environment, as well as the effects of those alterations upon human health and welfare. It was intended to give particular attention to those effects which are either global or shared in common by several nations. It serves as a non-governmental, interdisciplinary and international council of scientists, and as a source of advice for governments and inter-governmental agencies with respect to environmental problems.

SCOPE seeks to synthesize environmental information from diverse scientific fields, identifying knowledge gaps, and disseminating the results. During the last several years, the main emphasis has been on the following topics:

1. Biogeochemical cycles;
2. Dynamic changes and evolution of ecosystems;
3. Environmental aspects of human settlements;
4. Ecotoxicology;
5. Simulation modelling of environmental systems;
6. Environmental monitoring;
7. Communication of environmental information and societal assessment and response.

A number of publications have resulted, including *SCOPE 10: Environmental Issues*, which provides an overview of the environmental challenges of the next decade.

One result of SCOPE's activities in ecotoxicology is the publication of this report which gives a general overview of the subject, addresses the apparent difficulties in assessing dose–effect relationships of sublethal exposure to chemicals, and stresses the point at which research may provide better understanding of the processes involved. That research, however, requires international cooperation in order to ensure the comparability of data and the acceptance of test methods. Only with such assistance will international coordination and harmonization of regulations controlling the release of chemicals into the environment be possible. Too often, well-intentioned steps to cope with new chemicals are based upon incomplete and inconsistent information.

SCOPE expects to play an active role in ecotoxicology by promoting the establishment of an international forum of scientists which would seek to improve the scientific basis of test methods used. Such a new group would act as an international focal point for scientists concerned with the toxic effects of chemicals on ecosystems.

The officers of SCOPE are grateful to Dr. Gordon Butler for his leadership in editing this volume as well as to the preparatory committee which organized the workshops and to the authors whose contributions resulted in this publication. Financial contributions from the European Economic Community, the Deutsche Forschungsgesellschaft, the Gesellschaft fuer Strahlen- und Umweltforschung, the Holcomb Research Institute, and the Rockefeller Foundation are gratefully acknowledged.

GILBERT F. WHITE
President of SCOPE
University of Colorado
Boulder, Colorado USA

Contents

Foreword v
Preface ix
List of Contributors xi
List of Figures xiii
List of Tables xvii

Introduction xix

SECTION I: ENVIRONMENTAL BEHAVIOUR OF POLLUTANTS

1. General considerations 3
   D. R. Miller

2. Abiotic processes 11
   F. Korte

3. Biotic processes 37
   W. Klein and I. Scheunert

4. Models for total transport 71
   D. R. Miller

5. Estimation of doses and integrated doses 91
   G. C. Butler

SECTION II: THE STATISTICAL ANALYSIS OF DOSE–EFFECT RELATIONSHIPS

6. The statistical analysis of dose–effect relationships . . . 115
   C. C. Brown

SECTION III: EXPERIMENTAL TOXICOLOGY AND FIELD OBSERVATIONS RELATED TO ECOTOXICOLOGY

7. General aspects of toxicology 151
   A. Jernelöv, K. Beijer, and L. Söderlund

8. Terrestrial animals 169
   F. Moriarty

9. Aquatic animals 187
   P. D. Anderson and S. d'Apollonia

10. Terrestrial plants and plant communities 223
    M. Treshow

11. Ecotoxicology of aquatic plant communities 239
    C. Hunding and R. Lange

12. Toxic effects of pollutants on plankton 257
    G. E. Walsh

13. Toxic effects of pollutants on microorganisms 275
    R. R. Colwell

14. Physical and chemical changes in the environment with indirect
    biological effects 295
    R. E. Munn

SECTION IV: ECOSYSTEM RESPONSE TO POLLUTION

15. Ecosystem response to pollution 313
    P. Bourdeau and M. Treshow

SECTION V: CONCLUSIONS

16. Conclusions 333
    G. C. Butler

Index 337

Preface

This the twelfth in the series of SCOPE reports was prepared as one of the seven 'mid-term projects' under the guidance of the SCOPE Steering Committee chaired by Professor G. F. White. The work was planned and directed by a Preparatory Committee with the members:

Ph. Bourdeau, Belgium
G. C. Butler, Canada, Convener
A. G. Johnels, Sweden
F. Korte, West Germany
N. Nelson, U.S.A.
T. A. Tashev, Bulgaria
R. Truhaut, France

During the preparation two workshops were held, in Brussels in July, 1976, and in Neuherberg in December, 1976. Each of these was attended by the Preparatory Committee, the authors of chapters, and a number of observers. The workshops were financed by the European Economic Community, the Deutsche Forschungsgemeinschaft, the Gesellschaft für Strahlen- und Umweltforschung, and the Holcomb Research Institute. The preparation of the manuscript and final editorial meetings were financed by a grant from the Rockefeller Foundation.

The editor received valuable assistance from several members of the Division of Biological Sciences of the National Research Council of Canada, in particular Dr. D. R. Miller and Miss J. E. Marks. Miss E. Monson completed the manuscript.

G. C. BUTLER

List of Contributors

P. D. Anderson	*Department of Biological Sciences, Concordia University, 1455 de Maisonneuve Blvd. W. Montreal, P.Q. H3G 1M3*
S. d'Apollonia	*Department of Biological Sciences, Concordia University 1455 de Maisonneuve Blvd. W. Montreal, P.Q. H3G 1M3*
K. Beijer	*Swedish Water and Air Pollution Research Laboratory, Box 21060, 10031 Stockholm, Sweden.*
P. Bourdeau	*Directorate General for Research Science and Education, Commission of the European Communities, Brussels, Belgium.*
C. C. Brown	*Biometry Branch, National Cancer Institute, National Institutes of Health, Bethesda, Maryland 20014, U.S.A.*
G. C. Butler	*Division of Biological Sciences, National Research Council of Canada, Ottawa, Canada, K1A 0R6*
R. R. Colwell	*Department of Microbiology, University of Maryland, College Park, Maryland 20742, U.S.A.*
C. Hunding	*National Agency of Environmental Protection, The Freshwater Laboratory, 52 Lysbrogade, DK-8600 Silkeborg, Denmark.*
A. Jernelöv	*Swedish Water and Air Pollution Research Laboratory, Box 21060, 10031 Stockholm, Sweden.*
W. Klein	*Institut für Okologische Chemie Gesellschaft für Strahlen–und Umweltforschung mbH, München, FRG.*
F. Korte	*Institute of Chemistry, Technical University of Munich, 805 Freising-Weihenstephan, FRG*
R. Lange	*The Norwegian Marine Pollution Research and Monitoring Programme, 29 Munthes Gate, N-Oslo 2, Norway.*
D. R. Miller	*Division of Biological Sciences, National Research Council of Canada, Ottawa, Canada, K1A 0R6*
F. Moriarty	*Monks Wood Experimental Station, Institute of Terrestrial Ecology, Huntington, U.K.*
R. E. Munn	*Atmospheric Environment Service, Environment Canada, 4905 Dufferin Street, Downsview, Ontario, Canada M3H 5T4*

I. Scheunert	*Institut für Ökologische Chemie Gesellschaft für Strahlen–und Umweltforschung mbH, München, FRG*
L. Söderlund	*Swedish Water and Air Pollution Research Laboratory, Box 21060, 10031 Stockholm, Sweden.*
M. Treshow	*Department of Biology, University of Utah, Salt Lake City, Utah 84112, U.S.A.*
G. E. Walsh	*Environmental Research Laboratory, United States Environmental Protection Agency, Gulf Breeze, Florida 32561, U.S.A.*

List of Figures

1.1	Inputs to the environment associated with industrial and domestic use patterns	6
2.1	Fates of chemicals in the environment	13
2.2	Reactions of aldrin, chlordene, and 2,2'-dichlorobiphenyl with $O(^3P)$	21
2.3	Typical 'acceptor' molecules and oxygenated products for reaction with $^1O_2$	22
2.4	Photoisomerization of aldrin and dieldrin	23
2.5	$2\pi \to 2\sigma$ reaction	24
2.6	Hydrogen transfer reaction of methanoindenes	24
2.7	Photoreaction of 4,5,6,7,8,8-hexachloro-2,3,3a,4,7,7a-hexahydro-4,7-methano 1H-indene-1,3-dicarboxylic acid	25
2.8	Photodechlorination reaction of chlorinated methanoindene	26
2.9	Isomers of chlordene	26
3.1	Metabolism of cyclodiene insecticides by *Aspergillus flavus*	44
3.2	Metabolites of dieldrin from microorganisms	45
3.3	Metabolites of isodrin in cabbage	46
3.4	Metabolites of Lindane in insects	47
3.5	Formulae of aldrin, dieldrin, and metabolites in mammals	49
3.6	Degradation pathways of aldrin in the plant–soil system	51
3.7	Conversion products of buturon in the plant–soil system	52
3.8	Excretion of radioactivity by rats during long-term feeding with 2,4,6,2',4'-pentachlorobiphenyl-^{14}C, chloroalkylene-9-^{14}C, and 2,2'-dichlorobiphenyl-^{14}C	62
4.1	Pollutant transport model: mercury in an aquatic ecosystem	81
4.2	The dynamics of methylmercury production	82
4.3	Behaviour of mercury in an aquatic system	84
5.1	ICRP lung clearance model	94
5.2	Single uptake	100
5.3	Chronic uptake	101
5.4	Declining uptake resulting from initial contamination of the lungs or a wound	101
5.5	Several uptakes in a limited period	101
6.1	Example of relation between threshold distribution and dose–response curve	121
6.2	Estimated dose–response curve with 95% confidence limits (rotenone toxicity example)	131

List of Figures

6.3	Comparison of true dose–response curve having spontaneous occurrence with estimated curve assuming no spontaneous occurrence	135
6.4	Comparison of estimated and observed proportions of animals with skin tumours at four dose levels	139
8.1	A three-compartment model for the distribution of a pollutant within a vertebrate	170
8.2	Decrease in the concentration of dieldrin in rats' blood during the first 71 days after exposure	172
8.3	Increase in the concentration of dieldrin in sheeps' blood while ingesting 2 mg dieldrin/kg body weight/day	173
8.4	Linear regression for the steady-state concentration of p,p'-DDT in eggs of white leghorn hens on the concentration of p,p'-DDT in the diet	174
8.5	Changes in the concentration of dieldrin in sheeps' blood while ingesting 0.5 mg dieldrin/kg body weight/day	175
8.6	Changes from 1901–1969 of the eggshell index for the peregrine falcon in Great Britain	182
8.7	Relationship between mean clutch shell thickness and DDE residue of kestrel eggs collected in Ithaca, New York, during 1970 and same relationship experimentally induced with dietary DDE	183
9.1	Example of probit analysis applied to sublethal studies of avoidance behaviour in trout	190
9.2	Illustration of a sublethal toxicity curve derived by plotting EC_{50} values against the period of exposure	191
9.3	Relationship between per cent depression of olfactory response and the ambient concentration of discrete solutions of mercury and copper at 4 hours exposure	197
9.4	Toxicity curve showing the time to occurrence of hepatomas as a function of dose in rats administered p-dimethylaminoazobenzene	203
9.5	Phases of interactions between chemical constituents of pollutant mixtures	205
9.6	Linear regressions for discrete solutions and for mixtures of copper and nickel	209
9.7	Possible types of responses which can occur between two hypothetical toxicants, A and B, which have similar actions	209
9.8	Isobols for a mixture consisting of a pollutant A, an active toxicant if applied singly, and a pollutant B, a non-toxicant but which antagonizes or synergizes the response to pollutant A	211
11.1	The DDT concentrations in filtered pond water and in some plants as a function of time after labelling the pond	246
11.2	Interactions among PCBs, DDT, and DDE in *Thalassiosira pseudonana*	248
11.3	Response of *Skeletonema costatum* to various concentrations of pulp-mill effluent and to filtered sea-water controls	250
14.1	Schematic dose–response curve for frost damage to tobacco or oranges	297
14.2	Douglas Point generating station, Lake Huron	306

15.1 Diatom population in a stream not adversely affected by pollution (Ridley Creek, Pennsylvania, U.S.A., November 1951) 320
15.2 Diatom population in a polluted stream (Lititz Creek, Pennsylvania, November 1951) 321

List of Tables

2.1 Lifetimes of organophosphocompounds in aqueous solution 17
2.2 Rate constants for the reactions of ozone with olefins 19
2.3 Rate constants for the reactions of oxygen [O(^{3}P)] with olefins . . . 20
2.4 Photodegradation of photodieldrin 28
2.5 Results of ultraviolet irradiation of aromatic xenobiotics as solids in an oxygen stream 29
2.6 Results of ultraviolet irradiation of aromatic xenobiotics adsorbed on silica gel 31
3.1 Residues of environmental chemicals and metabolites in tissues and organs of rats after long-term feeding was discontinued 42
3.2 Radioactive material excreted within 10 days after single oral administration of 0.5 mg/kg dieldrin-^{14}C 50
3.3 Material balance in plants and soil five years after application of aldrin-^{14}C to soil 53
3.4 Residues of aldrin-^{14}C and conversion products in soil five years after treatment 54
3.5 Concentration of elements from sea water by plankton and brown algae . 58
3.6 DDT concentrations in samples collected from the pond, following a single addition of 1 μg/l 60
3.7 Behaviour of aldrin-^{14}C and di-2-ethylhexyl phthalate (DEHP) in a model ecosystem 61
4.1 Coefficients for the mercury transport model 83
5.1 Fraction of inhaled material absorbed 95
6.1 Expected per cent responding for various models over a range of dose levels 125
6.2 Expected per cent responding at low doses for models describing observed responses in the 5%–95% range equally well 125
6.3 Example of log-logistic model applied to quantal response data. Toxicity of rotenone 130
6.4 Example of log-logistic model applied to heterogeneous quantal data. Incidence of hepatomas in rainbow trout fed diets containing aflatoxin for 12 months 132
6.5 Maximum-likelihood iterative computation for rotenone toxicity example 147
8.1 Amounts of dieldrin in the blood, liver, and fat of rats fed on a diet containing 50 ppm dieldrin 177

List of Tables

8.2	The minimal percentage activity of seven acetylcholinesterase isozymes in the housefly after exposure to the LD_{50} dose of four insecticides	178
11.1	Chemical states of copper in natural waters	243
12.1	Size classes of plankton	258
12.2	Metabolic properties that can be used to characterize ecosystems	259
12.3	EC_{50} values for growth and photosynthesis by four genera of marine unicellular algae when exposed to Hill reaction inhibitors (ametryne, atrazine, and diuron) and non-Hill reaction inhibitors (silvex, diquat, and trifluralin)	261
12.4	Rates of phytoplankton production in oligotrophic and eutrophic lakes	262
12.5	Total primary productivity of phytoplankton per year in Lake Tahoe	262
12.6	Calculated and observed values of E (expected population density of algae as a percentage of the control) for three genera of marine unicellular algae exposed to nickel (as $NiCl_2$) and the technical acid of 2,4-D	265
13.1	Chemical contaminants found in aquatic environments	277
14.1	Estimated incidence of, and mortality from, skin cancers from steady-state release of selected chemicals	302

Introduction

G. C. Butler

Division of Biological Sciences, National Research Council of Canada, Ottawa, Canada, K1A 0R6

This report deals with pollutants and pollution, its processes and effects. Pollution of the environment results from the presence of agents in amounts that produce adverse changes. A pollutant is such an agent in such an amount.

Environmental contamination is not new but international efforts to do something about it are. They probably originated, in their current form, from the widespread concern caused by atmospheric testing of nuclear weapons in the nineteen fifties and sixties. The response of the United Nations to this situation was to create in 1955 the Scientific Committee on the Effects of Atomic Radiation which estimates levels of environmental radiation and their risks to human populations.

Soon after, individuals and organizations perceived and publicized the possible dangers of large-scale contamination by industrial chemicals released intentionally (pesticides) or unintentionally (PCB's, heavy metals) to the environment. In 1968 the United Nations decided to hold an international conference on the environment, later called the Stockholm Conference. Most of the present-day international environmental activities began with the preparations for that Conference.

In 1969 the Scientific Committee on Problems of the Environment (SCOPE) was formed by the International Council of Scientific Unions (ICSU). Since its inception SCOPE has provided a focus for environmental projects conducted by scientists acting independently of governments. The projects of SCOPE and their relations with other activities of ICSU are described in the publication 'Environmental Issues' (SCOPE, 1977).

Two of the recommendations from the Stockholm Conference are (United Nations, 1972):

Recommendation 73

It is recommended that Governments actively support, and contribute to, international programmes to acquire knowledge for the assessment of pollutant sources, pathways, exposures and risks and that those Governments in a position to do so provide educational, technical and other forms of assistance to facilitate broad participation by countries regardless of their economic or technical advancement.

Recommendation 80
It is recommended that the Secretary-General ensure:

(a) That research activities in terrestrial ecology be encouraged, supported and co-ordinated through the appropriate agencies, so as to provide adequate knowledge of the inputs, movements, residence times and ecological effects of pollutants identified as critical;

(b) That regional and global networks of existing and, where necessary, new research stations, research centres, and biological reserves be designated or established within the framework of the Man and the Biosphere Programme (MAB) in all major ecological regions, to facilitate intensive analysis of the structure and functioning of ecosystems under natural or managed conditions;

(c) That the feasibility of using stations participating in this programme for surveillance of the effects of pollutants on ecosystems be investigated;

(d) That programmes such as the Man and the Biosphere Programme be used to the extent possible to monitor: (i) the accumulation of hazardous compounds in biological and abiotic material at representative sites; (ii) the effect of such accumulation on the reproductive success and population size of selected species.

which form the directive for part of the action plan 'Earthwatch' to be carried out by the United Nations Environment Programme (UNEP). SCOPE Project 4, on Ecotoxicology, is a response to the challenge of those recommendations and we hope it may make a useful input to the planning and conduct of Earthwatch.

This report was prepared to provide an overview of an emerging subject in environmental science, namely, ecotoxicology. The history, scope, and needs of ecotoxicology have recently been reviewed by Truhaut (1977). The Preparatory Committee has agreed on the definition:

'Ecotoxicology is concerned with the toxic effects of chemical and physical agents on living organisms, especially on populations and communities within defined ecosystems; it includes the transfer pathways of those agents and their interactions with the environment.'

Since this is an emerging subject the report cannot be expected to deal adequately with all its aspects.

It was the ambition of the writers and the Preparatory Committee to provide a document that would be maximally relevant and useful to people dealing with environmental problems, e.g. researchers, teachers, research administrators, scientific advisers to decision makers and to authorities designing protocols for assessments of environmental impact. The subject matter of ecotoxicology is required for developing scientific criteria such as those of the environmental programme of the World Health Organization and for inclusion in data bases such as that being prepared by UNEP for its International Registry of Potentially Toxic Chemicals.

The SCOPE project on Ecotoxicology has concerned itself with chemical and radioactive pollutants but not with an exhaustive list. The biological effects dealt with are most of the important ones but are not confined to effects on humans. The report may contain more information about man than about any other single species but this simply reflects its greater availability. A complete treatment of the subject should contain a large chapter dealing with effects on man including clinical data and epidemiological considerations. It was, however, agreed to omit these since they are being studied and reported upon by the World Health Organization.

Not included are biological pollutants (e.g. parasites and infectious agents such as bacteria and viruses) and physical pollutants (e.g. heat, light, noise, vibration), but there is a brief description (Chapter 14) of some physical changes such as temperature with secondary biological effects. Also excluded are non-biological receptors such as structures erected by man. It was decided to omit consideration of activities which might have interactions with the environment, such as recreation.

The report is intended to be a source of ideas and principles rather than of data; it emphasizes the importance of quantitative information and describes, either implicitly or explicitly, the kinds of research needed in ecotoxicology and how to interpret the results of that research.

Since the report was prepared by about twenty specialists in different aspects of the subject, it may appear heterogeneous in style and approach. No effort was made to make it appear like a book written by a single author since it was hoped that the reader would find the variety refreshing and stimulating. An effort was made to limit the length of all chapters; exhaustive treatments of the subjects were not considered necessary because their purpose is mainly to stimulate the reader to think about the various subjects and to draw attention to the relevant literature for further study.

No effort was made to avoid overlaps in subject matter between several chapters since these were found to provide complementary rather than contradictory information; examples of this are to be found in comparing,

— Chapter 5 and Chapter 8
— Chapter 11 and Chapter 13
— Chapter 11 and Chapter 15

Finally, it should be pointed out that it will not be possible on first reading to assess how well the authors have achieved their purpose. Rather, this will be measured over the long term by how helpful the book proves to be in understanding and teaching the new science of ecotoxicology.

DEFINITIONS

A *population* consists of all the potentially interbreeding individuals of a species within a community.

A *community* consists of populations of all organisms (animals, plants, and microorganisms) in an ecosystem.

An *ecosystem* consists of communities of living organisms together with their habitat (or abiotic environment) and includes the interactions among these components.

*Exposure, the amount of a particular physical or chemical agent that reaches the target.

*Target (or receptor), the organism, population, or resource to be protected from specific risks.

*Risk, the expected frequency of undesirable effects arising from a specified (unit) exposure to a pollutant.

*Criteria, the quantitative relations between the exposure to a pollutant and the risk or magnitude of undesirable effects under specified conditions defined by environmental and target (receptor) variables.

Intake, the amount of substance entering the gastrointestinal or the respiratory tract.

Uptake, the amount of substance absorbed into the systemic circulation.

Mineralize, to convert to inorganic compounds, e.g. the oxidation of an organic chemical to carbonate and water.

Xenobiotic, not produced in nature — usually applied to man-made chemicals.

REFERENCES

SCOPE, 1977. *Environmental Issues* (Eds. M. W. Holdgate and G. F. White), *SCOPE Report 10*, John Wiley and Sons, London.

Truhaut, R., 1977. Ecotoxicology: objectives, principles and perspectives. In *The Evaluation of Toxicological Data for the Protection of Public Health* (Eds. W. J. Hunter and J. G. P. M. Smeets), Pergamon Press, Oxford, pp. 339–413; and *Ecotoxicol. Environ. Safety*, 1, 151–73.

United Nations, 1972. Draft document A/CONF. 48/14, 3 July 1972, for the United Nations Conference on the Human Environment held at Stockholm, 5–16 June 1972.

*Defined by the Stockholm Conference.

SECTION I
ENVIRONMENTAL BEHAVIOUR OF POLLUTANTS

CHAPTER 1

General Considerations

D. R. MILLER

Division of Biological Sciences, National Research Council of Canada, Ottawa, Canada, K1A 0R6

1.1 INTRODUCTION 3
1.2 SOURCES OF POLLUTANTS 4
1.3 NEED FOR ADVANCE ASSESSMENT 6
1.4 PREDICTING ENVIRONMENTAL BEHAVIOUR 7
1.5 PHYSICAL AND CHEMICAL PROPERTIES 7
1.6 MODELS AND MODEL ECOSYSTEMS 8
1.7 CONCLUSIONS 8
1.8 REFERENCES 9

1.1. INTRODUCTION

One of the main differences between classical toxicology and ecotoxicology is that the latter is a four-part subject. Any assessment of the ultimate effect of an environmental pollutant must take into account, in a quantitative way, each of the distinct processes involved (Truhaut, 1975).

First, a substance is released into the environment; the amounts, forms, and sites of such releases must be known if the subsequent behaviour is to be understood.

Second, the substance is transported geographically and into different biota, and perhaps chemically transformed, giving rise to compounds which have quite different environmental behaviour patterns and toxic properties. The nature of such processes is unknown for the majority of environmental contaminants, and the dangers arising from our ignorance of the ultimate fate of certain chemicals have been well documented in recent years.

The third part of the process is the exposure of one or more target organisms. For this to be assessed, one must first identify the nature of the target (man himself, livestock or similar resources, etc.) and the type of exposure that is to be examined.

Fourth, one has to assess the response of the individual organism, population or community to the specified (perhaps transformed) pollutant over the appropriate time scale.

In order for a proper ecotoxicological assessment to be made, this combination of steps must be examined in a quantitative and integrated way. Just as one must try not to allow environmental dangers to go unchecked, it is equally undesirable to impose restrictions on the use of some substance simply because it is 'potentially toxic', for example, if the above combination of processes has not been carefully examined (National Academy of Sciences, 1975).

Because of the requirement for quantitative precision, one might add a fifth facet to the procedure, and that is an estimate of the uncertainty and possible error in our current understanding of the other four. It is evident that our 'best' estimates of the final effect of releasing some substance should be based on our 'best' estimates of the various intervening steps. In the process of making such estimates, a natural by-product is an idea of the uncertainties in the various steps themselves, and the levels of confidence in the understanding of the various processes usually turn out to be quite different (NAS, 1975).

This is important for two reasons. First, any decisions based on an overall process, one or two parts of which are imperfectly understood, ought to incorporate correspondingly large safety factors, whereas if we are confident in predictions the margin for error may be reduced (Holcomb Research Institute, 1976).

The second, and equally important, reason is that we are naturally led to an ordering of priorities for further research. Problems given the highest priority should have to do with either highly uncertain pathways or those which preliminary study indicates involve considerable hazard (Patten, 1971).

The procedures for assessing some of these steps are, of course, just beginning to be developed. Also, the process of combining the conclusions into an overall judgment, including a judgment of the uncertainties involved, exists only in a preliminary form (Patten, 1976). However, some principles are available. It is the purpose of this Section to focus on several of these principles so that their further development may proceed more quickly.

1.2. SOURCES OF POLLUTANTS

There are more than 30,000 chemical substances in use in the world today, and this figure is increasing by 1,000–2,000 per year. Many are produced in quite large amounts; estimates vary and are changing regularly, but the following production figures are probably reasonable (Korte, 1976)

Amount produced (metric tons/year, world-wide)	Number of substances produced in excess of this level
500	1,500
50,000	100
1,000,000	50

It is worthwhile pointing out that these figures, although large, do not in themselves mean that concentrations at toxic levels are to be expected in most areas. Uniformly distributed over all the land area of the earth 500 metric tons amounts to a surface dose of 3.4×10^{-3} mg/m^2. Even if distributed over agricultural areas only, the dose would still be only 1.2×10^{-2} mg/m^2 (about 1.75×10^{-2} oz/acre). What this means is that, in general, concentrations of such substances are negligible; but, since local levels may easily rise to three orders of magnitude above average by natural concentrating factors alone, the potential for ecotoxic effects is very real for a wide range of substances.

An estimated growth rate of the chemical industry of even 2% per year on a world-wide basis, certainly conservative in the opinion of many, will mean more than a *sevenfold* increase in overall use within one hundred years, and concern for the environment must be a prime concern of mankind. It has been said that we have only perhaps two generations in which to change radically our handling of chemical substances in many countries.

It is convenient to begin with the chemicals themselves, and several characteristics of potentially hazardous substances can be set down. The first involves amounts, including amounts manufactured and the releases to the environment, both deliberate and accidental, which can be expected to accompany various use patterns (Freed and Haque, 1975). Such data, it must be admitted, are frequently quite difficult to obtain, since they provide considerable insight into the detailed operation of particular industries. Nonetheless, if we are adequately to assess ecotoxicological danger, a rather complete analysis must first be made of amounts which enter the various pathways, including those not released intentionally into the environment, along with an estimate of probable losses, avoidable and otherwise, from each. Those include mining or otherwise obtaining of raw materials, the production process itself, shipping, including imports and exports as well as transport to place of use, losses during application, and losses during the shipment and reprocessing. Finally, one must assess the release to the environment involved in deliberate application of the substance, or the discarding after use of products containing it (Figure 1.1).

Such an assessment of environmental inputs is a considerable undertaking. Nonetheless, only such an examination can provide reliable information on the total environmental consequences; public attention has frequently been focused on spectacular single or isolated cases of accidental release, while thoughtful examination of all sources may well indicate that widespread losses at much smaller local rates contribute more to the overall release into the biosphere. The petroleum industry is a case in point; the foundering of a tanker carrying tens of thousands of tons of petroleum is a matter of widespread press coverage, while estimates of completely unavoidable losses on an industry-wide basis of upwards of tens of millions of tons per year are deemed acceptable (Korte, 1976). It is to be hoped that such studies of losses associated with overall use patterns will soon be available for a considerable number of chemical substances.

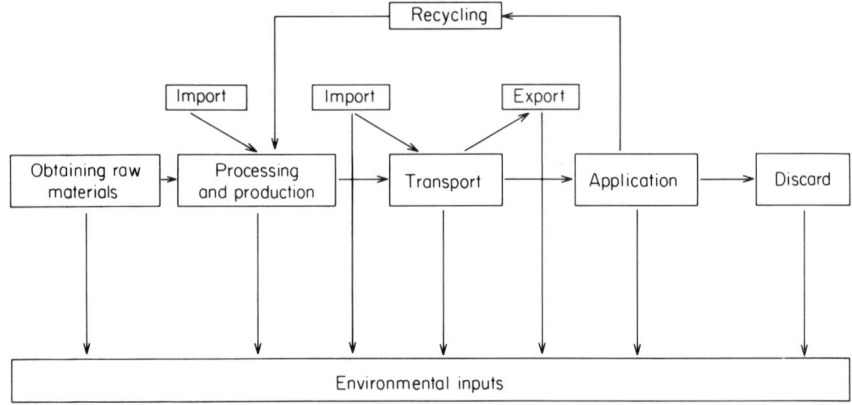

Figure 1.1 Inputs to the environment associated with industrial and domestic use patterns

It is also worthy of note that exposures, to man at least, involved in the production, shipping, and application of various substances have long been studied by workers in occupational health, pharmacology and so forth. Exposure resulting from accidental loss to the environment, or ultimate disposal, is an area less well understood (Goodman, 1974).

1.3. NEED FOR ADVANCE ASSESSMENT

The day is long past when one could introduce a chemical into common use and wait until events provided an accident to assess its effect upon release to the environment. Now, society demands that hazardous – in the sense of ecotoxic – substances be identified before being brought into use.

One may be able to find cases in the literature where accidental spillage of a similar or related substance has occurred and make a reasonable prediction for a new chemical. However, this approach presupposes an increasing supply of such 'accidents' – precisely what society, and regulatory agencies reflecting its wishes, will not permit. And it is not possible or desirable to contrive such an event on a real scale; for many environmentally dangerous substances, the ultimate effect is not expressed for many years, and a real-time experiment cannot be used to decide in advance whether the substance might be too dangerous to be used.

Thus we have an obviously impossible problem, namely that scientists must accurately predict the probability, and the nature, of an event that has never occurred. Furthermore, society is likely to demand improved levels of environmental safety, or at least increased reliability of predictions of ecotoxic effects, as time goes by. To an ever-greater extent, then, scientists will have to make assessments, on which real policy decisions will be based, but which are themselves based solely on extrapolation from a knowledge of the properties of the substance, or from small-scale, short-term experiments. Most important at this stage are the particular principles and experiments to be selected.

1.4. PREDICTING ENVIRONMENTAL BEHAVIOUR

As mentioned above, ecotoxicological assessment of a chemical requires information about its behaviour in the environment and in the receptor. The present discussion deals only with the former.

The essential data on environmental behaviour include those for pathways and rates (van Dobben and Lowe-McConnell, 1975). Some general predictions for these subjects are possible from a knowledge of the chemical and physical properties of a substance and the mechanisms of its reactions and transformations. As more quantitative information on pathways becomes available there is correspondingly less need for research on properties and mechanisms as a basis for prediction.

1.5. PHYSICAL AND CHEMICAL PROPERTIES

There have been several attempts recently to identify those physical and chemical characteristics of a chemical which would make it suspect as an ecotoxic substance in the wide sense or would help to predict its environmental behaviour (Goodman, 1974). These examples, in fact, serve as excellent illustrations of two approaches to the question.

Goodman (1974) concentrates on what might be called functional properties. From the standpoint of environmental behaviour, he identifies (i) persistence in the environment, (ii) environmental mobility, and (iii) failure to form inert compounds, as key properties of ecotoxic substances (in addition to various others such as toxicity, the property of sequestering in lipid or bone, etc., which are perhaps best addressed from the standpoint of toxic effects).

Accepting such characteristics as indicative of potential hazard, we can move to the chemical and physical properties of the substance that would enable a prediction to be made. In such a list would be included:

solubility (in water)
partition coefficient (between solid and liquid, polar and non-polar media)
dissociation constants
formation of chemical complexes
degradation, hydrolysis or photolysis
volatilization
leaching and dissipation characteristics

For an understanding of the behaviour of most chemicals in an ecosystem the properties listed above are the most important. Usually, biological behaviour contributes very little to the large-scale transport or transformation of chemicals and is important mainly for an understanding of routes to, and effects on, receptors.

1.6. MODELS AND MODEL ECOSYSTEMS

If the various characteristics in the above list, of the substance, are well understood, it is possible to predict the overall distribution that should be observed in a typical simplified ecosystem. The most important characteristics that must be understood in an aquatic system, for example, are the partitioning between water and bottom sediment, between water and suspended materials, rates of volatilization and degradation, and availability for uptake by biota. Most of these can be estimated on the basis of the chemical and physical properties.

It might be useful to point out that refining numerical estimates of parameters, once the natures of the dominant mechanisms have been determined, is not a difficult process. Programming of a given model on a digital computer is a familiar business, and the application of such techniques as sensitivity analysis has become routine (Patten, 1976).

This statement unfortunately applies only to simple, quasi-static ecosystems. In more unstable situations, such as would be expected when organic chemicals are exposed to sunlight, or metabolized on a large scale by bacteria, the mechanisms would be more complicated (van Dobben and Lowe-McConnell, 1975).

What is needed, then, is progress in two directions. The first is a refinement of techniques of discussing quantitatively the chemical and physical parameters in the above list, and the second is a way of identifying mechanisms that might be dominant in ecosystems of a slightly different type. The two chapters to follow deal primarily with the second subject area.

1.7. CONCLUSIONS

By way of providing a summary of this section, we may identify the following conclusions:

(1) The technology of predicting environmental behaviour of pollutants has developed substantially in recent years, but needs to be further refined.

(2) Such refinement will be accomplished most effectively by concentrating on overall pathways and movements and general physical and chemical properties, not on detailed chemical or biochemical mechanisms.

(3) A list of those physical and chemical properties most important in predicting environmental behaviour can be compiled.

(4) Predictive mathematical models will be increasingly useful, but more attention must be paid to quantitative assessment of errors and uncertainties of their predictions.

(5) Examination of mathematical models can materially assist in identifying those pathways and mechanisms most in need of further investigation.

1.8. REFERENCES

van Dobben, W. H. and Lowe-McConnell, R. H. (Eds.), 1975. Unifying concepts in ecology. *Proc. First Internat. Cong. Ecology*, The Hague, Sept. 8–14, 1975.

Freed, V. H. and Haque, R. (Eds.), 1975. *Environmental Dynamics of Pesticides*, Plenum Press, New York, 384 pp.

Goodman, G. T., 1974. How do chemical substances affect the environment? *Proc. Roy. Soc. Lond.*, **B185**, 127–48.

Holcomb Research Institute, 1976. *Environmental Modelling and Decision Making*. Report prepared by U.S. National Committee on SCOPE Project 5, Simulation Modelling, Praeger Publishers, New York, 152 pp.

Korte, F., 1976. Personal communication, December 1976.

National Academy of Sciences, 1975. *Decision Making for Regulating Chemicals in the Environment*. Report of the Environmental Studies Board, NAS, 2101 Constitution Avenue, N.W., Washington, D.C., 232 pp.

Patten, B. C., 1971 to 1977. *Systems Analysis and Simulation in Ecology*, 4 Vols., Academic Press, New York.

Truhaut, R., 1975. Ecotoxicology – A new branch of toxicology. In *Ecological Toxicology Research* (Eds. A. D. McIntyre and C. F. Mills). *Proc. NATO Science Comm. Conf.*, Mt. Gabriel, Quebec, May 6–10, 1974, Plenum Press, New York, 323 pp.

CHAPTER 2

Abiotic Processes

F. KORTE

Institute of Chemistry, Technical University of Munich,
805 Freising-Weihenstephan, FRG

2.1 PHYSICAL PROCESSES	11
(i) Local, regional, and global transport of chemicals	12
(ii) Leaching of ions and organic compounds in soil and landfills	14
(iii) Evaporation of organic chemicals from soil and surface waters	14
(iv) Atmospheric washout of organic chemicals	15
(v) Dry deposition from the atmosphere	15
(vi) Sedimentation of organic chemicals	15
2.2 CHEMICAL PROCESSES	16
(i) Sedimentation of inorganic chemicals	16
(ii) Hydrolysis of organic chemicals	16
(iii) Oxidations	17
(iv) Photochemically induced processes	18
(v) Photoreactions of cyclodienes	22
(vi) Photoreactions of some chlorinated compounds	26
(vii) Evaluation of the reactions of organic chemicals	30
(viii) Laboratory models for testing abiotic degradability	33
2.3 REFERENCES	33

2.1. PHYSICAL PROCESSES

When considering the effects of industrial chemicals in the environment, physical-chemical processes have a special importance since they are responsible, first, for the dispersion and, second, for the chemical changes which occur under abiotic conditions. The significance of biotic processes has, at times, been overemphasized, since degradation by this pathway is often quantitatively a minor factor by comparison with the physical-chemical processes.

The importance of abiotic degradation under atmospheric conditions has to be emphasized. In the past, research has been focused on the photochemical changes and degradation of organic chemicals, including investigations of the reaction mechanisms and the kinetics of such reactions. These studies involved laboratory experiments that did not attempt to simulate actual atmospheric conditions. It has been known for a long time that organic molecules are subject to isomerization, conversion, and incomplete degradation reactions through ultraviolet irradiation. It has not generally been recognized that mineralization* may take place even in

---

*As used in this section 'mineralization' means the complete degradation of organic compounds to inorganic products, e.g. carbonates and chlorides.

diffuse daylight. So far, only some organic chemicals such as methane, propane, and unsubstituted arenes which participate in the formation of the so-called photochemical smog, have been studied in great detail. The studies by Gäb *et al.* (1974a,b) demonstrate that even persistent chemical substances (e.g. photodieldrin) can be mineralized within relatively short periods if irradiated with light of wavelengths between 230 and 300 nm, as well as above 300 nm, the latter being a wavelength found in diffuse daylight.

Considering the fact that an instant availability of energy exists in the atmosphere, the abiotic (photochemical) degradation is probably a more important process than biodegradation which, in most cases, only leads to conversion products easily excreted by the living organism. Therefore, the atmosphere can be regarded as a large sink for persistent organic chemicals. Comparative studies of today's global concentrations of some persistent chemicals such as DDT and dieldrin and the total amounts released suggest that the bulk may be mineralized.

Moreover it may be possible to estimate permissible emission levels by determining the rates of photochemical mineralization reactions in the atmosphere.

(i) Local, Regional, and Global Transport of Chemicals

The mechanisms leading to global dispersion of industrial chemicals are to some extent complex and can only be described approximately by scientific methods. A good correlation is found between the dispersion of ^{90}Sr from atmospheric atomic bomb tests and the dispersion of DDT, where in both cases higher concentrations are present in the northern hemisphere. In order to achieve a better understanding of the transport phenomena, the differences between the local, regional, and global transport possibilities should be mentioned. Local transport mechanisms include those that change the environmental quality within a limited area intentionally, as in the case of pesticides and fertilizers. The pathways of these substances leading to the contamination of food are depicted in Figure 2.1.

Regional and global transport mechanisms involve an undesired dispersion of chemicals outside the area being treated, which leads to an occurrence of the corresponding chemicals in the global system. Although it is of greater importance for the understanding of the dispersion of industrial chemicals to deal with transport phenomena within the troposphere, the required three-dimensional models (which simulate the troposphere) can only be developed with great difficulties. For this reason, there are to date no satisfactory calculations for the lower atmosphere. The results of the model reported in the works of Pressman and Warneck (1970) and Bolin *et al.* (1963, 1970) seem to be better suited to the treatment of this problem by categorizing the dispersion mechanisms of the troposphere into vertical, longitudinal, intrahemispherical and interhemispherical transport pathways. The relevant references for these four aspects of transport are listed below.

Abiotic Processes

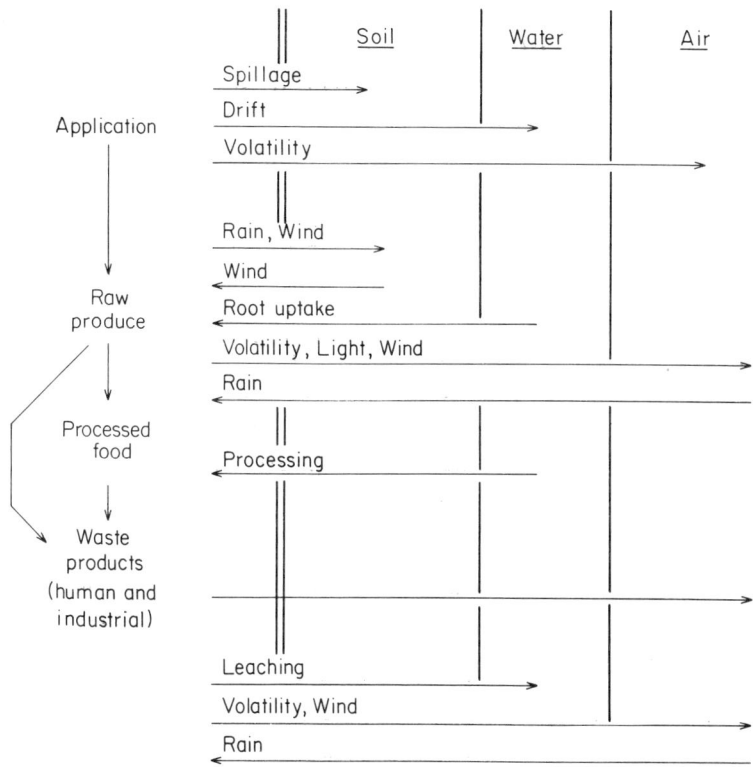

Figure 2.1 Fates of chemicals in the environment

(a) Vertical and longitudinal mixing:
 Jakobi and Andre (1963)
 Junge (1962)
 Kroenig and Ney (1962)
 Lettau (1951)

(b) Intrahemispherical and interhemispherical mixing:
 Junge (1962)
 Lal and Rama (1966)
 Levy (1974)
 Nydal (1968)

(c) Troposphere — stratosphere exchange:
 Junge (1962)
 Pressman and Warneck (1970)
 Reiter et al. (1967)
 Smith (1968)

(ii) Leaching of Ions and Organic Compounds in Soil and Landfills

In the case of the washing out of organic compounds from the ground, the affinity of the soil for the chemicals must be considered. In general, and as expected, polar organic compounds adsorbed onto the soil can be more easily washed out than non-polar compounds. Therefore the classes of substances have increasing affinities in the following order: hydrocarbons < ethers < tertiary amines < nitro compounds < esters < ketones < aldehydes < amides < alcohols < acids.

The behaviour of pesticides serves as a practical example. It has been found experimentally that the affinity for the soil depends mainly on the structure of the chemical in question, on the physical properties of the soil, and finally on the chemical composition of the soil.

Transport phenomena are also strongly influenced by the accumulation of detergents or inorganic salts in the soil. Lichtenstein and Schultz (1965) have reported on these effects and have attempted to correlate the results of laboratory experiments with phenomena occurring in natural soils. The experiments with non-polar substances such as cyclodiene insecticides seem to verify the significance of these adsorption effects.

Due to their low water solubility, the insecticides aldrin and dieldrin are not transported deeply into the soil and do not enter into the ground-water. Recent studies show however, that after treating a 10 cm deep soil layer with ^{14}C-aldrin, small amounts of aldrin, dieldrin, and photodieldrin penetrate up to 60 cm into the ground but that some degradation products are so strongly adsorbed that they cannot be extracted by organic solvents. The main product is dihydrochlordendicarboxylic acid, which occurs at a depth of 60 cm in ground-water in concentrations as high as 0.05 ppm (Moza et al., 1972).

(iii) Evaporation of Organic Chemicals from Soil and Surface Waters

The removal of organic chemicals from soil is mainly dependent on the physical properties of the lower troposphere. Such factors as wind erosion and air extraction resulting from various types of air currents or soil agitation caused by living organisms and urbanization have a significant influence on the evaporation of organic chemicals from the soil. Organochlorine insecticides, such as dieldrin and heptachlor, have been shown to volatilize from soil into the atmosphere under field conditions. Five months after incorporation of these insecticides into the soil, 2.8 and 3.9%, respectively, had volatilized into the air (Caro et al., 1971). Studies carried out in the laboratory indicate that soil-incorporated residues of chlordane volatilize at a rate faster than that of dieldrin. Thus it would be expected that part of the chlordane applied to soil would be released into air, although the degree of volatilization would depend on the type of soil, soil moisture content, and weather conditions (Edwards, 1966). In the case of evaporation from surface water, physical and mathematical methods are often employed since the entire process can be formulated as a type of steam distillation.

Generally, if two components are insoluble in each other (e.g. water and non-polar organic compounds), then each does not influence the vapour pressure of the other. Therefore, the total vapour pressure of a heterogeneous mixture is simply the sum of the vapour pressures of its individual constituents. Since the total vapour pressure is higher than the vapour pressure of the individual constituents, the boiling point of such a mixture must be lower than that of its lowest boiling constituent. This phenomenon takes place also for very dilute solutions of organic substances in water, since the composition of the distillate is independent of the absolute amount of the components.

(iv) Atmospheric Washout of Organic Chemicals

When considering the washout of organic chemicals in the atmosphere one of the main problems is to determine in what form the organic substances occur. The question of whether they exist in a gas phase, are adsorbed on solid or liquid aerosols or exist as clusters in the troposphere has been answered only for some chemicals. Previous work has generally been limited to the washout of radioactive particles and aerosols and it has been estimated that these particles have a lifetime of approximately 10 days (Francis *et al.*, 1970) although some authors have reported lifetimes ranging from 20 to 50 days. It should be noted that rainout and washout usually occur in only the lowest 5 km of the troposphere.

(v) Dry Deposition from the Atmosphere

Although very little information is available on dry deposition, it could in some cases lead to the deposition of amounts comparable with those resulting from wet deposition.

(vi) Sedimentation of Organic Chemicals

Sedimentation of organic chemicals is especially noticeable in calm-water bodies where the process is often indirect, since organic chemicals are introduced attached to solid aerosols. Although there are a few reports of the sedimentation of organic chemicals in flowing waters, not much is known of the process or mechanisms.

Large amounts of chlordane residues were found in the sediments of the lower Mississippi River and several of its tributaries (Barthel *et al.*, 1969). Thus, 0.12 to 1.28 mg/kg (dry weight) of chlordane residues were found in the mud samples from Wolf River/ Cypress Creek in Memphis, Tennessee. The high levels were attributed to a manufacturing operation in Memphis (highest levels were found at the factory site). Chlordane residues of 0.25 to 1.55 mg/kg of sample (dry weight) were found in sediments near formulating plants in Mississippi. Agricultural use of pesticides was not regarded as a significant contributor to the pollution of these waterways.

2.2. CHEMICAL PROCESSES

The abiotic changes which environmental chemicals undergo in nature can be classified into two groups, depending on whether or not ultraviolet irradiation from the sun occurs. The processes occurring in the absence of ultraviolet irradiation are limited to sedimentation by complexing with inorganic chemicals, hydrolysis and oxidation of organic chemicals.

(i) Sedimentation of Inorganic Chemicals

Sedimentation of inorganic chemicals by the formation of complexes with organic or inorganic compounds and pH-dependent equilibrium reactions all fall into this category. Accumulations of metals in river and sea sediments can also be included here, since they depend on the chemical and physical composition of the water. The precipitation reactions and pH dependence of equilibrium reactions have been reported in numerous publications. Although of interest, there have been few investigations on the sedimentation processes of biologically important metals such as mercury in rivers and in the ocean.

Hasselrot (1968) found 34–168 ng/g (dry weight) of mercury in river sediments upstream of factories using mercury. Downstream of a paperboard mill using a mercurial fungicide the sediment was found to contain 18,400 ng/g at 1.6 km, 4,300–10,100 ng/g at 5 km, and 3,500–8,000 ng/g at 7 km, mostly in the upper 2 cm of sediment. Similarly, below a chloralkali plant, he found 11,600–26,000 ng/g in the upper 4 cm of sediment 550 m downstream and 1,200 ng/g in the top 1 cm of sediment 750 m downstream. Even 7 km downstream 440 ng/g were found. Similar analyses have been reported from the United States where contaminated sediments below a chloralkali plant contained 5,400–86,000 ng/g. Other areas, less contaminated, contained 600–4,000 ng/g (Anon, 1970). Saito (1967) reported 350–3,730 ng/g in river muds near Japanese industrial plants (processes unspecified), the overlying water generally having less than 10 ng/g.

Marine sediments near the Swedish and Japanese coasts have been examined. Sediments sampled in the sound between Denmark and Sweden, adjacent to the Swedish coast, were found to contain high mercury levels, up to 2,000 ng/g (dry weight), apparently due to pollution (Ackefors *et al.*, 1970). Other marine sediments examined near the Swedish coast contained 1,000–1,500 ng/g. Sediments in the region of the discharge of mercury-containing effluents in Minamata Bay, Japan, were found to contain from 7,160–801,000 ng/g (Saito, 1967).

(ii) Hydrolysis of Organic Chemicals

It is known that many pesticides lose their toxic properties through hydrolysis in the environment; thus the reactivity of a pesticide in aqueous solutions can be used as an important criterion for its ecotoxicological behaviour.

Table 2.1 Halflives ($t_{1/2}$) of Organophosphorus Compounds in Aqueous Solution. (From Mühlmann, V. R., and Schrader, G. (1957). Hydrolysis of the phosphoric acid ester insecticide (in German), *Z. Naturforsch.*, **126**, 196–208. Reproduced by permission of the publishers)

Compound	Temperature (°C)						
	0	10	20	30	40	50	60
	Days						
Parathion	13,800	3,000	690	180	50	15	5
Methyl-parathion	3,600	760	175	45	12.5	4	1.3
Chlorthion	2,900	600	138	36	10	3	1
Metasystox Rl	4,800	970	236	62	18	5	1.7
Disulfoton	23,200	4,830	1,110	290	78	24	8
Azinphosmethyl	5,200	1,070	240	62	18	5.5	2
Trichlorphon	11,600	2,400	526	140	41	11	3

In Table 2.1 (Steller *et al.*, 1960), the half-times of persistence of several phosphocompounds is presented, where the half-time is the time required for a pesticide in aqueous solution to be degraded to one half its original concentration.

It can be seen that the stability of all the chemicals rapidly decreases with increasing temperature. Stability also decreases with extremes of pH.

Hydrolysis is also an important fate for chlorinated hydrocarbons. Heptachlor in aqueous solution is hydrolysed to hydroxychlordene, in which only the *exo* isomer is formed (Parlar *et al.*, 1975a).

(iii) Oxidations

Oxygen can react with certain organic compounds, giving a hydroperoxide

$$RH + O_2 \longrightarrow ROOH$$

These reactions, which can take place under natural conditions, are referred to as autooxidation, and have the following free-radical mechanism:

$$RH \xrightarrow{-H^\bullet} R^\bullet$$
$$R^\bullet + {}^\bullet O-O^\bullet \longrightarrow R-O-O^\bullet$$
$$R-O-O^\bullet + HR \longrightarrow R-O-OH + R^\bullet$$

The chain reaction is stopped (suppressed) by, for example, the reaction of the radical initiators ROO· or R· with each other, and accelerated by peroxides, u.v. irradiation from the sun, and traces of heavy metals. The reaction is autocatalytic, since peroxides are formed in the course of the reaction.

Heavy metals have a catalytic effect on the reaction and are able to convert peroxides into radicals according to the following steps:

$$ROOH + Me^+ \longrightarrow RO^\bullet + OH^- + Me^{2+}$$
$$ROOH + Me^{2+} \longrightarrow ROO^\bullet + H^+ + Me^+$$

Due to its lower reactivity, the peroxide radical is selective, and attacks more reactive CH bonds.

It has long been known that such reactions occur in nature. They have been observed in the spoiling of oils and fats and in the ageing of rubber and other polyenes. In the case of pesticides, it is known that aldrin reacts with oxygen from the air to give dieldrin in small yields. The oxidation of P=S bonds (found in phosphoric acid insecticides) to P=O groups probably results from an autooxidation process.

(iv) Photochemically Induced Processes

The atmosphere can be considered as a large chemical reactor in which chemicals react under the influence of irradiation from the sun and of catalysts in the form of trace elements. The reaction steps involved consist of parallel, sequential, and competitive reactions. To determine the individual mechanisms, conditions must be developed which mimic the atmosphere and permit, at the same time, a clear interpretation of the results. In this case complications arise due to the various reactions which occur simultaneously. Another approach would be the investigation of individual reaction steps. This can be achieved for example by deactivating the more reactive primary compounds and isolating and characterizing the resulting products.

(a) Photochemical Smog

The characteristic symptoms of photochemical air pollution were first encountered in the mid 1940's in Southern California. This type of 'smog' was characterized by the presence of organic compounds not found in the 'London smog', which consisted of SO_2 and aerosols. Several years later, Haagen-Smit (1964) established that it was indeed a new kind of air pollution, caused by the action of u.v. light on the exhaust emitted by the motor vehicles in the Los Angeles Basin. Haagen-Smit demonstrated that a mixture of nitrogen oxides with gasoline or olefins in the presence of sunlight, reproduced the smog damage on crops and showed that ozone was produced when individual organics such as olefins, alcohols, paraffins, and carbonyl compounds were irradiated in the presence of nitrogen dioxide. The same results could be demonstrated by the irradiation of automobile exhaust mixed with NO_2.

Experiments under laboratory conditions with a mixture of nitrogen oxides and propylene in ppm concentrations show that NO is rapidly converted to NO_2.

Table 2.2 Rate Constants for the Reactions of Ozone with Olefins. (Reproduced by permission of Verlag Chemie, Weinheim, Germany from Pitts and Finlayson, 1975)

Compound	$k \times 10^{-3}$ ($1\,\text{mol}^{-1}\,\text{s}^{-1}$)
Ethylene	0.93 ± 0.09
Propylene	7.5 ± 0.6
Isobutene	8.2
trans-2-Butene	165 ± 14
Toluene	7.2×10^{-3}
Acetylene	52×10^{-3}
Acetaldehyde	20×10^{-3}
Methane	0.72×10^{-3}
Carbon monoxide	0.6×10^{-3}
2-Methylbutene	296 ± 10
2,3-Dimethylbutene	906 ± 408

Simultaneously, the olefin concentration decreases and acetaldehyde appears. When the nitrogen oxide concentration is low, peracetyl nitrate and ozone build up while the nitrogen dioxide and C_3H_6 concentrations fall. The major light-absorbing compound in this mixture is NO_2, which dissociates to NO and ground-state oxygen atoms, i.e.

$NO_2 \rightarrow NO + O\cdot$

In air this is rapidly followed by

$O + O_2 + M \rightarrow O_3 + M$

where M is a 'third body'. In the presence of C_3H_6:

$O + C_3H_6 \rightarrow$ Products

$O_3 + C_3H_6 \rightarrow$ Products.

Table 2.2 gives typical rate constants for the reactions of ozone with some olefins and with other species present in polluted urban atmospheres (Pitts and Finlayson, 1975).

Using these rates and commonly encountered pollutant concentrations, one calculates that the only important losses of ozone occur by reaction with unsaturated compounds. The rapid reaction of ozone with NO ($O_3 + NO \rightarrow NO_2 + O_2$) is the reason that ozone does not begin to accumulate until the NO concentration has decreased to a low value.

(b) Reactions of Organic Compounds with Active Oxygen

Table 2.3 gives the values of selected rate constants for the reactions of $O(^3P)$ with some olefins and arenes (Pitts and Finlayson, 1975). These rates, which show

Table 2.3 Rate Constants for the Reactions of Oxygen [O(^{3}P)] with Olefins. (Reproduced by permission of Verlag Chemie, Weinheim, Germany from Pitts and Finlayson, 1975)

Olefin	$k \times 10^{-8}$ (1 mol^{-1} s^{-1})	Arene	$k \times 10^{-8}$ (1 mol^{-1} s^{-1})
Ethylene	4.3 ± 0.5	Benzene	0.144 ± 0.02
Propylene	20 ± 1.7	Toluene	0.45 ± 0.045
1-Butene	24 ± 3.7	o-Xylene	1.05 ± 1.11
cis-2-Butene	92 ± 15	m-Xylene	2.12 ± 0.21
2-Methyl-2-butene	313 ± 30	p-Xylene	1.09 ± 0.11
2,3-Dimethylbutene	425 ± 46	1,2,3-Trimethylbenzene	6.9 ± 0.7
		1,2,4-Trimethylbenzene	6.0 ± 0.6
		1,3,5-Trimethylbenzene	16.8 ± 2.0

an electrophilic trend, represent addition to the double bond. While it was suggested that abstraction may occur 15% of the time at 300°K in the O(^{3}P)–1-butene reaction, ·OH was not detected as an intermediate in this reaction, supporting previous assumptions that hydrogen abstraction is too slow to compete with addition at room temperature.

The model environmental chemicals aldrin (I), chlordene (II), and 2,2'-dichlorobiphenyl (III) react with O(^{3}P) to give oxygenated compounds (Saravanja-Bozanic et al., 1977), as shown in Figure 2.2.

(c) Reactions of Singlet Molecular Oxygen (Pitts, 1969)

It is known that the photooxygenation reactions of chemicals in photochemical smog fall into two groups: (a) those in which the photoexcited sensitizer molecule interacts directly with the substrate to produce free-radical species which react with the ground-state molecular oxygen present to give products, and (b) those in which the excited sensitizer transfers its energy to molecular oxygen producing singlet molecular oxygen, which in turn reacts with the substrate to give oxygenated products.

Briefly, the photosensitized oxidation by singlet oxygen involves the following sequence of reactions: it is now known to be general for gas-phase systems.

$$\text{Sens (S}_0\text{)} + h\nu \xrightarrow{\text{Absorption}} \text{Sens (S}_1\text{)}$$

$$\text{Sens (S}_1\text{)} \xrightarrow{\text{Intersystem crossing}} \text{Sens (T}_1\text{)}$$

$$\text{Sens (T}_1\text{)} + {}^3\text{O}_2 \xrightarrow{\text{Energy transfer}} \text{Sens (S}_0\text{)} + {}^1\text{O}_2$$

$${}^1\text{O}_2 + \text{Acceptor} \xrightarrow{\text{Chemical reaction}} \text{Product}$$

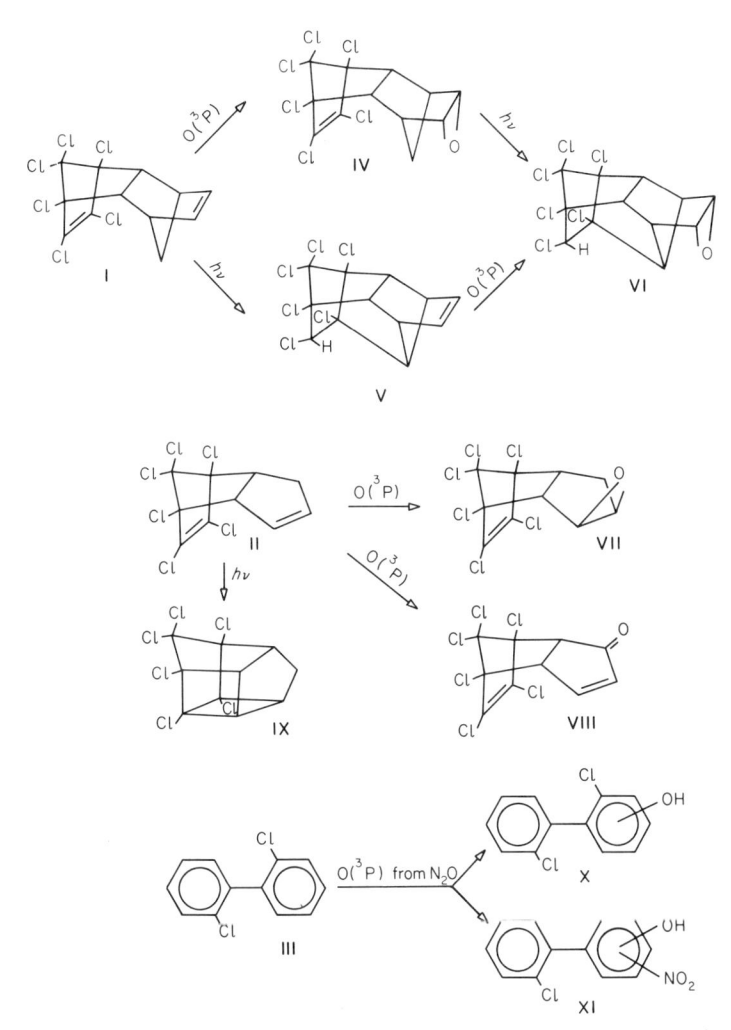

Figure 2.2 Reactions of aldrin (I), chlordene (II), and 2,2'-dichlorobiphenyl III with O(^{3}P). (Reprinted with permission from *Chemosphere*, **6**, 21, Saravanja–Bozanic, V., Gäb, S., Hustert, K., and Korte, F., Ecological chemistry, CXXXIII. Reactions of aldrin, chlordene and 2,2'-dichlorobiphenyl with O(^{3}P). © 1977, Pergamon Press, Ltd.)

There are two general reactions of $^1O_2$ with organic compounds:

(1) The oxygenation of olefins containing allylic hydrogen atoms resulting in the shift of the double bond and formation of an allylic hydroperoxide; this is analogous to the so-called 'ene' reaction.

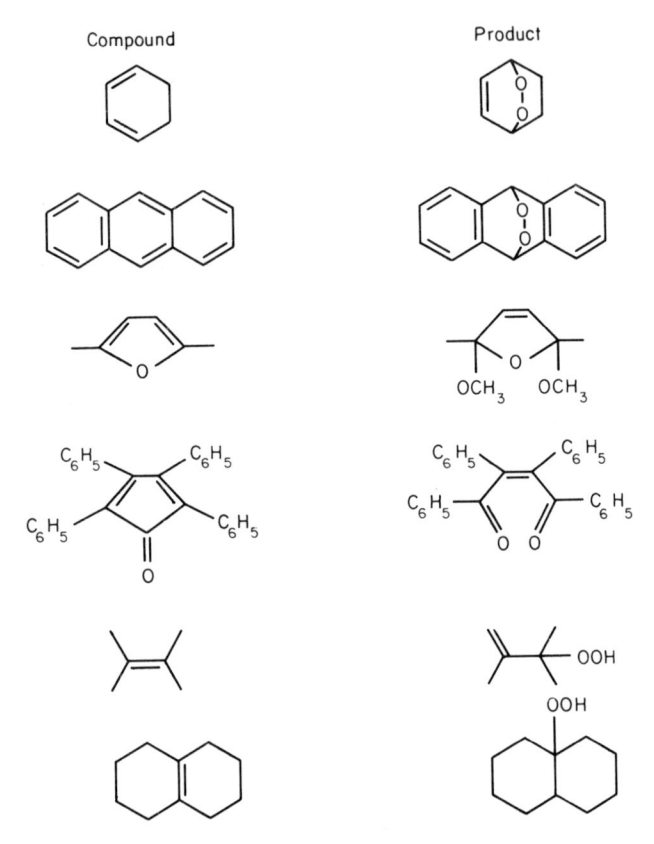

Figure 2.3 Typical acceptor molecules and oxygenated products for reaction with $^1O_2$. (Reproduced by permission of John Wiley and Son, Inc., from Pitts and Metcalf, 1969)

(2) The oxygenation of polycyclic aromatic hydrocarbons such as cyclopentadienes and heterocycles which give endoperoxides; this is analogous to the Diels—Alder reaction (Figure 2.3).

Singlet molecular oxygen may also be formed in several reactions involving neutral oxygen atoms, molecular oxygen, and ozone, as well as reactions of other simple inorganic molecules. Many of these are important upper-atmosphere processes and some may be significant in the lower atmosphere.

(v) Photoreactions of Cyclodienes

The photoreactions of cyclodienes can be divided into six classes of reaction:

(a) photoisomerization by $\pi\sigma \rightarrow 2\sigma$ reaction

(b) photoisomerization by $2\pi \to 2\sigma$ reaction

(c) photoreversible and irreversible hydrogen transfer reactions

(d) dechlorination reactions

(e) photochemically induced radical reactions

(f) dimerization reactions.

(a) Photoisomerization by $\pi\sigma \to 2\sigma$ Reactions

Cyclodiene insecticides contain a chlorinated double bond, which can be excited by light of wavelengths greater than 300 nm. By interaction with the methylene bridge in the non-chlorinated part of the molecule, the corresponding photoisomeric product can be formed by a $\pi\sigma \to 2\sigma$ reaction. The excited double bond abstracts the opposite H atom, whereby a new σ bond is formed (bridging) (Figure 2.4).

It has been determined from the irradiation of these compounds in D_6-acetone, that deuterium is not incorporated into the corresponding isomerization product. The reaction is thus intramolecular (Fischler and Korte, 1969).

(b) Photoisomerization by $2\pi \to 2\sigma$ Reaction

All cyclodiene derivatives possessing a double bond on the unchlorinated ring at the *endo* position react according to this mechanism (Figure 2.5).

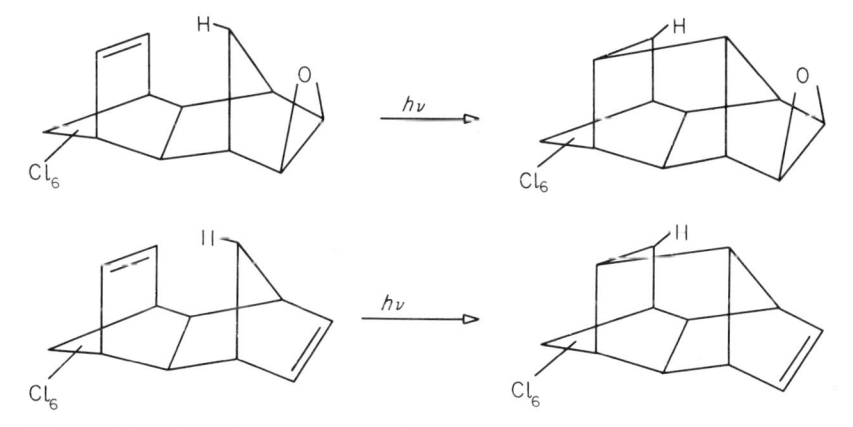

Figure 2.4 Photoisomerization of aldrin and dieldrin. When sensitized, these reactions proceed almost quantitatively. However they can also be observed unsensitized in the solid and gas phases. (Reprinted with permission from *Tetrahedron Letters*, 32, 2793–2796, Fischler, H. M., and Korte, F. Sensitized and unsensitized photoisomerization of cyclodiene insecticides. © 1969, Pergamon Press, Ltd.)

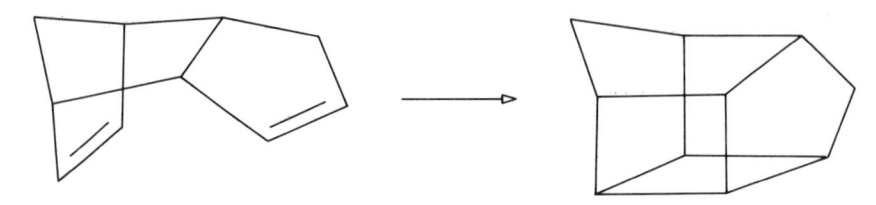

Figure 2.5 $2\pi \rightarrow 2\sigma$ reaction

(c) Hydrogen Transfer Reactions

Besides the $\pi\sigma \rightarrow 2\sigma$ reactions resulting from the irradiation of dihydrochlordene derivatives there are reversible and irreversible hydrogen shifts. This type of suprafacial shift to two hydrogen atoms is referred to as a synchronous reaction. In the case of an unconcerted process, the radical formed in the first step can either recombine to form a bridged alkane or transfer a second hydrogen atom to form another alkene. A reverse reaction forming the corresponding starting compound should also be expected (Figure 2.6).

It could be shown that, during sensitized irradiation of 4,5,6,7,8,8-hexachloro-2,3,3a,4,7,7a-hexahydro-4,7-methano-1*H*-indene-1,3-dicarboxylic acid dimethyl ester (**1**), intramolecular reversible hydrogen shifts are possible as well as the

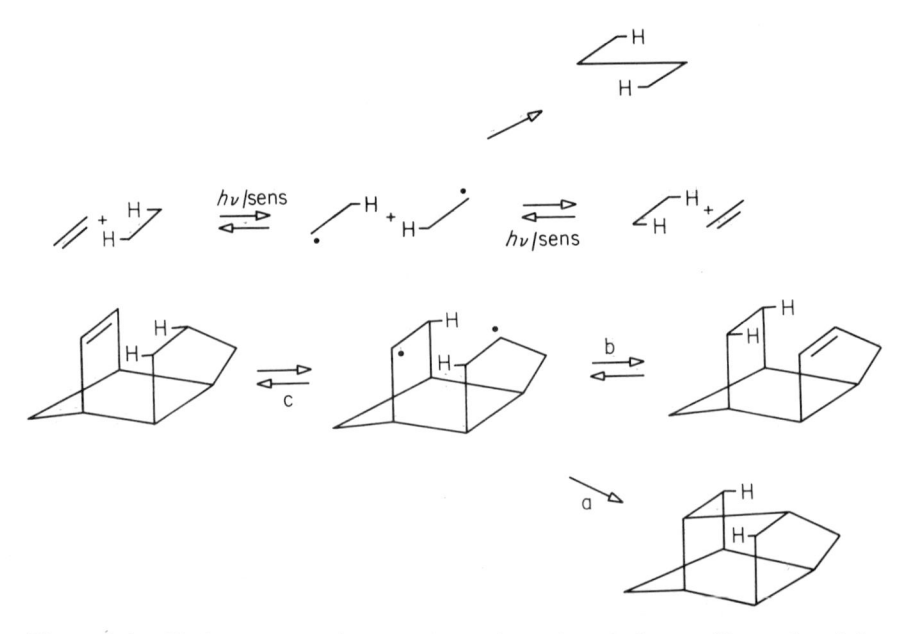

Figure 2.6 Hydrogen transfer reaction of methanoindenes. (Reproduced by permission of Verlag Chemie, Weinheim, Germany from Parlar *et al.*, 1975b).

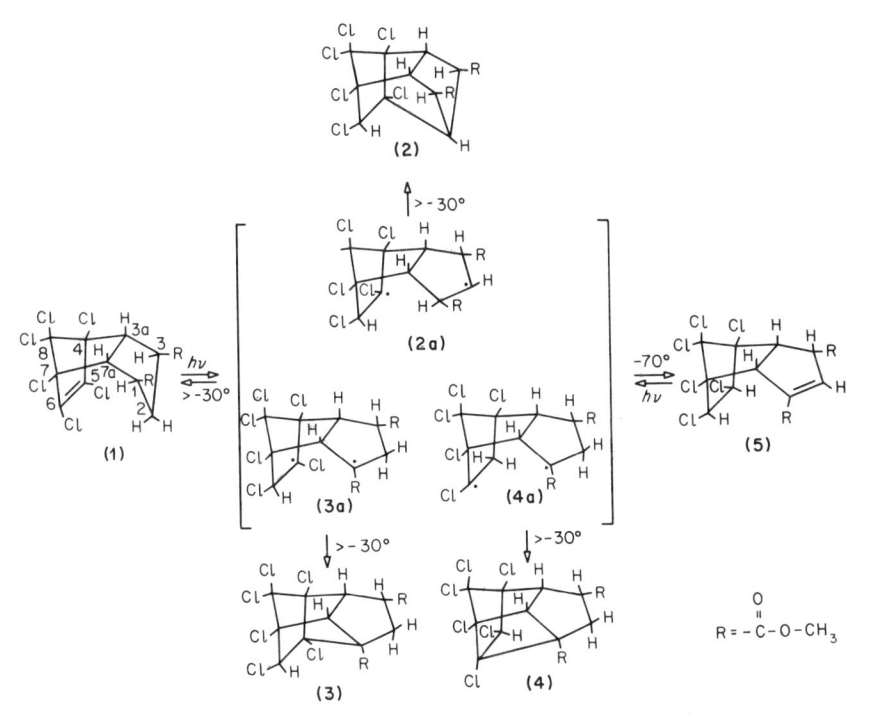

Figure 2.7 Photoreaction of 4,5,6,7,8,8-hexachloro-2,3,3a,4,7,7a-hexahydro-4,7-methano 1H-indene-1,3-dicarboxylic acid. (Reproduced by permission of Verlag Chemie, Weinheim, Germany from Parlar et al., 1975b)

$\pi\sigma \to 2\sigma$ reactions typical for this class of substances. Irradiation of (1) at low temperatures showed that below $-30°C$ (5) is the only photoproduct. The photoisomerization products (2), (3) and (4), which arise by a $\pi\sigma \to 2\sigma$ reaction only above $-30°C$ suggest that the reaction proceeds through the biradical intermediates (2a), (3a), and (4a). The parent compound (1) and the photoisomerization products (2), (3) and (4) are obtained by irradiation of (5) (Parlar et al., 1975b) (Figure 2.7).

(d) Dechlorination Reactions

In protonated solvents as well as in the solid phase, cyclodienes are photochemically dechlorinated at the chlorinated double bond. It may be assumed that the dechlorination reactions proceed from the singlet state of the molecule. In contrast to the above described photoisomerization reactions, in this case an intermolecular reaction also occurs whereby, after the photolysis of the C—Cl compound, the abstraction of a nearby H· (from solvent) occurs (Parlar and Korte, 1973) (Figure 2.8).

Figure 2.8 Photodechlorination reaction of chlorinated methanoindene

(e) Photochemically Induced Radical Reactions

It is evident from the formation of higher chlorinated compounds that a Cl radical is formed, which can chlorinate the starting compound. Therefore, heptachlor and isoheptachlor are formed from the irradiation of chlordene (Parlar and Korte, 1972). The formation of α-, β-, and γ-chlordene probably results from the attack on chlordene by a chlorine radical at the unchlorinated double bond (Gäb *et al.*, 1975) (Figure 2.9).

(f) Dimerization Reactions

Chlordene derivatives with the structural requirement to undergo a $2\pi \rightarrow 2\sigma$ reaction react at wavelengths above 280 nm in concentrated solutions to give dimers. Experiments have shown that this reaction requires a lower activation energy than the intramolecular $2\pi \rightarrow 2\sigma$ reaction (Parlar and Korte, 1972).

(vi) Photoreactions of Some Chlorinated Compounds

(a) DDT

It was found that DDT in the vapour phase in the presence of a large excess of air is converted very slowly to 1,1-dichloro-2,2-bis(*p*-chlorophenyl)ethane (DDD) and about 15 times more rapidly to 1,1-dichloro-2,2-bis(*p*-chlorophenyl)ethylene (DDE) by ultraviolet light with wavelength 290–310 nm. DDD is stable to further

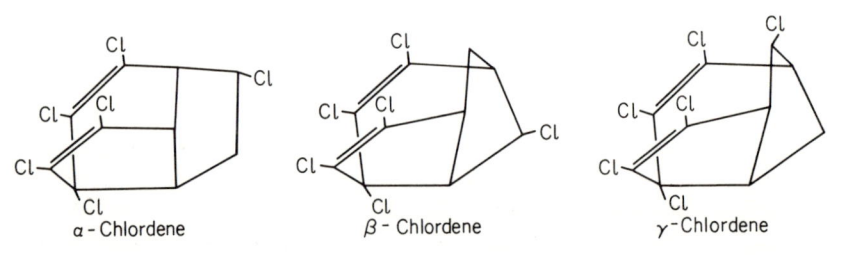

Figure 2.9 Isomers of chlordene

irradiation, but additional breakdown products are derived from DDE by the action of the ultraviolet radiation. In this case, 4,4'-dichlorobenzophenone (DDCO) and the intermediate 1-chloro-2,2-bis(*p*-chlorophenyl)ethylene (DDMU) are formed. DDCO is relatively stable to ultraviolet light. It has been shown, however, that it is transformed at a finite rate into 4,4'-dichlorobiphenyl (Maugh, 1973). Because all biphenyls are resistant to further irradiation in the vapour phase, one could expect that they should accumulate in the biosphere. This may in part explain the wide distribution of the PCBs which has been demonstrated in isolated areas far from the regions of original application.

(b) PCBs (Hustert and Korte, 1972, 1974)

The abiotic degradation of polychlorinated biphenyls was investigated using ultraviolet light of various wavelengths. In such experiments with 2,2'-, 4,4'-, 6,6'-hexachlorobiphenyl in non-polar solvents, a step-wise displacement of chlorine atoms, with the final production of unsubstituted biphenyl, could be demonstrated. Polymerizations and isomerizations are possible side reactions. In polar solvents, oxygen-containing compounds are also formed, for example, hydroxylated products. It is possible that the extremely toxic polychlorodibenzofurans, which have been detected as impurities in various industrial products, are also formed. Irradiation in the solid or liquid phase without solvent produces higher chlorinated products such as hepta- or octachlorobiphenyl.

(c) Dieldrin

During irradiation of dieldrin in solution, adsorbed or in the gas phase, photodieldrin is produced. This substance was chosen for photolysis studies in order to determine the conditions that cause a complete decomposition (mineralization) because it has up to now been considered to be one of the most persistent chemicals under normal atmospheric conditions.

During irradiation of adsorbed photodieldrin with light of wavelengths greater than 300 nm, photoaldrin-chlorhydrin and both photoaldrin ketone isomers could be isolated in pure condition. Besides these compounds, a mixture of substances, which could be separated by gas chromatography, was isolated in small yield. It was found that during chromatography compounds of lower chlorine content and low molecular weight developed. Parallel to these fragmentation products, however, higher molecular products may also be formed (Gäb *et al.*, 1974a).

Irradiation of photodieldrin adsorbed on silica gel with wavelengths below 300 nm showed that more than 70% of the applied photodieldrin was changed under such conditions. Besides photoaldrin-chlorhydrin, the same products were identified as those which develop during adsorbed irradiation. These irradiations show that photodieldrin loses HCl in a primary step which is responsible for the opening of the epoxide ring.

Table 2.4 Photodegradation of Photodieldrin. (Reprinted with permission from *Chemosphere,* 3(5), Gäb, S., Parlar, H., and Korte, F., Ecological chemistry. LXXXII. Ultraviolet-irradiation reactions of photodieldrin as a solid on glass and adsorbed to silica gel. © 1974, Pergamon Press, Ltd.)

Compound	Yield (mg)	Mass number	Number of Cl Atoms
Photodieldrin	130	378	6
Photoaldrinchlorhydrin	140	414	7
Photoaldrin ketone I	45	378	6
Photoaldrin ketone II	25	378	6
I*	10	432	8
II	5	398	7
III	5	358	6
IV	2	352	6
V	5	344	5
VI	5	310	4
VII	5	324	4
VIII	2	294	3
IX	2	270	3

*Compounds I–IX are photoproducts which could not be identified by spectroscopic methods.

Irradiation of adsorbed photodieldrin in an oxygen stream with wavelengths below 300 nm revealed up to 95% degradation of photodieldrin to CO_2 and HCl.

Tests in the gas phase, however, revealed that isomeric photodieldrin is produced in large yield. This can be explained by a much longer duration of the electronically stimulated state of the dieldrin molecule in the gas phase. These conditions produce a larger yield from the bridging reaction which is favoured in comparison to the bimolecular reactions with oxygen species. Tests with pure ozone produced a hydroxy product in smaller quantities. Irradiation of adsorbed dieldrin in an oxygen stream at wavelengths below 300 nm revealed that dieldrin is almost quantitatively degraded to CO_2 and HCl. Small amounts of photodieldrin were found as well.

(d) Various Chlorinated Aromatics

The photochemical behaviour of hexachlorobenzene, pentachlorobenzene, pentachlorophenol, 1,1,1-trichloro-2,2-bis(*p*-chlorophenyl)ethane (DDT), 1,1-dichloro-2,2-bis(*p*-chlorophenyl)ethylene (DDE), 2,4,5,2′,4′,5′-hexachlorobiphenyl and 2,5,2′,5′-tetrachlorobiphenyl was investigated in the presence of a large excess of oxygen exposed to u.v. light with wavelengths longer than 230 nm (quartz glass) as well as with wavelengths longer than 290 nm ('Pyrex' glass). It was observed that the conversion rates of the substances adsorbed on particulate matter were far

Table 2.5 Results of Ultraviolet Irradiation of Aromatic Xenobiotics as Solids in an Oxygen Stream

Compound	Irradiation conditions	Initial solid material (mg)	Recovered solid material (mg)	Mineralization products CO_2	HCl	Cl_2
2,4,5,2',4',5'- Hexachlorobiphenyl	Quartz (7 days)	73	36	53 mg	19 mg	n.d.
	'Pyrex' (7 days)	70	68	n.d.	n.d.	n.d.
2,5,2',5'- Tetrachlorobiphenyl	'Pyrex' (7 days)	101	99	n.d.	n.d.	n.d.
Hexachlorobenzene	Quartz (2 days)	63	23	46 mg	13 mg	13 mg
	'Pyrex' (7 days)	82	81	n.d.	n.d.	n.d.
Pentachlorobenzene	'Pyrex' (7 days)	64	62	n.d.	n.d.	n.d.
Pentachlorophenol	'Pyrex' (7 days)	80	69	13 mg	6 mg	n.d.
DDT	'Pyrex' (7 days)	54	89	12 mg	2 mg	n.d.
DDE	'Pyrex' (7 days)	98	85	10 mg	8 mg	n.d.

n.d. = not detected.

higher than with those deposited as solids or thin films on a glass surface (Gäb *et al.*, 1974b).

These differences can be attributed to the bathochromic shift, changes in the relative extinction, or appearance of new absorption bands as a consequence of the adsorption on silica gel, as well as to the greater dispersion of pesticide molecules in the adsorbed phase resulting in a higher pesticide-oxygen contact. The disappearance of the applied substances can be explained neither by their vapour pressures nor by formation of photoproducts. Since CO_2 and HCl are formed during 'Pyrex' irradiation of pentachlorophenol, DDT and DDE as solids on glass and the u.v. absorption bands of these compounds are located in nearly the same region as those substances adsorbed on silica gel, it is evident that in this case mineralization products are evolved as well.

Ultraviolet irradiation of chlorinated olefins (e.g. vinyl chloride, dichloroethylene, trichloroethylene and tetrachloroethylene) in the presence of excess oxygen also results in products of low molecular weight such as CO_2, $COCl_2$, and HCl. The primary step of this photooxidation process is the formation of the respective epoxides. Irradiation of ethyl acetate and ethanol in the presence of NO_2, SO_2, and H_2O in concentrations which can be found in polluted areas resulted in a degradation of these compounds under simulated tropospheric conditions.

(vii) Evaluation of the Reactions of Organic Chemicals

(a) Significance of Dechlorination Reactions

Approximately 40% of insecticides contain chlorine, which is in most cases responsible for the efficiency and toxicity of the compound. A biological transformation or u.v. irradiation in the atmosphere can dechlorinate an environmental chemical, which often results in a loss of its toxicity.

Experiments have shown that a pesticide remains for only a relatively short time after application in a treated area and is quickly dispersed into the atmosphere. An understanding of the following factors is necessary to predict the rate of dechlorination in the atmosphere:

(a) The amount produced, the area treated, and the dispersion of environmental chemicals.

(b) Intensity of irradiation from the sun depending on the height in the atmosphere.

(c) Reactivity and reaction pathways after adsorption on surfaces (liquid and solid aerosols).

(d) Kinetics and quantum yields of the photochemical reactions of chemicals under simulated atmospheric conditions.

Table 2.6 Results of Ultraviolet Irradiation of Aromatic Xenobiotics Adsorbed on Silica Gel

Compound	Initial quantity adsorbed[a] (mg)	Quantity remaining adsorbed after 4 days' irradiation[b] (mg)	Quantity remaining adsorbed after 7 days' irradiation[c] (mg)	Photoproducts detected after 7 days' irradiation[b,c]
Pentachlorophenol	102 (100%)	24 (25%)	12 (12%)	None
DDT	185 (100%)	298 (77%)	255 (66%)	None
DDE	162 (100%)	91 (25%)	69 (19%)	Dichlorobenzophenone 38 mg; trichlorobenzophenone 7 mg

[a] In each case the substance was adsorbed on 100 g silica gel.
[b] Quantitation was by means of gas chromatography (BDC) and comparison with standard solutions of known concentration as well as by measurements of the u.v. extinction.
[c] The possibility that more polar compounds arise, which are not eluted under these conditions, and consequently are not detected, cannot be excluded.

Reported results provide sufficient information about the above-mentioned points, although comparative investigations with other classes of substance are still lacking.

(b) Significance of Photoisomerization Reactions

Photoisomerizations are specific reactions which have a lower activation energy in comparison to other photoreactions. For this reason it is important to investigate the toxic and ecotoxicological properties of the photoisomers formed as primary products under atmospheric conditions. All photoisomers of the cyclodienes are more toxic than the starting compounds. The extent of damage to the environment caused by these stable photoproducts, as well as their direct side effects, can be estimated by determining the lifetime under atmospheric conditions of the corresponding chemical.

(c) Significance of Photooxidation Reactions

The rearrangement of chemicals under atmospheric conditions is strongly dependent on the type of oxidants present during the reaction. For example, the oxygen molecule possesses two unpaired electrons in the ground state and is in the triplet state ($^3O_2$) and paramagnetic. Two electron configurations are possible for the excited singlet state of oxygen, the so-called $^1\Delta g$ and $^1\Sigma g^+$ configurations, although their formation through the absorption of irradiation from the sun is not significant. In the case of widening of the lines of O_2 due to collisions, the excited singlet $O_2(^1\Delta g)$ can be formed in such concentrations and be so long-lived as to react (favourably) with olefins and aromatic compounds in the atmosphere.

Oxygen molecules can also be converted into oxygen atoms by u.v. light of shorter wavelength. This process is found in the higher layers of the atmosphere (outside the troposphere).

Oxygen atoms in atmospheric layers near ground level are formed from the photodissociation of nitrogen dioxide in concentrations estimated at approximately 10^{-8} ppm. Ozone also plays a special role here. It is therefore interesting in an ecological sense to study the reactions of environmental pollutants in the presence of O_2, O_3, NO_2, O, and $O_2(^1\Delta g)$, which in many cases can be carried out under simulated conditions. These studies furnish necessary data for estimating the quality of the environment and information about the primary photoreactions leading to the buildup of smog.

(d) Significance of Photomineralization Processes

It has been shown that the irradiation of several chemically stable compounds (aldrin, dieldrin, chlordene, photodieldrin, hexa- and tetrachlorobiphenyl, hexachlorobenzene, pentachlorobenzene, DDT, and DDE) in their solid states and at wavelengths smaller than 290 nm leads to a mineralization of those compounds into CO_2 and HCl. In the case of aldrin, dieldrin, DDT, DDE, and pentachlorophenol, photomineralization is also achieved at wavelengths greater than 290 nm. Similar results are found for the irradiation of these compounds when absorbed on silica gel. It can be concluded from the experiments that the u.v. irradiation from the sun possesses sufficient energy to decompose these stable compounds. A systematic study of other specific classes of compounds is needed to determine whether those reactions also apply to other environmental chemicals.

(viii) Laboratory Models for Testing Abiotic Degradability

In order to investigate the reactions of environmental chemicals under natural conditions, it is necessary to develop experimental conditions which correspond to atmospheric conditions. There are generally five types of reactions:

(1) photochemical reactions of environmental chemicals in solution;

(2) photoreaction of environmental chemicals on solids;

(3) photoreactions of environmental chemicals in the gas phase;

(4) photoreactions of environmental chemicals in the gas phase with other gases (e.g. NO_2, SO_2, O_3, O);

(5) photocatalysed reaction of environmental chemicals in the gas phase at the surface of solid or liquid phases (this type of reaction simulates the reaction of chemicals on the surface of dust particles and on solid or liquid aerosols found in the atmosphere).

Standardized values have been found from these experiments which can be used as parameters in the evaluation of environmental chemicals.

2.3. REFERENCES

Ackefors, R., Löfroth, G., and Rosen, G. G., 1970. A survey of the mercury pollution problem in Sweden with special reference to fish. *Oceanog. Mar. Biol. Ann. Rev*, 8, 203–24.

Anonymous, 1970. U.S. Dept. of Interior, 108.

Barthel, W. F., Hawthorne, J. C., Ford, J. H., Bolton, G. C., McDowell, L. L., Grissinger, E. H., and Parsons, D. A., 1969. Pesticides in water. Pesticide residues in sediments of the lower Mississippi River and its tributaries. *Pestic. Monit. J.*, 3, 8–66.

Bolin, B. and Bischof, W., 1970. Variations of the carbon dioxide content of the atmosphere in the northern hemisphere. *Tellus*, 22(4), 431–2.

Bolin, B. and Keeling, C. D., 1963. Large-scale atmospheric mixing as deduced from the seasonal and meridional variations of carbon dioxide. *J. Geophys. Res.*, 68, 3899–920.

Caro, H. J., Taylor, A. W., and Lemon, E. R., 1971. Measurement of pesticide concentrations in the air overlying a treated field. *Proc. Internat. Symp. on Identification and Measurement of Environmental Pollutants*, Ottawa, June 14–17.

Edwards, C. A., 1966. Insecticide residues in soils. *Residue Rev.*, **13**, 83–132.

Fischler, H. M. and Korte, F., 1969. Sensitized and unsensitized photoisomerization of cyclodiene insecticides. *Tetrahedron Letters*, **32**, 2793–6.

Francis, C. W., Chesters, G., and Haskin, L. A., 1970. Determination of lead-210 mean residence time in the atmosphere. *Environ. Sci. Technol.*, **4(7)**, 586–9.

Gäb, S., Cochrane, W. P., Parlar, H., and Korte, F., 1975. Photochemical reactions of chlordene isomers of technical chlordane. *Z. Naturforsch.*, **30B (3–4)**, 239–44.

Gäb, S., Parlar, H. and Korte, F., 1974a. Ecological chemistry. LXXXII. Ultraviolet-irradiation reactions of photodieldrin as a solid on glass and adsorbed to silica gel. *Chemosphere*, **3(5)**, 187–92.

Gäb, S., Parlar, H., Nitz, S., Hustert, K., and Korte, F., 1974b. Ecological chemistry. LXXXI. Photochemical degradation of aldrin, dieldrin, and photodieldrin as solids in a current of oxygen. *Chemosphere*, **3(5)**, 183–6.

Haagen-Smit, A. J., 1964. The control of air pollution. *Scientific American*, **210**, 25–31.

Hasselrot, T. B., 1968. Report on current field investigations concerning the mercury content in fish, bottom sediments, and water. *Sweden Institute of Freshwater Research Reports*, **48**, 101–11.

Hustert, K. and Korte, F., 1972. Ecological Chemistry. XXXVIII. Synthesis of polychlorinated biphenyls and their reactions under UV-irradiation. *Chemosphere*, **1(1)**, 7–10.

Hustert, K. and Korte, F., 1974. Ecological chemistry. LXXVIII. Reactions of polychlorinated biphenyls during UV-irradiation. *Chemosphere*, **3(4)**, 153–6.

Jakobi, W. and Andre, K., 1963. The vertical distribution of radon, thoron, and their decay products in the atmosphere. *J. Geophys. Res.*, **68(13)**, 3799–814.

Junge, C. E., 1962. Global ozone budget and exchange between stratosphere and troposphere. *Tellus*, **14**, 363–77.

Kroenig, J. L. and Ney, E. P., 1962. Atmospheric ozone. *J. Geophys. Res.*, **67**, 1867–75.

Lal, D. and Rama, D., 1966. Characteristics of global tropospheric mixing based on man-made ^{14}C, ^3H and ^{90}Sr. *J. Geophys. Res.*, **71(12)**, 2865–74.

Lettau, H., 1951. *Compendium of Meterology* (Ed. T. F. Malone), Am. Meteorological Society, Boston, p. 320.

Levy, H., 1974. Photochemistry of the troposphere. Smithsonian Institution Astrophysical Observatory, Cambridge, Massachusetts.

Lichtenstein, E. P. and Schultz, K. R., 1965. Residues of aldrin and heptachlor in soils and their translocation into various crops. *J. Agr. Food Chem.*, **13**, 57–63.

Maugh, T. H., 1973. DDT: an unrecognized source of polychlorinated biphenyls. *Science*, **180**, 578–9.

Moza, P., Weisgerber, I., and Klein, W., 1972. Leaching of water-soluble carbon-14-labelled decomposition products of aldrin from soils. *Chemosphere*, **1(5)**, 191–5.

Nydal, R., 1968. Transfer of radiocarbon in nature. *J. Geophys. Res.*, **73(12)**, 3617–35.

Parlar, H., Gäb, S., Lahaniatis, E. S., and Korte, F., 1975a. Ecological chemistry. XCIII. Synthesis and analytical behaviour of 1-*endo*-hydroxychlordene and 1-*exo*-hydroxychlordene. *Chemosphere*, **4(1)**, 15–20.

Parlar, H., Gäb. S., Lahaniatis, E. S., and Korte, F., 1975b. The photoreversible hydrogen shift: a competitive reaction for the $\pi\sigma \to 2\sigma$ isomerization of bridged chlorinated hydrocarbons. *Chem. Ber.*, **108**(12), 3692–9.

Parlar, H. and Korte, F., 1972. Ecological chemistry. XLIV. Reaction of UV-irradiated chlordene in solution and in gas phase. *Chemosphere*, **1**(3), 125–8.

Parlar, H. and Korte, F., 1973. Ecological chemistry. LX. Photochemistry of chlordane derivatives. *Chemosphere*, **2**(4), 169–72.

Pitts, J. N. Jr., 1969. Photochemical air pollution: singlet molecular oxygen as environmental oxidant. *Advan. Environ. Sci.*, **1**, 289–337.

Pitts, J. N. Jr. and Finlayson, B. J., 1975. Mechanisms of photochemical air pollution. *Angew. Chem.*, **87**, 18–33.

Pitts, J. N. Jr. and Metcalf, R. L. (Eds.), 1969. *Advances in Environmental Sciences and Technology*, Vol. 1, Wiley–Interscience.

Pressman, J. and Warneck, P., 1970. Stratosphere as a chemical sink for carbon monoxide. *J. Atmos. Sci.*, **27**(1), 155–63.

Reiter, E. R., Glasser, M. E., and Mahlmann, J. D., 1967. Atmospheric Science Paper, No. 107, Colorado State University, Fort Collins.

Saito, N., 1967. *Levels of Mercury in Environmental Materials*, Expert Meeting on Mercury Contamination in Man and his Environment, IAEA Tech. Rept. Ser., No. 137.

Saravanja-Bozanic, V., Gäb, S., Hustert, K., and Korte, F., 1977. Ecological chemistry. CXXXIII. Reactions of aldrin, chlordene and 2,2′-dichlorobiphenyl with $O(^3P)$. *Chemosphere*, **6**, 21–6.

Smith, M. R., 1968. A preview on the determination of mass return flow of air and water vapor into the stratosphere using tritium as a tracer. *Tellus*, **20**, 76–81.

Steller, W. A., Klotsas, K., Kuchar, E. J., and Norris, M. V., 1960. Colorimetric estimation of dodecylguanidine acetate residues. *J. Agr. Food Chem.*, **8**, 460–4.

CHAPTER 3

Biotic Processes

W. KLEIN AND I. SCHEUNERT

Institut für Okologische Chemie Gesellschaft für Strahlen — und Unweltforschung mbH München, FRG

3.1 GENERAL PRINCIPLES OF STRUCTURAL CHANGES OF ENVIRONMENTAL
    CHEMICALS 37
3.2 ORGANOTROPISM 39
    (i) Time-course studies 40
    (ii) Effects of chemical structure on distribution between organs 40
    (iii) Differences between animal species 41
    (iv) Sex differences of organ distribution 43
3.3 METABOLIC BALANCE OF ENVIRONMENTAL CHEMICALS 43
    (i) *In vitro* 43
    (ii) *In vivo* 44
3.4 LABORATORY SCREENING TESTS 55
    (i) Tests of biodegradation 55
    (ii) Tests of bioconcentration 56
    (iii) Significance of screening tests 56
3.5 FOOD CHAINS 57
    (i) Environmental data 57
    (ii) Laboratory test models 60
3.6 PREDICTABILITY OF THE BEHAVIOUR OF XENOBIOTICS FROM
    STRUCTURAL CHARACTERISTICS 61
    (i) Influence of chlorine content 61
    (ii) Influence of substituents (epoxy groups) 63
    (iii) Influence of conformation 63
3.7 CONCLUSION 64
3.8 REFERENCES 64

3.1. GENERAL PRINCIPLES OF STRUCTURAL CHANGES OF ENVIRONMENTAL CHEMICALS

In evaluating the enzymatic attack of organisms on environmental organic chemicals, the following situations should be considered: the chemical is easily digested and mineralized without persistent or biologically active intermediates, or it is degraded to low molecular weight fragments circulating in the carbon pool, or it is chemically altered by co-metabolism, without significant breakdown to products joining the natural carbon pool. This chapter is concerned mainly with the last situation since, besides the effects of the parent compounds on organisms and ecosystems, the potential effects of the conversion products have also to be considered.

Evaluation is required for metabolic pathways as well as for the amounts and persistence of the metabolites formed. In general, the principles of structural changes in xenobiotics by enzymes are the same as those of natural compounds.

As examples for enzymatic transformations of inorganic compounds, the metals mercury, lead, and tin are discussed. The transformation of inorganic mercury to methylmercury in biological and related systems (Wood et al., 1968; Jensen and Jernelov, 1969; Neujahr and Bertilsson, 1971; Imura et al., 1971; Landner, 1971) including man (Edwards and McBride, 1975) is well documented. The formation of the volatile dimethylmercury (Wood et al., 1968; Jensen and Jernelov, 1969; Imura et al., 1971), the formation of diphenylmercury from phenylmercury by several soil and aquatic microorganisms (Matsumura et al., 1971), the decomposition of methyl-, ethyl-, and phenylmercury to elemental mercury and the corresponding hydrocarbon by mercury-resistant pseudomonas bacteria (Furukawa et al., 1969), and the vaporization of mercury by activated sludge when treated with mercuric chloride and phenylmercury (Yamada et al., 1969) have all been reported (Fishbein, 1974). For methanogenic bacteria, it was shown that the methylation proceeds via the transfer of a methyl group from a Co^{3+} atom bound in a complex organic molecule, methylcobalamine (a methylated form of vitamin B12; Wood et al., 1968). The methylation of lead has also been observed (Wong et al., 1975). The formation of monomethylmercury is accompanied by an increase of toxicity ('activation'). Furthermore, this derivative may enter biological cycles (Gavis and Ferguson, 1972; Wood, 1974) and may be accumulated in food chains (see also Chapter 5). For lead (Zeman et al., 1951; Cremer, 1959) and tin (Cremer, 1957, 1958), the conversion of the tetraethyl to the triethyl metal has been shown, and also results in an increase of toxicity.

For organic chemicals, primary changes, mainly oxidative, hydrolytic, or reductive, may be accompanied by an increase in toxicity ('activation') or by a decrease ('detoxication') (Klein and Korte, 1970). They are often followed by secondary changes of the primary conversion products, e.g. by alkylation, acetylation, conjugation, or binding with biological molecules. In animals, the secondary processes are frequently accompanied by detoxication through conjugation which results in water-soluble molecules easily eliminated from the body. In plants, detoxication is achieved by fixation of the xenobiotic substances within natural macromolecular structures like cell wall components (Kaufman et al., 1976). This results in the so-called 'unextractable residues', i.e. the incorporated xenobiotic cannot be extracted from the tissue without destruction of the macromolecular bonds. In soil, even more complex reactions occur; the xenobiotic may be involved in humus formation and replace natural constituents in the humic acid macromolecule (Hsu and Bartha, 1976). Oxidative processes are the commonest enzymatic changes of xenobiotics. Mixed-function oxidases are the enzymes generally involved (Mason, 1957; Brooks, 1972). Hydroxylation of aliphatics and aromatics, epoxidation and oxidative cleavage of double bonds, oxidation of phosphorothionates to phosphates, of hydroxyl to keto groups, of thioethers to sulphoxides and sulphones, dehydrogenations, etc., are well known. Some of these processes, like the epoxidation of cyclodienes, the oxidation of phosphorothionates to phosphates, or the hydroxylation of polychlorinated biphenyls (Yamamoto and Yoshimura, 1973), result in a biological activation. Plant tissues are rich in peroxidase, and the abundance of phenolic products in higher plants suggests generally high oxidase activity (Brooks, 1972), with the result that the occurrence of metabolites with increased toxicity in plants must be taken into account; on the other hand, oxidative processes may lead to a stepwise degradation of the foreign compound. For instance, for cyclodienes (aldrin, isodrin, heptachlor), epoxidation or hydroxylation of the double bond is followed by ring cleavage and by loss of carbon atoms upon decarboxylation (see also paragraph 3.3. iib). Certain microorganisms achieve a ring cleavage of chlorinated aromatics (Furukawa and Matsumura, 1976).

Hydrolytic conversions, e.g. by esterases and amidases which are widely distributed in nature, are also well known, as, for example, organophosphorus compounds, carbamates, aliphatic esters (e.g. 'kelevan'; Sandrock et al., 1974), cyclodiene epoxides, or urea derivatives. The recently discovered degradation of phenylurea herbicides (monolinuron, buturon) to carbamates in plant cultures could be due to the enzyme urease. Since, in hydroponic plant experiments, this conversion was observed to be higher in nitrogen-deficient systems than in normal nutrient media, there arises the question whether plants or microorganisms are able to consume the side-chain nitrogen of the herbicides and excrete the remaining moiety of the molecule as carbamate into the nutrient medium (Haque et al., 1977a).

Reductive conversions are known for xenobiotics containing nitro groups (e.g. the fungicide pentachloronitrobenzene) which are reduced to amines, containing aldehydes and ketones, and containing chloro groups (reductive dechlorination, e.g. DDT → DDD).

The secondary processes in animals, especially conjugations with sulphuric acid, glucuronic acid, or glutathione, which lead to the excretion of the foreign compound, may be regarded as a pathway to elimination for the individual organism but not for the ecosystem of which it is a member, since the xenobiotic continues to exist there and may be taken up by other members of the system. This applies also to the 'unextractable residues' in plants, but very little is known of their biological availability, i.e. whether they are taken up by animals. Even when the xenobiotic is fixed in the soil as a constituent of the polymerizate of humic acids, the possibility that it may become available again for plant uptake cannot be excluded, since the humic acids are constantly undergoing biosynthesis and degradation. Further work on this topic is urgently needed (see also 3.3. iie).

3.2. ORGANOTROPISM

The uptake of a chemical and its distribution within the animal body, its accumulation, remobilization, and excretion are strongly dependent on chemical and biological factors, one of which is the structure of the chemical. Since chemical alterations may occur by metabolism, not only the organotropism of the parent compound, but also that of the metabolites which often have different physical-chemical properties, must be considered.

(i) Time-course Studies

In order to investigate the time-course of organ distribution, three ^{14}C-labelled model compounds are discussed, 2,2'-dichlorobiphenyl was taken as a lower chlorinated constituent of the commercial PCB mixtures. Dieldrin was chosen as a representative of the cyclodiene group, and hexachlorobenzene as a representative of the chlorinated benzenes and as one of the chemicals most persistent in the environment. The experiments were conducted by oral administration to rats (Iatropoulos et al., 1975).

2,2'-Dichlorobiphenyl is rapidly absorbed from the upper gastrointestinal tract and is transported to the liver mainly by the portal venous system with some participation of lymphatic transport. In the liver it is metabolized rapidly (Milling et al., 1975) and excreted into the intestinal tract without ever reaching the storage depot of adipose tissue.

Dieldrin, like dichlorobiphenyl, is absorbed rather rapidly from the intestinal tract and is transported to the liver mainly through the portal venous system, reaching a peak concentration within the first hour. As the metabolic conversion of dieldrin is much slower than that of dichlorobiphenyl, only part of it can be metabolized and excreted, the major amount being redistributed into the storage depot of adipose tissue. During this redistribution process, the lymphatic system seems to be a major transport pathway; the parallel increases of the contents of lymph nodes and adipose tissue indicate an equilibrium between lymph and depot fat.

Hexachlorobenzene is absorbed more slowly than dichlorobiphenyl or dieldrin; the portal venous transport to the liver seems to be a minor pathway because in spite of its extremely slow metabolic conversion (Rozman et al., 1975), hexachlorobenzene never builds up to high concentrations in the liver. The major part of the ingested hexachlorobenzene is absorbed by the lymphatic system in the region of the duodenum and jejuno-ileum and deposited in the fat, bypassing the systemic circulation and the excretory organs. As with dieldrin, the comparison of lymph nodes and adipose tissue contents indicates an equilibrium between lymph and fat.

(ii) Effects of Chemical Structure on Distribution Between Organs

As examples of the influence of chemical structure on the distribution of residues between organs, the results of the following studies with six radiolabelled chemicals are presented. The chemicals were given daily orally to rats in long-term feeding experiments. 2,2'-Dichlorobiphenyl, 2,5,4'-trichlorobiphenyl, and

2,4,6,2',4'-pentachlorobiphenyl, three PCB isomers, were chosen to study the influence of chlorine content on organ distribution. Chloroalkylene-9, an isopropylated 2,4-dichlorobiphenyl which was developed as a PCB substitute, was included to observe possible additional effects of the side-chain. Endrin was selected as a representative of the cyclodiene insecticide group, and 'Imugan', a dichloroaniline-derived fungicide, as a so-called 'non-persistent' chemical.

The daily dose was about $1-2\ \mu g/g$ in the diet, corresponding to $32-75\ \mu g/$animal. The application was carried out until a plateau level of accumulation was reached in the body, i.e. until the daily administered dose was daily excreted. Tissue analysis was performed by counting of total radiocarbon 1–3 days after the application was discontinued (exception: in the experiment with 2,4,6,2',4'-pentachlorobiphenyl, the feeding was discontinued before a plateau level was reached since the formation of a plateau would have taken a very long time). Table 3.1 shows the results of tissue analysis (Klein *et al.*, 1968; Lay *et al.*, 1975; Begum *et al.*, 1975; Kamal *et al.*, 1976b; Lay, 1976; Viswanathan *et al.*, 1976).

The table indicates the variation of tissue concentration with chemical structure. All substances applied are lipophilic and, therefore, should be more concentrated in the fatty tissues. However, this fact was observed only for the medium or higher chlorinated biphenyls and for chloroalkylene (columns 1–3). These three substances show, besides their tendency to be accumulated in the abdominal and subcutaneous fat, also generally a higher level in the other organs than the substances in columns 5 and 6. These results are a consequence of increased chemical stability which prevents catabolism to more hydrophilic products and results in an increased general body concentration. First of all, 2,4,6,2',4'-pentachlorobiphenyl which is one of the less degradable PCBs (Schulte and Acker, 1974), has the highest tissue concentrations although the experiment was discontinued before the saturation level was reached. The radioactivity in the organs was found to be the unchanged parent compound (exception: liver and kidneys). Additionally, the low excretion rate as well as the relatively low percentage of metabolites in the excreta (Lay *et al.*, 1975) demonstrate that this substance is only slowly metabolized.

Endrin (column 4) is susceptible to metabolic attack by hydroxylation which results in metabolites of increased water solubility. Consequently, the highest residues are not in the abdominal fat but in blood and spleen. In this respect, endrin is not typical of the cyclodiene group. 2,2'-Dichlorobiphenyl and 'Imugan' also give the highest residues in blood and liver respectively. This is in good agreement with their excretion rate and the high percentage of metabolites in the excreta (nearly 100%).

(iii) Differences Between Animal Species

Hexachlorobenzene-^{14}C was administered orally to rats and to rhesus monkeys (about 0.5 mg/kg body weight; Rozman *et al.*, 1975).

Table 3.1 Residues of Environmental Chemicals and Metabolites in Tissues and Organs of Rats after Long-term Feeding (1–2 ppm) was Discontinued. (Normalized to unit concentration of 'Imugan'-^{14}C in muscle, 0.003)

Organ	2,4,6,2',4'-Pentachloro-biphenyl-^{14}C	2,5,4'-Trichloro-biphenyl-^{14}C	Chloro-alkylene-9-^{14}C	Endrin-^{14}C	2,2'-Dichloro-biphenyl-^{14}C	'Imugan'-^{14}C
Liver	2,040	150	85	90	37	60
Lungs	4,000	120	45	100	40	44
Kidneys	4,000	230	150	100	77	18
Skin and subcutaneous fat	1,900	100	100	250	10	14
Blood			30	370	87	9
Abdominal fat	8,700	330	170	130	27	7
Stomach + duodenum		150	35	220	40	6
Spleen		100	45	1,000	50	6
Heart	2,100	130	30	190	80	4
Genitals	4,200	130	90	320	20	3
Brain		20		83	10	1
Muscles	530	60	20	20	10	1

In this experiment, the differences of total radiocarbon content between tissues of rats and rhesus monkeys were small. In some tissues, the concentration was slightly higher for the rhesus monkey. This corresponds to the lower excretion rate of this animal as compared to the rat (Rozman *et al.*, 1975).

However, there may exist differences in the chemical nature of organ residues for different species. For instance, dieldrin is preferably converted to 12-hydroxydieldrin in rats and primates including man (Richardson and Robinson, 1971), but preferably to aldrin-*trans*-diol in mice and rabbits (Müller *et al.*, 1975). The rate of excretion of the diol was by far the highest in the mouse, which must necessarily result in relatively high concentrations of the diol in the mouse liver.

(iv) Sex Differences of Organ Distribution

For the data listed in Table 3.1 and in Rozman *et al.* (1975) differences between male and female animals were small. However, larger differences due to different accumulation capacities are possible. In a long-term feeding experiment with aldrin-^{14}C, females reached the plateau level of body concentration only after 200 days, males after 53 days (Ludwig *et al.*, 1964). This results in a much higher concentration in the abdominal fat at plateau time (females 3.5 μg/g, males 0.29 μg/g, when fed 0.2 μg/g of diet).

3.3. METABOLIC BALANCE OF ENVIRONMENTAL CHEMICALS

(i) *In vitro*

Whereas *in vivo* experiments give information on the balance of distribution and conversion for the whole organism, *in vitro* experiments give information on the possible primary enzymatic attacks on xenobiotics. *In vitro* experiments may be carried out with cell cultures or with various cell fractions, preferably from rat liver or kidney. Experiments with perfused rat liver may accomplish these studies and, when compared with *in vivo* studies, help to localize the site of metabolic attack for the substance in question.

The metabolites detected *in vitro* may not in all cases be the same as those formed *in vivo*. In the isolated cell fractions, metabolic reactions may occur which are not possible in the intact organism. There, competitive reactions may reduce or even prevent the reaction observed *in vitro*. Additionally in the intact organism the primary product formed by a cell fraction may be altered immediately by secondary reactions. The following examples demonstrate the non-applicability of *in vitro* experiments to living animals.

For 2,2'-dichlorobiphenyl, all four theoretically possible monohydroxy derivatives and four dihydroxy derivatives were found *in vitro* in rat cell fractions (Greb *et al.*, 1975a); *in vivo*, however, only three monohydroxy isomers were detected one of which occurred only in very small amounts, and only three dihydroxy isomers (Kamal *et al.*, 1976a). The only metabolite formed *in vitro* from the cyclodiene insecticide α-(*trans*)-chlordane was dehydrochlordane (Spitzauer, unpublished); *in vivo*, however, the epoxide and hydroxylated products were the major conversion products (Schwemmer *et al.*, 1970; Poonawalla and Korte, 1971; Barnett and Dorough, 1974). In human fat, the epoxide was detected (Biros and Enos, 1973).

These examples show that *in vitro* experiments are suitable for rapid preliminary tests, but that their interpretation should be cautious and that the evaluation of their significance should be made only in conjunction with *in vivo* experiments.

(ii) *In vivo*

(a) Microorganisms

The conversion and degradation of xenobiotics by microorganisms has been the topic of laboratory investigations for many years. Figure 3.1 shows the comparative conversion of model pesticides by the fungus *Aspergillus flavus* (Korte, 1968).

A large number of such studies has been and is being done, and they are usually part of registration procedures for the use of pesticides (Sanborn *et al.*, 1976). For dieldrin, Matsumura and Boush (1967; Matsumura *et al.*, 1968) have found several soil microbial strains capable of degrading dieldrin significantly (Figure 3.2).

However, these results could not be confirmed by using cultures of normal or of dieldrin-contaminated soils, nor by any other microorganisms tested (Vockel and Korte, 1974).

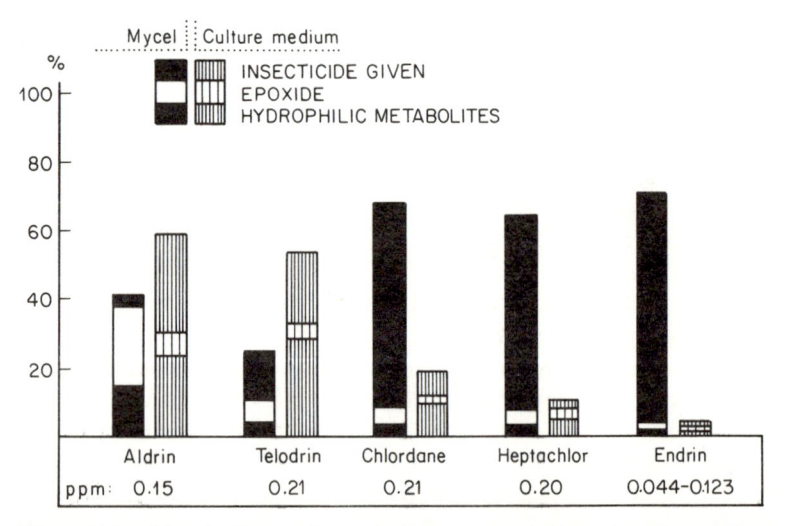

Figure 3.1 Metabolism of cyclodiene insecticides by *Aspergillus flavus*. (Reproduced by permission of Schweizerischer Chemiker-Verband from Korte, 1968)

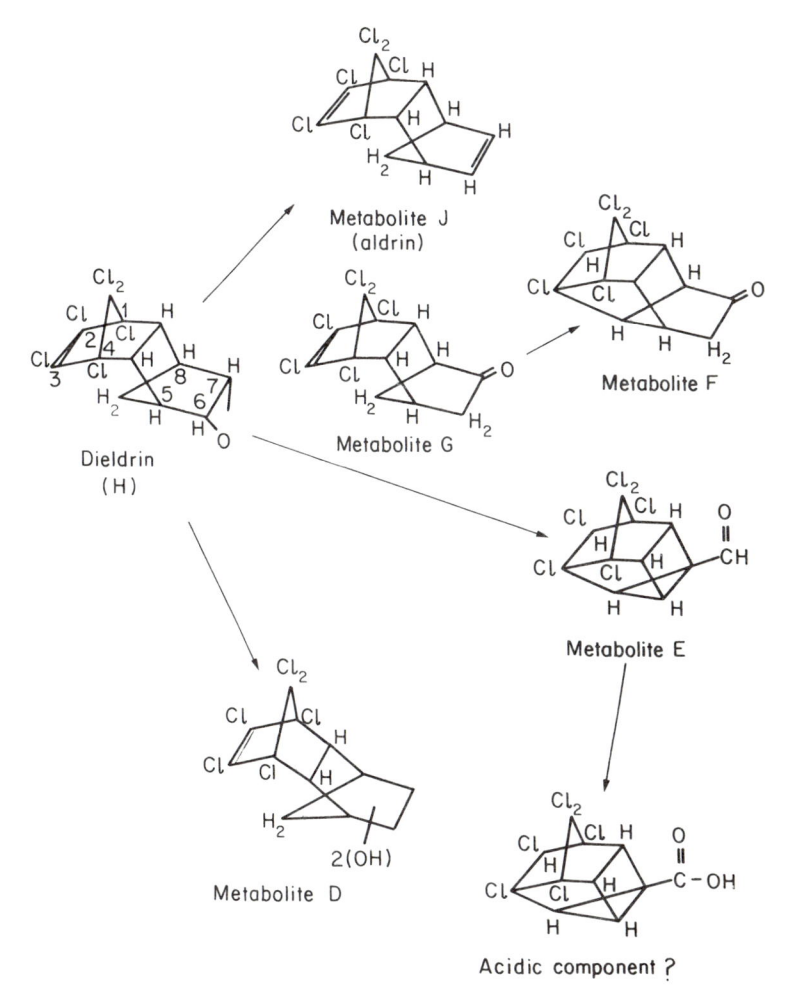

Figure 3.2 Metabolites of dieldrin from microorganisms. Reproduced by permission of Macmillan Journals Ltd., from Matsumura et al., 1968)

Two types of research into microbial degradation of pesticides are of practical importance.

(a) The development of adapted microbial strains that can use pesticides as the sole source of carbon. These can be used in industrial sewage treatment to remove pesticides such as chlorinated phenols which kill the normal bacterial populations.

(b) The identification of chemical intermediates that should be analysed for in environmental samples.

Regulatory agencies frequently require the testing of a pesticide and whole soil reacting together. Under properly controlled conditions such tests yield qualitative information about the nature of metabolites but little quantitative data about the capacity for degradation. They are, however, useful for estimating the relative persistence of pesticides in soil. More information about the environmental impact of pesticides should be obtained from field experiments under outdoor conditions (see 3.3. iie).

The capability of thermophilic microorganisms involved in composting to degrade so-called persistent environmental chemicals present in waste has been investigated by laboratory simulation. It was shown that, during the composting process, most of the organic chemicals tested remained largely unchanged (Müller and Korte, 1974; Müller et al., 1974; Müller and Korte, 1975). Among these were dieldrin and various polychlorinated biphenyls, although degradation of PCB to benzoic acids by *Alkaligenes* sp. was demonstrated to occur (Furukawa and Matsumura, 1976).

(b) Plants

Since in the normal environment, plants grow in soil and form an ecosystem with the soil organisms, it is difficult to investigate plant metabolism in isolation. *In vitro* studies with plant enzymes may give preliminary information on possible

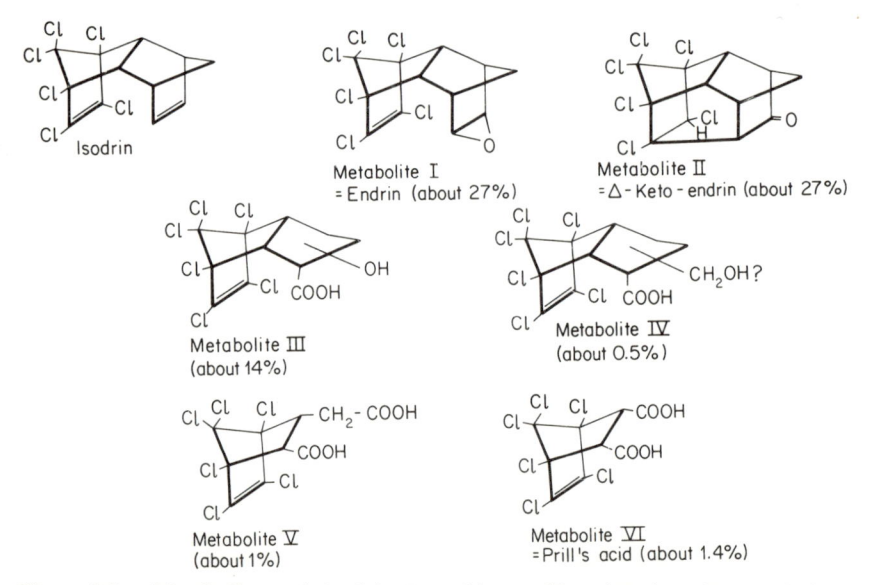

Figure 3.3 Metabolites of isodrin in cabbage. (Reprinted with permission from *Chemosphere*, 4(2), 99–104. Weisgerber, I., Tomberg, W., Klein, W., and Korte, F. Ecological chemistry. XCV. Isolation and structure elucidation of hydrophilic carbon-14-labelled isodrin metabolites from common cabbage. © 1975, Pergamon Press, Ltd.)

metabolic reactions, but for their interpretation, the same problems are encountered as for animal enzyme investigations (see 3.3. i). In hydroponic culture, the influence of microorganisms can be eliminated by running a blank with the nutrient medium without plants. Plants grown in normal soil can also give good information about plant metabolism when the xenobiotic is applied on the leaves, since the transport from the leaf surface to the soil via roots is very small within the normal laboratory test time.

Figure 3.3 shows the conversion of the cyclodiene isodrin by cabbage plants after foliar treatment. This substance shows how, by means of stepwise oxidation, the molecule is degraded to Prill's acid, a substance with three carbon atoms less than the parent compound (Weisgerber *et al.*, 1975b). This example demonstrates the importance of plant oxidases, mentioned in paragraph 1, for the degradation of xenobiotics in the environment.

Similar experiments with heptachlor showed also, besides epoxidation and hydroxylation, oxidative ring cleavage leading to Prill's acid (Weisgerber *et al.*, 1974a).

(c) Insects

Since insects are the target organisms of insecticides, their metabolism has been widely investigated. All reaction mechanisms discussed in the first paragraph of this chapter are known for insects.

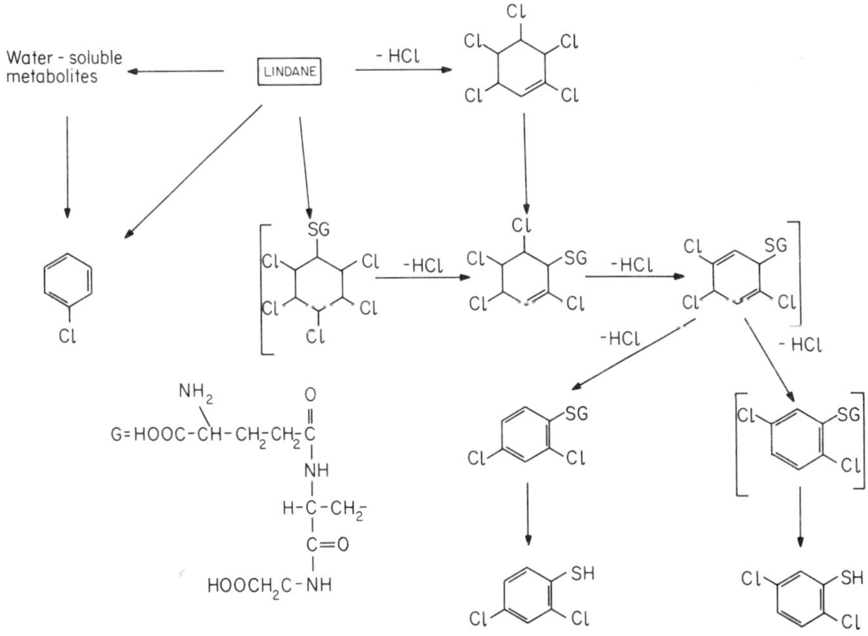

Figure 3.4 Metabolites of Lindane in insects. (Reproduced by permission of Springer-Verlag from Klein and Korte, 1970)

Figure 3.4 presents, as an example, the fate of Lindane in insects (Klein and Korte, 1970). It shows a stepwise dehydrochlorination of the molecule through the intermediacy of glutathione conjugation and results in aromatic compounds.

The problem of resistance to insecticides and its causes has been investigated in many studies (e.g. Klein and Korte, 1970) and will not be discussed here in detail. From the ecotoxicological point of view, it should be mentioned that this phenomenon will lead to an increase in the number and quantity of insecticides used, and, therefore, of effects on the ecosystem.

(d) Animals

Using dieldrin as an example, the dependence of metabolism and excretion by mammals on the animal species and its consequences for the effects of chemicals and for the extrapolation of animal experimental data to man will be discussed.

Dieldrin-^{14}C was given as a single oral dose of 0.5 mg/kg body weight to five animal species: mice, rats, rabbits, rhesus monkeys, and chimpanzees. Besides dieldrin, 12-hydroxydieldrin and 4,5-aldrin-*trans*-diol were found in the excreta; the formulae are shown in Figure 3.5, along with other, minor mammalian metabolites of aldrin. Table 3.2 shows the amounts of the three substances excreted by each animal species within 10 days, in per cent of administered dose (Müller *et al.*, 1975). In all species the faecal excretion of unchanged dieldrin was high in the first 48 hours and then declined rapidly, probably due to the completed excretion of unabsorbed dieldrin. The urine samples contained only metabolites and no dieldrin.

In all five species 12-hydroxydieldrin and 4,5-aldrin-*trans*-diol were the major metabolites. Regarding the ratio of the two metabolites, the rat seems to be comparable to the primates; direct oxidation resulting in the monohydroxy metabolite is their common main metabolic pathway. There is strong evidence that this is also true for man (Richardson and Robinson, 1971). In the mouse and the rabbit, on the other hand, the opening of the epoxide to the diol is the predominant reaction. These findings also demonstrate that, because of metabolism similar to that in man, the rat is the suitable experimental animal for dieldrin and not the mouse.

In fish, the occurrence of methylated metals like methylmercury is well known. However, information is limited on metabolic conversions of organic compounds. Enzymatic hydroxylation seems to be of minor importance in comparison with other vertebrates. For example, the hydroxylation of polychlorinated biphenyls detected for many organisms (summary: Klein and Weisgerber, 1976) has been reported to occur in fish only slowly (Melancon and Lech, 1976; Herbst *et al.*, 1976), or not at all (Hutzinger *et al.*, 1972).

(e) Ecosystems

The most important ecosystems involved in transformation and degradation of environmental chemicals are the plant—soil and the aquatic ecosystems.

Biotic Processes 49

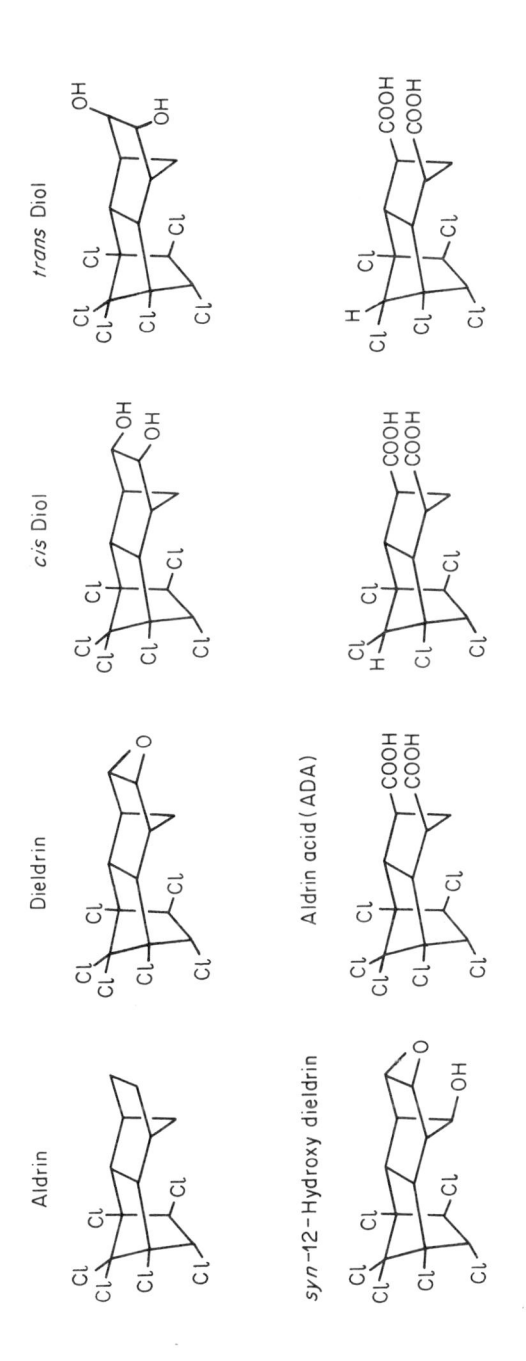

Figure 3.5 Formulae of aldrin, dieldrin, and metabolites in mammals

Table 3.2 Radioactive Material Excreted Within 10 Days After Single Oral Administration of 0.5 mg/kg Dieldrin-^{14}C (per cent of administered dose). (Reprinted with permission from *Chemosphere*, **4**(2), Müller, W., Nohynek, G., Woods, G., Korte, F., and Coulston, F. Comparative metabolism of Carbon-14 labelled dieldrin in mouse, rat, rabbit, rhesus monkey and chimpanzee. © 1975, Pergamon Press, Ltd.)

	Mice male	Mice female	Rats male	Rats female	Rabbits male	Rabbits female	Rhesus male	Chimpanzee female
Dieldrin	5.5	3.2	0.8	2.8	0.3	0.5	9.0	3.2
12-Hydroxy-dieldrin	13.0	7.5	8.8	4.6		0.2	9.4	2.0
Aldrin-*trans*-diol	20.0	26.0	2.3	2.4	1.5	2.0	2.0	1.1
Total	38.5	36.7	11.9	9.8	1.8	2.7	20.4	6.3
Faeces	36.6	35.0	11.3	9.3	0.3	0.5	16.0	5.0
Urine	1.9	1.7	0.6	0.5	1.5	2.2	4.4	1.3

Biotic Processes

For the plant–soil system the metabolism of environmental chemicals is characterized by uptake of chemicals by plants from soil, distribution within plants and soil, conversion reactions in soil and plants, residue loss by evaporation, leaching, mineralization, or assimilation, and residue fixation in plant or soil macromolecules.

Laboratory tests on the uptake of chemicals from soil by plants are not predictive for the environment, since the uptake is much higher than in the environment when small laboratory pots are used. Under open-air conditions, the uptake of residues by plants is less than one per cent of residue present for the cyclodiene insecticides and other chemicals tested. Although there is strong evidence that plants actively contribute to the disappearance of residues from soil, the reduction of soil burden by harvesting is negligible.

When lysimeters of 60 × 60 × 60 cm size are used under outdoor conditions with ^{14}C-labelled model substances, the quantitative data on uptake and residues in plants and soil are comparable to field data and, therefore, relevant for practical evaluation.

For the conversion of environmental chemicals by the plant–soil system, two examples are presented: the degradation of aldrin in Figure 3.6 (Klein *et al.*, 1973;

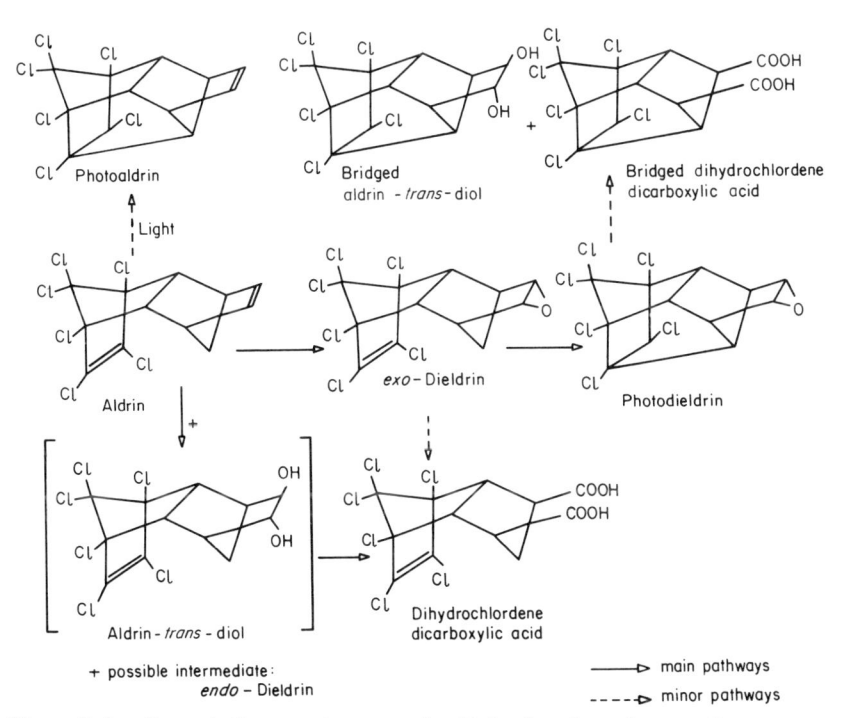

Figure 3.6 Degradation pathways of aldrin in the plant–soil system. (Reproduced by permission of Academic Press, Inc. from Scheunert *et al.*, 1977)

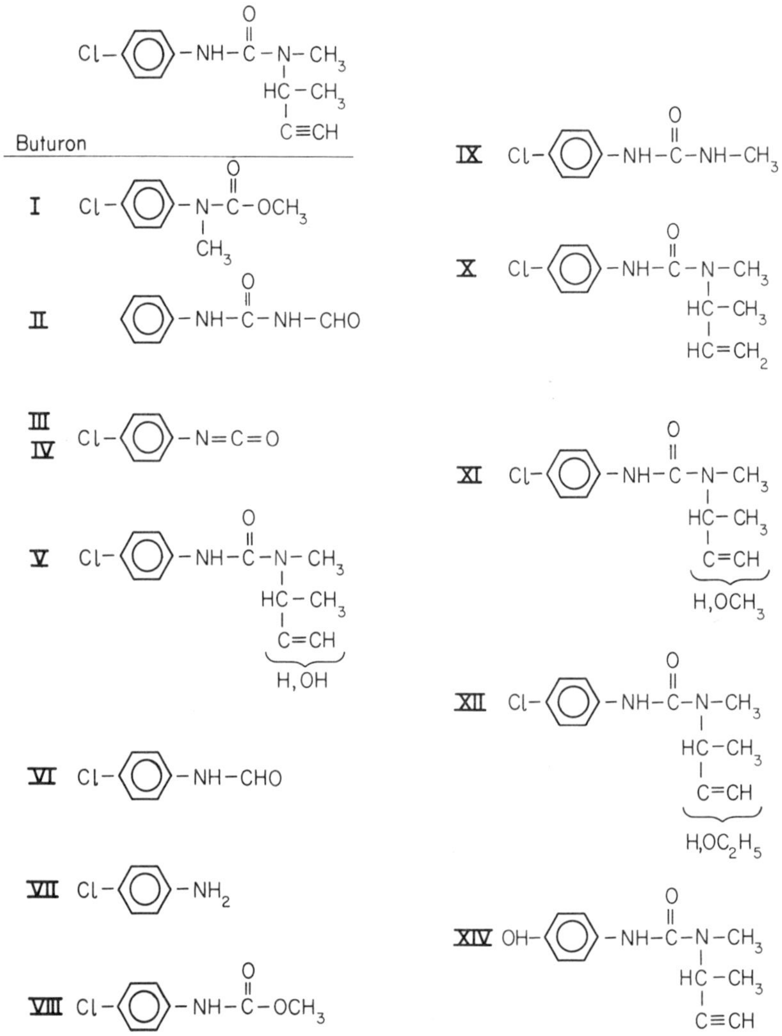

Figure 3.7 Conversion products of buturon in the plant–soil system. (Substances I–IX in wheat plants, I, V, VII–XII in soil, I, VI, XIV in leaching water.) (Reprinted from Hague et al., 1976, by courtesy of Marcel Dekker, Inc.)

Weisgerber et al., 1974b; Kilzer et al., 1974; Weisgerber et al., 1975a), and the conversion products of the phenylurea herbicide buturon in Figure 3.7 (Haque et al., 1976; Haque et al., 1977b).

The substances shown in the two figures are the products of complex combinations of agents like plant enzymes, soil microorganisms, and abiotic effects.

The conversion products in plants may originate from plant metabolism as well as be formed in soil by microorganisms or by abiotic reactions and be taken up by plants. Although the formation mechanisms of individual conversion products cannot be elucidated in detail, the results reported here are significant as they give a realistic description of the nature of residues that must be expected under field conditions and, above all, in human food.

When looking at the material balance of the system after five years, in the case of aldrin the significance of soluble metabolites seems to decrease. After this time, the only important soluble residue is dieldrin; the 'unextractable residues' have increased and are of higher importance than in the first years after application (Tables 3.3 and 3.4).

Furthermore, Table 3.3 demonstrates that within five years half of the radioactivity applied was lost by volatilization. The identity of the volatilized radioactivity was not investigated in this study (Weisgerber et al., unpublished). Thus volatilization is a major pathway for residue loss in ecosystems, and the uptake by plants and the leaching by water are of minor importance. For chemicals other than aldrin, the trend seems to be comparable (e.g. polychlorinated biphenyls; Moza et al., 1976). Little is known of the chemical nature and the fate of volatilized residues; they may be the unchanged compounds, conversion products, or carbon dioxide resulting from total mineralization. Also, information on the 'unextractable residues' is very limited (see also paragraph 1). They may be conversion products bound to macromolecules (lignin, humic acids), or copolymerized with the natural monomers of macromolecules, or they may be ^{14}C reassimilated from $^{14}CO_2$ or its low molecular weight precursors. Further work is required on the possibility of remobilization of these residues and their uptake by plants.

Table 3.3 Material Balance in Plants and Soil Five Years After Application of Aldrin-^{14}C to Soil

Applied in 1969:	103 mg (100%)
Residue in soil 1973:	34.5%
Taken up by plants 1969:	0.1%
Taken up by plants 1970:	0.2%
Taken up by plants 1971:	0.1%
Taken up by plants 1972:	0.1%
Taken up by plants 1973:	0.1%
Leaching water 1969:	1.6%
Leaching water 1970:	4.9%
Leaching water 1971:	2.4%
Leaching water 1972:	1.3%
Leaching water 1973:	0.4%
Recovery after 5 years:	45.7%
Volatilization within 5 years:	54.3%

Table 3.4 Residues of Aldrin-^{14}C and Conversion Products in Soil Five years After Treatment (% of applied radioactivity)

Sample	Aldrin	Metabolite X (unidentified)	Dieldrin	Photo-dieldrin	Extracted hydrophilic metabolites	Unextractable residue	Total
Soil depth 0–10 cm	0.8	<0.4	11.7	0.4	1.9	4.3	19.4
Soil depth 10–20 cm	0.4	<0.4	5.0	<0.4	1.2	1.9	8.9
Soil depth 20–30 cm	<0.4	<0.4	1.5	<0.4	0.4	1.2	3.1
Soil depth 30–40 cm	<0.4	<0.4	1.5	<0.4	0.4	1.2	3.1
Soil total	1.2	0.4	19.7	0.7	3.9	8.6	34.5

When the balance of metabolism of xenobiotics is to be studied in aquatic ecosystems, the simulation of stable aquatic ecosystems in the laboratory is indispensable. Systems consisting of water and microorganisms, of water, microorganisms, and sediment, of water, microorganisms, sediment, and plants, and also systems including higher animals and fish have been described. In tests with 2,2'-dichlorobiphenyl-^{14}C, 2,5,4'-trichlorobiphenyl-^{14}C, 2,4,6,2',4'-pentachlorobiphenyl-^{14}C, and chloroalkylene-9-^{14}C, an isopropylated 2,4'-dichlorobiphenyl, the bioaccumulation was highest for chloroalkylene-9 (53.1% of the applied radioactivity in the biomass). This substance also had the highest conversion rate, in the biomass as well as in the water. The toxicity, also, was highest for chloroalkylene-9, especially for Daphnia.

From 2,2'-dichlorobiphenyl, three hydroxylated metabolites were identified in water and biomass, from 2,5,4'-trichlorobiphenyl, one hydroxylated metabolite. From chloroalkylene-9, a number of metabolites were detected in small amounts, six of which could be characterized. These were hydroxylated or methoxylated products or products with partially degraded side-chains. No metabolite could be isolated from the experiment with 2,4,6,2',4'-pentachlorobiphenyl.

Mixed terrestrial-aquatic model ecosystems including sand with sorghum plants and a 'pond' with water, microorganisms, plankton, Daphnia, snails, algae, mosquito larvae, and fish, have been developed by several authors such as Metcalf (1974). When radiolabelled model chemicals were used, the determination of material balance was possible.

There are only a few reports on studies with radiolabelled compounds in natural-water bodies (see also paragraph 5; Salonen and Vaajakorpi, 1974). Since the radioactivity used has to be very low, an identification of metabolites is not possible; only the concentration of radioactivity is determined in water, sludge, and various organisms.

3.4. LABORATORY SCREENING TESTS

The above procedures to study balances and conversions of chemicals are rather complex and demanding. Therefore, a number of laboratory test methods have been elaborated to overcome such problems. Such test methods may involve the determination of physicochemical characteristics, biochemical determinations of specific results with a chemical, and predictive environmental experiments. Although many of these methods are quick and feasible and appropriate for certain kinds of environmental technology, their value for ecotoxicological evaluations, even on a chemical-to-chemical comparative basis, remains to be proven (see also 3.3. iia). Consequently they cannot be used at present to predict quantitatively the environmental fate of chemicals.

(i) Tests of Biodegradation

Biodegradation tests have been developed for sewage treatment and are related to the analysis of water pollutants by biological oxygen demand (BOD) and chemical oxygen demand (COD) determinations. The OECD detergent test is widely used and gives results that can be extrapolated to sewage treatment conditions. More feasible flask methods, when used under standardized conditions, may, however, be as good (Zahn and Wellens, 1974). These tests normally include adaptation of the activated sludge for the investigated chemical, thus allowing prediction of degradation upon continuous or long-term exposure. For less polluted aquatic environments, rivers and lakes, these tests are less applicable and feasible, and direct analysis of sources and disappearance of chemicals by BOD, COD, total oxygen demand (TOD), or single compound analysis seems to be more appropriate. Whether laboratory tests, with trace concentrations of pollutants and all the difficulties of nutrients and blanks, allow a quantitative correlation to watersheds remains to be proven. In order to provide metabolic data for aquatic systems comparable to those given in section 3.3, more complicated experiments have to be carried out.

The same applies to standardized degradation tests with soil. Simple laboratory tests, frequently carried out with 100 g of soil, permit measurement of conversion and identification of metabolites but do not give data that are quantitatively related to terrestrial systems (see also 3.3. iia).

(ii) Tests of Bioconcentration

The difficulties of testing for biodegradation also apply to bioaccumulation testing. In addition there is the possible influence of other contaminants. For this reason it would be appropriate to measure the bioaccumulation factor of a certain fish species using river water and not 'clean water' to which was added one chemical only.

(iii) Significance of Screening Tests

For a comparative screening of environmental chemicals for biodegradation and accumulation, tests under standardized conditions and the determination of physicochemical characteristics such as partition coefficients are valuable but it must be emphasized that the sequence of chemicals found in these tests may not be identical to that in the environment.

Due to the limited significance of screening tests for quantitative ecotoxicological investigations of the behaviour of pollutants, no detailed discussion of procedures and results is given here.

One extreme example should be mentioned to underline this statement: In all laboratory degradation tests phthalate plasticizers show only low persistence. In the environment, however, the major amount of these chemicals after use is buried in landfills where they persist for decades and may be slowly released with leaching water.

3.5. FOOD CHAINS

Although the mass of the biota is minute as compared to the mass of the abiotic parts of the biosphere, the biota contain a proportionally greater amount of lipophilic foreign compounds, since these are concentrated in organisms. This bioconcentration in organisms is important for man since it serves as a pathway to human food. For DDT it has been roughly estimated that the biota contain about 0.2% of all DDT ever produced or about 3% of the annual production in the mid-sixties (Woodwell *et al.*, 1971). Provided this estimate is correct within an order of magnitude and also valid for other persistent environmental chemicals it may be concluded that transport and movement in the biota does not play a significant role in the global long-term dispersion of these chemicals. Consequently the global dispersion of organic and inorganic chemicals quantitatively is mainly based on physical and physicochemical factors and it may be easily understood that the global 'distribution patterns' of persistent organics and strontium-90 are alike (Appleby, 1970).

Bioconcentration of nutrients and trace elements from their environment is a general and basic activity of living cells and organisms (Table 3.5).

Nutrient elements are used for the biosynthetic production of the organisms so that this type of bioconcentration sustains life. In contrast bioaccumulation of xenobiotics in higher organisms may be explained as a type of elimination of non-metabolized lipophilic chemicals into internal sinks.

Table 3.5 represents concentration factors of nutrients as well as xenobiotic elements from sea water by plankton and brown algae (Bowen, 1966). The bioconcentration factor of xenobiotics depends on the one hand on the partition coefficient between the releasing (generally aqueous) and the uptaking (generally lipophilic) medium. On the other hand, species-specific factors are also included such as active absorption (penetration through membranes), and metabolic conversion and excretion.

More recent investigations into the accumulation in food chains revealed that higher concentrations of chemicals in organisms at higher levels of the food chains (both aquatic and terrestrial) may result from the slower rate of elimination in the higher levels of the chains (Moriarty, this volume, Chapter 8). On the other hand, there are natural toxins like the compound causing the ciguatera fish disease which is formed by blue-green algae and reaches an edible fish as the tertiary step in a food chain. This chemical is not concentrated in the fish from the aquatic environment (Russell, 1965).

(i) Environmental Data

Data from environmental monitoring of food chains for pollutants are mainly available for mercury, methylmercury, and organochlorine compounds. Most original publications, however, include only one step of the respective chains. As far

Table 3.5 Concentration of Elements from Sea Water by Plankton and Brown Algae[a]. (Reproduced by permission of Academic Press Inc., London – New York – San Francisco from Bowen, 1966)

Element	State in sea water	Plankton CF 1	Plankton CF 2	Brown algae CF 1	Brown algae CF 2
Ag	Anion	210		240	
Al	Particulate?	25,000		1,550	
As	Anion			2,500	200–6,000
Au	Anion?			270	
B	Molecule			6.6	
Ba	Cation?	120		260	
Be	Particulate?				1,500
Br	Anion			2.8	
Ca	Cation	5	10	7.2	0.5–10
Cd	Cation	910		890	
Cl	Anion	1		0.062	
Co	Cation?	4,600		650	450
Cr	Particulate?	17,000	1–5	6,500	300–10,000
Cs	Cation			33	1–100
Cu	Cation	17,000		920	100
F	Anion			0.86	
Fe	Particulate?	87,000	2,000–140,000	17,000	20,000–35,000
Ga	Particulate?	12,000		4,200	
Ge	Particulate?				15–200
Hg	Anion			250	
I	Anion	1,200		6,200	3,000–10,000
K	Cation			34	3–50
La	Particulate?			8,300	
Li	Cation			8?	

Mg	Cation	0.59		0.96	
Mn	Cation?	9,400	750	6,500	6,500
Mo	Anion	25		11	
N	Variable	19,000		7,500	
Na	Cation	0.14		0.78	1
Nb	Particulate?				450–1,000
Ni	Cation	1,700		140	500
P	Anion + organic	15,000		10,000	10,000
Pb	Cation?	41,000		70,000	
Ra	Cation	4,500	2,750	370	100
Rb	Cation			15	5–50
Ru	Anion?		600–3,000		15–2,000
S	Anion	1.7		3.4	10
Sb	Anion?		50		
Sc	Particulate?				1,500–2,600
Si	Particulate?	17,000		120	
Sn	Particulate?	2,900		92	
Sr	Cation	8	9	44	1–40
Th	Particulate?				10
Ti	Particulate?	20,000		3,000	
U	Anion?				10
V	Anion?	620		250	
W	Anion?			87	
Y	Particulate?				100–1,000
Zn	Cation	65,000	1,000	3,400	100–13,000
Zr	Particulate?		1,500–3,000		350–1,000

[a]CF = Concentration factor = ppm in fresh organism/ppm in sea water for the element concerned. CF 1 refers to Bowen (1966), while CF 2 refers to the compilation by Mauchline and Templeton (1964).

Table 3.6 DDT Concentrations in Samples Collected from the Pond, Following a Single Addition of 1 µg/l (ppm in organic matter[a]; seston = plankton + organic matter). (Reproduced by permission of the authors from Salonen and Vaajakorpi, 1974)

	30 days	59 days
Filtered water	0.00004	0.00001
Seston, 8 µm sample	0.0	0.0
Seston, 3 µm sample	0.0	0.0
Sediment	0.69 ± 0.02	0.71 ± 0.02
Bladderwort	1.98 ± 0.04	
Moss	0.38 ± 0.01	0.22 ± 0.01
Dragonfly larvae	0.20 ± 0.02	0.17 ± 0.01
Caddisfly larvae	0.43 ± 0.02	0.38 ± 0.02
Backswimmer	0.24 ± 0.02	
Bivalve mollusc	0.24 ± 0.03	0.17 ± 0.03
Newt	0.58 ± 0.03	
Perch: muscle	0.44 ± 0.05	0.24 ± 0.03
gills	3.73 ± 0.19	2.15 ± 0.08
liver	2.79 ± 1.02	0.52 ± 0.02
mes. adip.	17.20 ± 4.60	23.80 ± 6.60
Crucian carp: muscle	0.10 ± 0.01	0.37 ± 0.04
gills	0.33 ± 0.02	0.83 ± 0.03
liver	0.09 ± 0.01	0.20 ± 0.01
Crucian carp (whole)	2.06 ± 0.06	2.02 ± 0.05

[a] For most samples the dry weight and the weight of organic matter were identical.

as wildlife is concerned, the work of the Patuxent Wildlife Research Center should be cited (e.g. Anderson and Hickey, 1976; Clark and Lamont, 1976). Further data are given by, for example, Moore and Walker (1964) and Walker *et al.* (1967). Only a few conclusive open-air experiments have been carried out so far. Salonen and Vaajakorpi (1974) have treated a small pond with DDT-^{14}C (1 nCi/l) and analysed the concentrations of ^{14}C as DDT equivalents in water, sediment, many animal species, fish, invertebrates, and plants. Table 3.6 gives a summary of the results. A total balance has not been attempted in this experiment.

(ii) Laboratory Test Models

Several laboratory test models, microcosms or micro-ecosystems have been developed which yield results difficult to interpret for food chain accumulation.

The laboratory model ecosystem of Metcalf (1974) which has been designed for balance studies, including metabolism, bioconcentration, biodegradability and effects on the organisms used, at least permits comparison of chemicals. The experimental procedure has been described in Metcalf *et al.* (1971) with a summary

Table 3.7 Behaviour of Aldrin-^{14}C and Di-2-ethylhexyl Phthalate (DEHP) in a Model Ecosystem (E.M. = ecological magnification). (Reproduced by permission of the author from Metcalf, 1974)

	Concentration (ppm equivalents) in				
	Water	Algae	Snail	Mosquito	Fish
Aldrin					
Total ^{14}C	0.0117	19.70	57.20	1.13	29.21
Aldrin	0.00005	1.95	2.23		0.157
Dieldrin	0.0047	16.88	52.40	1.1	28.00
9-OH-Dieldrin	0.00052	0.12	0.17		0.322
9-C=O Dieldrin	0.0004	0.079	0.217		0.088
Unknown	0.00039	0.585	2.05		0.612
Polar metabolites	0.0040	0.015	0.097		0.004
E.M. values					3,140
DEHP					
Total ^{14}C	0.0078	19.105	20.325	36.609	0.206
DEHP	0.00034	18.322	7.302	36.609	0.044
MEHP	0.00099	0.325	2.541		0.021
Phthalic anhydride	0.00363	0.180	5.772		0.113
Phthalic acid	0.00077	0.094	2.724		0.018
Unknown	0.00190	0.029	0.768		
Polar metabolites	0.00016	0.155	1.218		0.010
E.M. values		53,890	21,480	107,670	130

of data in 1974. Table 3.7 gives an example of the data for aldrin and di-2-ethylhexyl phthalate with this system (from Metcalf, 1974).

3.6. PREDICTABILITY OF THE BEHAVIOUR OF XENOBIOTICS FROM STRUCTURAL CHARACTERISTICS

As examples of the predictability of the environmental behaviour of xenobiotics from structural characteristics of the molecule, the influence of three different structural characteristics is discussed: chlorine content, substituents (epoxy group), and stereochemical configuration.

(i) Influence of Chlorine Content

Among the first organic substances known to be environmental pollutants were chlorinated compounds. It has also been known for many years that substitution with chlorine increases the chemical stability of a compound. When the first polychlorinated biphenyls were detected in the environment and when the gas chromatograph pattern of environmental samples was compared to that of the technical PCB mixtures used, it was observed that the higher chlorinated components of the mixtures were present. It was concluded that the lower chlorinated components were more easily degraded.

This negative correlation between chlorine content and degradability or convertability to hydrophilic derivatives has been confirmed by many studies (e.g. Matthews and Anderson, 1975).

The urinary excretion of PCB's by mammals may be regarded as a measure of the metabolism to hydrophilic compounds and, consequently, is strongly dependent on the chlorine content. Figure 3.8 shows the excretion of radioactivity during long-term feeding of three selected radiolabelled PCB's by rats: 2,4,6,2',4'-pentachlorobiphenyl, 2,2'-dichlorobiphenyl, and chloroalkylene-9, an isopropylated 2,4'-dichlorobiphenyl (Begum et al., 1975).

The figure shows that a plateau level of body concentration is reached very slowly for pentachlorobiphenyl. The two less chlorinated compounds, on the other hand, reach the saturation level in a relatively short time and at a relatively low level. The tissue concentrations pertaining to these studies are discussed in section 3.2 and are, as expected, high for the pentachlorobiphenyl and relatively low for the other two compounds. 2,5,4'-Trichlorobiphenyl which is not shown in this figure but included in Table 3.1 on tissue concentrations, is placed between pentachlorobiphenyl and the dichlorobiphenyl.

However, the quantitative excretion pattern is dependent not only on the chlorine content but also on the chemical nature of the metabolites excreted. 2,4'-Dichlorobiphenyl is excreted by rhesus monkeys mainly in monohydroxylated

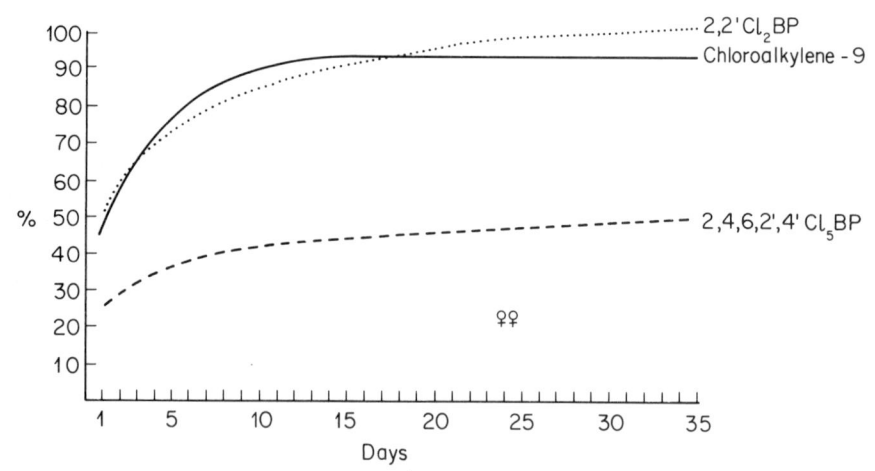

Figure 3.8 Excretion of radioactivity by rats during long-term feeding with 2,4,6,2',4'-pentachlorobiphenyl-^{14}C, chloroalkylene-9-^{14}C, and 2,2'-dichlorobiphenyl-^{14}C. (Reprinted with permission from *Chemosphere*, 4, 241. Begum, S., Lay, J. P., Klein, W., and Korte, F. Ecological chemistry. CIII. Elimination, accumulation, and distribution of chloroalkylene-9-^{14}C. © 1975, Pergamon Press, Ltd.)

form (66.5% of excreted radioactivity), 2,2',5-trichlorobiphenyl, however, is mainly in dihydroxylated form (8% of excreted radioactivity). The higher the chlorine content, the lower is the water solubility; therefore, the higher chlorinated isomers need more hydroxy groups to be excreted. When, for highly chlorinated isomers, the introduction of several hydroxy groups is impossible, the excretion becomes more and more difficult, and the substance is accumulated in the body (Greb et al., 1975b).

From these examples it is evident that the metabolism and, consequently, the excretion of PCB's is dependent on the number of chlorine atoms in the molecule. Furthermore, the position of the chlorine atoms has an influence on degradability. Schulte and Acker (1974) have demonstrated that PCB's can be metabolized by mammals, including man, provided that on at least one ring two neighbouring positions are unsubstituted. When these atoms are in *para* and *meta* position, the molecule is metabolized relatively easily, when they are in *ortho* and *meta* position, the degradation takes place more slowly. The preference of the *meta* or *para* position for enzymatic hydroxylation was also reported by Goto et al. (1975) and by Sugiura et al. (1976).

(ii) Influence of Substituents (Epoxy Groups)

As an example of the influence of the substituents on environmental behaviour, the epoxy group in cyclodienes is discussed.

When, in the cyclodiene insecticides aldrin, isodrin, and heptachlor, the double bond of the non-chlorinated ring is replaced by an epoxy group, the biological activity is increased. Additionally, an *exo*-epoxy group results in stabilization of the molecule and in resistance to enzymatic attack, especially against cleavage of the ring system. In the case of mammals where ring cleavage is of minor importance in the metabolism of these chemicals, the difference between metabolism of olefin and that of epoxide is small. The difference between olefin and epoxide is highest for plants whose metabolism is mainly oxidative and leads to ring cleavage. Thus aldrin (Klein et al., 1973) and isodrin (Weisgerber et al., 1975b) are metabolized by plants to hydrophilic products much faster than dieldrin (Kohli et al., 1973) and endrin (Weisgerber et al., 1968).

(iii) Influence of Conformation

For the influence of stereochemical conformation on the environmental behaviour two examples are selected: the cyclodiene insecticides and BHC isomers.

Aldrin and isodrin are conformational isomers; dieldrin and endrin are the corresponding epoxides. These differences in configuration result in significant differences in metabolic behaviour. The *endo—endo* structure of isodrin and endrin is more susceptible to enzymatic attack than the *endo—exo* structure of aldrin and dieldrin (Klein et al., 1968). In the mammal, endrin is metabolized more rapidly than aldrin, which means that the stereochemical structure has a greater influence on the rate of metabolism than the substitution with the epoxide. In plants, on the other hand, the percentage of hydrophilic metabolites is higher for aldrin and dieldrin (Weisgerber et al., 1970) than for isodrin (Weisgerber et al., 1975b) and endrin (Weisgerber et al., 1968); however, when looking at the formulae of metabolites, it is evident that isodrin (Figure 3.3, paragraph 3.3. iib) is degraded to smaller molecules than aldrin (Figure 3.6, paragraph 3.3. iie). Isodrin is broken down to substances containing up to three carbon atoms less than the parent compound, aldrin undergoes only ring-cleavage to a relatively stable dicarboxylic acid which still contains the same number of carbon atoms as the parent compound. Thus minute structural changes lead to significant changes in the metabolic behaviour.

Also, for the BHC isomers, there exists a significant difference in environmental behaviour. The fact that the β-isomer is detected in environmental samples in greater concentrations than those corresponding to its occurrence in the technical mixture, indicates that it is more persistent than the α- or γ-isomer. For bacteria as model organisms, it was shown that the adsorption–diffusion mechanism is also different for the three isomers (Sugiura et al., 1975).

3.7. CONCLUSION

It may be concluded from this chapter that, for the ecotoxicological evaluation of a foreign compound, not only the unchanged parent compound but also each conversion product must be considered. The numbers and the quantitative amounts of conversion products are higher for the so-called 'non-persistent' compounds (e.g. buturon, 3.3. iie) than for the persistent ones. The 'unextractable' residues bound in tissues or soil, which have been overlooked for a long time, are also higher for non-persistent compounds. The same applies to the residues which are volatilized, i.e. which escape from the target ecosystem. Therefore, besides the indispensable investigation of persistent xenobiotics, the study of non-persistent compounds should also be stressed.

3.8. REFERENCES

ACS Symposium Monograph, 1975. *Conjugated and Bound Particle Residues*, 1976.

Anderson, D. W. and Hickey, J. J., 1976. Dynamics of storage of organochlorine pollutants in herring gulls. *Environ. Pollut.*, **10**(3), 183–200.

Appleby, W. G., 1970. IUCN 11th Technical Meeting, New Delhi, 1969.

Barnett, J. R. and Dorough, H. W., 1974. Metabolism of chlordane in rats. *J. Agric. Food Chem.*, **22**(4), 612–9.

Begum, S., Lay, J. P., Klein, W., and Korte, F., 1975. Ecological chemistry. CIII. Elimination, accumulation, and distribution of chloroalkylene-9-^{14}C. *Chemosphere*, **4**(4), 241–6.

Biros, F. J. and Enos, H. F., 1973. Oxychlordane residues in human adipose tissue. *Bull. Environ. Contam. Toxicol.*, 10(5), 257–60.

Bowen, H. J. M., 1966. *Trace Elements in Biochemistry*, Academic Press, London–New York, p. 86.

Brooks, G. T., 1972. In *Environmental Quality and Safety*, Vol. I, Georg Thieme Publishers Stuttgart, Academic Press Inc., New York, p. 106.

Clark, D. R. Jr. and Lamont, T. G., 1976. Organochlorine residues in females and nursing young of the big brown bat. *Bull. Environ. Contam. Toxicol.*, 15(1), 1–8.

Cremer, J. E., 1957. Metabolism *in vitro* of tissue slices from rats given triethyltin compounds. *Biochem. J.*, 67, 87–96.

Cremer, J. E., 1958. The biochemistry of organotin compounds. *Biochem. J.*, 68, 685–92.

Cremer, J. E., 1959. Biochemical studies on the toxicity of tetraethyl lead and other organo-lead compounds. *Br. J. Ind. Med.*, 16, 191–9.

Edwards, T. and McBride, B. C., 1975. Biosynthesis and degradation of methylmercury in human feces. *Nature*, 253(5491), 462–4.

Fishbein, L., 1974. Mutagens and potential mutagens in the biosphere. II. Metals, mercury, lead, cadmium and tin. *Sci. Total Environ.*, 2(4), 341–71.

Furukawa, K. and Matsumura, F., 1976. Microbial metabolism of polychlorinated biphenyls. Relative degradability of polychlorinated biphenyl components by *Alkaligenes* species. *J. Agric. Food Chem.*, 24(2), 251–6.

Furukawa, K., Suzuki, T., and Tonomura, K., 1969. Decomposition of organic mercurial compounds by mercury-resistant bacteria. *Agric. Biol. Chem. (Tokyo)*, 33(1), 128–30.

Gavis, J. and Ferguson, J. F., 1972. In *Water Research*, Vol. 6, Pergamon Press, p. 989.

Goto, M., Hattori, M., and Sugiura, K., 1975. Ecological chemistry. VI. Metabolism of pentachloro- and hexachlorobiphenyls in the rat. *Chemosphere*, 4(3), 177–80.

Greb, W., Klein, W., Coulston, F., Golberg, L., and Korte, F., 1975a. Ecological chemistry. LXXXIII. *In vitro* metabolism of carbon-14-labelled polychlorinated biphenyls. *Bull. Environ. Contam. Toxicol.*, 13(4), 424.

Greb, W., Klein, W., Coulston, F., Golberg, L., and Korte, F., 1975b. Ecological chemistry. LXXXIV. Metabolism of lower carbon-14-labelled polychlorinated biphenyls in the Rhesus monkey. *Bull. Environ. Contam. Toxicol.*, 13(4), 471–6.

Haque, A., Weisgerber, I., and Klein, W., 1977a. Absorption, efflux and metabolism of the herbicide (^{14}C) buturon as affected by plant nutrition. *J. Exp. Bot.*, 28, 468–79.

Haque, A., Weisgerber, I., Kotzias, D., and Klein, W., 1977b. Conversion of (^{14}C) buturon in soil and leaching under outdoor conditions. *Pestic. Biochem. Physiol.*, 7, 321–31.

Haque, A., Weisgerber, I., Kotzias, D., Klein, W., and Korte, F., 1976. Contributions to ecological chemistry. CXII. Balance of conversion of buturon-^{14}C in wheat under outdoor conditions. *J. Envrion. Sci, Health.*, B11(3), 211–23.

Herbst, E., Weisgerber, I., Klein, W., and Korte, F., 1976. Contributions to ecological chemistry. CXVIII. Distribution, bioaccumulation and transformation of carbon-14-labelled 2,2'-dichlorobiphenyl in goldfish. *Chemosphere*, 5(2), 127–30.

Hsu, T.-S. and Bartha, R., 1976. Hydrolyzable and nonhydrolyzable 3,4-dichloroaniline humus complexes and their respective rates of biodegradation. *J. Agric. Food Chem.*, **24(1)**, 118–22.

Hutzinger, O., Nash, D. M., Safe, S., de Freitas, A. S. W., Norstrom, R. J., Wildish, D. J., and Zitko, V., 1972. Polychlorinated biphenyls: Metabolic behaviour of pure isomer in pigeons, rats, and brook trout. *Science*, **178**, 312–314.

Iatropoulos, M. J., Milling, A., Müller, W., Nohynek, G., and Rozman, K., 1975. International Centre of Environmental Safety, Holoman AFB, New Mexico, U.S.A., unpublished.

Imura, N., Sukegawa, E., Pan, S.-K., Nagao, K., Kim, J.-Y., Kwan, T., and Ukita, T., 1971. Chemical methylation of inorganic mercury with methylcobalamin, a vitamin B-12 analog. *Science*, **172(3989)**, 1248–9.

Jensen, S. R. and Jernelov, A., 1969. Biological methylation of mercury in aquatic organisms. *Nature*, **223(5207)**, 753–4.

Kamal, M., Klein, W., and Korte, F., 1976a. Ecological chemistry. CXXIX. Isolation and characterization of metabolites after long-term feeding of 2,2'-dichlorobiphenyl-^{14}C to rats. *Chemosphere*, **5(5)**, 349–56.

Kamal, M., Weisgerber, I., Klein, W., and Korte, F., 1976b. Contributions to ecological chemistry. CXIV. Fate of 2,2'-dichlorobiphenyl-^{14}C in rats upon long-term feeding. *J. Environ. Sci. Health*, **B11**, 271–87.

Kaufman, D. D., Still, G. G., Paulson, G. D., and Bandal, S. K., 1976. Bound and conjugated pesticide residues. ACS Symposium Ser., No. 29, American Chemical Society, Washington, D.C., U.S.A.

Kilzer, L., Detera, S., Weisgerber, I., and Klein, W., 1974. Ecological chemistry. LXXVII. Distribution and metabolism of aldrin–dieldrin metabolite *trans*-4,5-dihydroxy-4,5-dihydroaldrin-^{14}C in lettuce and soil. *Chemosphere*, **3(4)**, 143–8.

Klein, W., Kohli, J., Weisgeber, I., and Korte, F., 1973. Fate of carbon-14-labelled aldrin in potatoes and soil under outdoor conditions. *J. Agric. Food Chem.*, **21(2)**, 152–6.

Klein, W. and Korte, F., 1970. In *Chemie der Pflanzenschutz- und Schädlingsbekämpfungsmittel*, Vol. 1, Springer-Verlag, Berlin–Heidelberg–New York, p. 199.

Klein, W., Müller, W. and Korte, F., 1968. Insecticides in metabolism. XVI. Excretion, distribution and metabolism of endrin-^{14}C in rats. *Justus Liebigs Ann. Chem.*, **713**, 180–5.

Klein, W. and Weisgerber, I., 1976. In *Environmental Quality and Safety*, Vol. V, Academic Press, New York, p. 237.

Kohli, J., Weisgerber, I., Klein, W., and Korte, F., 1973. Ecological chemistry. LIX. Residue retention and transformation of carbon-14-labelled dieldrin in cultivated plants, soils and ground water after soil application. *Chemosphere*, **2(4)**, 153–6.

Korte, F., 1968. Ecological chemistry. Metabolism of ^{14}C-chlorinated hydrocarbon insecticides. *Chimia*, **22**, 399–401.

Landner, L., 1971. Biochemical model for the biological methylation of mercury suggested from methylation studies *in vivo* with *Neurospora crassa*. *Nature*, **230(5294)**, 452–4.

Lay, J. P., 1976. Unpublished.

Lay, J. P., Klein, W., and Korte, F., 1975. Ecological chemistry. C. Excretion, accumulation, and metabolism of 2,4,6,2',4'-pentachlorobiphenyl-^{14}C after long-term feeding to rats. *Chemosphere*, **4(3)**, 161–8.

Ludwig, G., Weis, J., and Korte, F., 1964. Excretion and distribution of aldrin-^{14}C and its metabolites after oral administration for a long period of time. *Life Sci.*, **3(2)**, 123–30.

Mason, H. S., 1957. Mechanisms of oxygen metabolism. *Science*, **125**, 1185–8.

Matsumura, F. and Boush, G. M., 1967. Dieldrin; degradation by soil microorganisms. *Science*, **156**(3777), 959–61.
Matsumura, F., Boush, G. M., and Tai, A., 1968. Breakdown of dieldrin in the soil by a microorganism. *Nature*, **219**(5157), 965–7.
Matsumura, F., Gotoh, Y., and Boush, G. M., 1971. Phenylmercuric acetate. Metabolic conversion by microorganisms. *Science*, **173**(3991), 49–51.
Matthews, H. B. and Anderson, M. W., 1975. Effect of chlorination on the distribution and excretion of polychlorinated biphenyls. *Drug Metab. Dispos.*, **3**(5), 371–80.
Mauchline, J. and Templeton, W. L., 1964. Artificial and natural radioisotopes in the marine environment. *Oceanogr. Mar. Biol. Ann. Rev.*, **2**, 229–79.
Melancon, M. J. Jr. and Lech, J. J., 1976. Isolation and identification of a polar metabolite of tetrachlorobiphenyl from bile of rainbow trout exposed to ^{14}C-tetrachlorobiphenyl. *Bull. Environ. Contam. Toxicol.*, **15**(2), 181–8.
Metcalf, R. L., 1974. A laboratory model ecosystem for evaluating the chemical and biological behaviour of radiolabelled micropollutants. In *Comparative Studies of Food and Environmental Contamination*, IAEA, Wien, STI/PUB/348, pp. 49–63.
Metcalf, R. L., Sangha, G. K., and Kapoor, I. P., 1971. Model ecosystem for the evaluation of pesticide biodegradability and ecological magnification. *Environ. Sci. Technol.*, **5**(8), 709–13.
Milling, A., Müller, W., Korte, F., and Coulston, F., 1975. Unpublished.
Moore, N. W. and Walker, C. H., 1964. Organic chlorine insecticide residues in wild birds. *Nature*, **201**, 1072–3.
Moza, P., Weisgerber, I., and Klein, W., 1976. Fate of 2,2′-dichlorobiphenyl-^{14}C in carrots, sugar beets and soil under outdoor conditions. *J. Agric. Food Chem.*, **24**(4) 881–5.
Müller, W. P. and Korte, F., 1974. Microbial transformation of xenobiotics in waste compost. *Naturwissenschaften*, **61**(7) 326.
Müller, W. P. and Korte, F., 1975. Ecological chemistry. CII. Microbial degradation of benzo(a)pyrene, monolinuron and dieldrin in waste composting. *Chemosphere*, **4**(3), 195–8.
Müller, W., Nohynek, G., Woods, G., Korte, F., and Coulston, F., 1975. Comparative metabolism of carbon-14-labelled dieldrin in mouse, rat, rabbit, rhesus monkey and chimpanzee. *Chemosphere*, **4**(2) 89–92.
Müller, W., Rohleder, H., Klein, W., and Korte, F., 1974. Model studies of the behaviour of xenobiotic substances in waste composting. GSF-Berichte Ö 104, Published by the Gesellschaft für Strahlen- und Umweltforschung mbH, München, 8042 Neuherberg, Ingolstater Landstrasse 1, FRG (in German).
Neujahr, H. Y. and Bertilsson, L., 1971. Methylation of mercury compounds by methylcobalamin. *Biochemistry*, **10**(14), 2805–8.
Poonawalla, N. H. and Korte, F., 1971. Metabolism of *trans*-chlordane-^{14}C and isolation and identification of its metabolites from the urine of rabbits. *J. Agric. Food Chem.*, **19**(3), 467–70.
Richardson, A. and Robinson, J., 1971. Identification of a major metabolite of HEOD (dieldrin) in human feces. *Xenobiotica*, **1**(3), 213–9.
Rozman, K., Müller, W., Iatropoulos, M., Coulston, F., and Korte, F., 1975. Separation, body distribution, and metabolism of hexachlorobenzene after oral administration to rats and rhesus monkeys. *Chemosphere*, **4**(5), 289–98.
Russell, F. E., 1965. Marine toxins and venomous and poisonous marine animals. In *Advances in Marine Biology* (Ed. F. S. Russell), Vol. 3, Academic Press, London–New York, pp. 307–14.

Salonen, L. and Vaajakorpi, H. A., 1974. Bioaccumulation of ^{14}C-DDT in a small pond. In *Comparative Studies of Food and Environmental Contamination*, IAEA, Wien, STI/PUB/348, pp. 130–40.
Sanborn, J. R., Francis, B. M., and Metcalf, R. L., 1976. The degradation of selected pesticides in soil: a review of the published literature. U.S. Environmental Protection Agency, Office of Research and Development, Cincinnati, Ohio 45268.
Sandrock, K., Bieniek, D., Klein, W., and Korte, F., 1974. Ecological chemistry. LXXXVI. Isolation and structural elucidation of kelevan-^{14}C metabolites and balance in potatoes and soil. *Chemosphere*, 3(5), 199–204.
Scheunert, I., Kohli, J., Kaul, R., and Klein, W., 1977. Fate of ^{14}C aldrin in crop rotation under outdoor conditions. *Ecotox. Environ. Safety*, 1, 365–85.
Schulte, E. and Acker, L., 1974. Identification and metabolizability of polychlorinated biphenyls. *Naturwissenschaften*, 61(2), 79–80.
Schwemmer, B., Cochrane, W. P., and Polen, P. B., 1970. Oxychlordane, animal metabolite of chlordane: isolation and synthesis. *Science*, 169(3950), 1087.
Sugiura, K., Sato, S., and Goto, M., 1975. Ecological chemistry. VIII. Adsorption–diffusion mechanisms of BHC residues consideration based on bacteria experiments as models. *Chemosphere*, 4(3), 189–94.
Sugiura, K., Tanaka, N., Takeuchi, N., Toyoda, M., and Goto, M., 1976. Ecological chemistry. IX. Accumulation and elimination of PCB's in the mouse. *Chemosphere*, 5(1), 31–7.
Viswanathan, R., Klein, W., and Korte, F., 1976. Contributions to ecological chemistry. CXXX. Balance of the fate of 'Imugan'-C-14 in rats upon long-term feeding. *Chemosphere*, 5(5), 357–62.
Vockel, D. and Korte, F., 1974. Ecological chemistry. LXXX. Microbial degradation of dieldrin and 2,2'-dichlorobiphenyl. *Chemosphere*, 3(5), 177–82.
Walker, C. H., Hamilton, G. A., and Harrison, R. B., 1967. Organochlorine insecticide residues in wild birds in Britain. *J. Sci. Food Agric.*, 18(3), 123–9.
Weisgerber, I., Bieniek, D., Kohli, J., and Klein, W., 1975a. Isolation and identification of three unreported carbon-14-labelled photodieldrin metabolites in soil. *J. Agric. Food Chem.*, 23(5), 873–7.
Weisgerber, I., Detera, S., and Klein, W., 1974a. Ecological chemistry. LXXXVIII. Isolation and identification of some heptachlor-^{14}C metabolites from plants and soil. *Chemosphere*, 3(5), 221–6.
Weisgerber, I., Klein, W., Djirsarai, A., and Korte, F., 1968. Insecticides in metabolism. XV. Distribution and metabolism of endrin-^{14}C in cabbage. *Justus Liebigs Ann. Chem.*, 713, 175–9.
Weisgerber, I., Klein, W., and Korte, F., 1970. Ecological chemistry. XXVI. Transformation and residue behaviour of aldrin-^{14}C and dieldrin-^{14}C in cabbage, spinach and carrots. *Tetrahedron*, 26(3), 779–89.
Weisgerber, I., Kohli, J., Kaul, R., Klein, W., and Korte, F., 1974b. Fate of aldrin-^{14}C in maize, wheat, and soils under outdoor conditions. *J. Agric. Food Chem.*, 22(4), 609–12.
Weisgerber, I., Tomberg, W., Klein, W., and Korte, F., 1975b. Ecological Chemistry. XCV. Isolation and structure elucidation of hydrophilic carbon-14-labelled isodrin metabolites from common cabbage. *Chemosphere*, 4(2), 99–104.
Wong, P. T. S., Chau, Y. K., and Luxon, P. L., 1975. Methylation of lead in the environment. *Nature*, 253(5489), 263–4.
Wood, J. M., 1974. Biological cycles for toxic elements in the environment. *Science*, 183(4129), 1049–52.

Wood, J. M., Kennedy, F. S., and Rosen, C. G., 1968. Synthesis of methylmercury compounds by extracts of a methanogenic bacterium. *Nature,* **220(5163)**, 173–4.

Woodwell, G. M., Craig, P. P., and Johnson, H. A., 1971. DDT in the biosphere. Where does it go? *Science,* **174(4014)**, 1101–7.

Yamada, M., Dazai, M. and Tonomura, K., 1969. Changes of mercurial compounds in activated sludge. *Hakko Kogazu Zasshi (J. Ferment. Technol.),* **47(2)**, 155–60.

Yamamoto, H. and Yoshimura, H., 1973. Metabolic studies on polychlorinated biphenyls. CXI. Complete structure and acute toxicity of the metabolite of 2,3',4,4'-tetrachlorobiphenyl. *Chem. Pharm. Bull.* **21(10)**, 2237–42.

Zahn, R. and Wellens, H., 1974. Simple procedure for testing the biodegradability of products and sewage components. *Chem. Ztg.,* **98(5)**, 228–32.

Zeman, W., Gadermann, E., and Hardebeck, K., 1951. The genesis of disturbances of circulatory regulation. Toxic effect of tin peralkyls. *Dtsch. Arch. Klin. Med.,* **198**, 713–21.

CHAPTER 4

Models for Total Transport

D. R. MILLER

*Division of Biological Sciences, National Research Council of Canada,
Ottawa, Canada, K1A 0R6*

4.1 GENERAL 71
4.2 GEOGRAPHICAL TRANSPORT 73
4.3 TROPHIC-LEVEL AND SPECIES DISTRIBUTION 75
4.4 MODEL CONSTRUCTION: GENERAL 76
4.5 COMPARTMENT MODELS 78
4.6 AN EXAMPLE: MERCURY IN WATER 80
4.7 USES OF MODELS 85
4.8 FINAL THOUGHTS 87
4.9 CONCLUSIONS 88
4.10 REFERENCES 88

4.1. GENERAL

This chapter deals with the problem of setting up a quantitative and predictive mathematical model for pollutant distribution. It will be based on information of the type discussed in the previous chapters and should make possible the dose calculations described in the chapter to follow.

First, let us make clear that we mean to address the broad question of how a pollutant is partitioned in various parts of the environment, how this partition changes with time, and how the pollutant may be converted to different chemical forms as part of the process. When we use the word distribution we imply a description involving all these facets.

It seems (Goodall, 1974)[*] that there are four principal factors that must be considered. While they are all interrelated, we can try to separate them for simplicity. The first is geographical; different pollutants are typically transported over different distances, and local, regional, and global movements usually involve different combinations of transport mechanisms.

---

[*]There are many references describing the process of modelling in an ecological context; for general treatments see Jeffers (1972), Smith (1974), or Waide and Webster (1976). References dealing with pollutant transport in general terms are rather fewer, although the field is well developed for radionuclide transport and gross pollution of potentially potable water supplies (for review see Argentisi *et al.*, 1973; for the most recent general treatment see Holcomb Research Institute, 1976). Most other references cited in this chapter deal specifically with pollutant movement.

The second is time scale. It is clear that movement of a pollutant over a time scale of years is quite different from that over a scale of minutes or hours; also it involves different distance scales. In fact, distances and time are so interrelated that they need to be discussed together (Goodall, 1974), as they are in what follows.

The third important thing to be described is distribution of a pollutant among trophic levels and species (Goodman, 1974). Much work has been devoted to this subject; if a certain amount of aggregation is permitted, a reasonable description can often be put together on the basis of information on uptake of pollutants by individual species, accumulation in the different trophic levels and so forth. Much of the material presented later in this volume relates to this step.

The fourth aspect is the chemical form of the pollutant. This aspect is all-pervasive, and strongly influences the other three (Wood, 1974). A dominant consideration is that the chemical form may change in various ways during movement of the pollutant, both by physical mechanisms, particularly in the atmosphere, and in association with biota, by various processes of biotransformation and biodegradation. Thus this aspect has profound importance both for mechanisms of transport and for influence on uptake, retention, and toxicity in biota. These have been the main subjects addressed in the preceding two chapters; our approach here is not to duplicate what has been said, but only to restate some of it in the form most suitable for use in the discussion of model formulation to follow.

Before we proceed, we should admit that any such approach is bound to be a simplification. This, of course, is true of any scientific approach; no experimental or theoretical analysis ever includes every possible interaction between elements of the system (Rapoport, 1972). As applied to this discussion, it means that not every possible interaction can be included in the models presented. The whole idea of modelling involves identification not of every possible pathway but only of those that account for the largest part of the phenomena observed (Smith, 1974). The reader will undoubtedly notice that some interconnections have been omitted; it should not be assumed that they have been forgotten or ignored.

Connected with this point is another aspect, or purpose, of mathematical models, namely their use in identifying pathways for which our understanding is relatively imprecise (relative, that is, to their importance in the system) as well as those which, although they may be biologically or chemically interesting, are not worthy of further refinement for the quantitative study of pollutant transfer under consideration. Indeed, it may well be said that at the present stage of development of the modelling art, the main advantage of formulating a mathematical model is not the making of precise predictions of the future state of the system. Rather, the value lies in the very exercise of stating precisely how much is really known about each pathway, and in the possibility of allocating future research effort in such a way as to contribute most effectively to quantitative knowledge of the overall behaviour of the system (Miller *et al.*, 1976).

4.2. GEOGRAPHICAL TRANSPORT

The continuum of distances over which a pollutant can be transported is traditionally, if somewhat arbitrarily, divided into levels such as local, regional, and global. By these we mean, respectively, distances of a few kilometres, of a hundred to perhaps a thousand kilometres, and of many thousands of kilometres. In most cases, geographical transport by physical factors greatly dominates that due to movement of or by biota (Bolin, 1976). Atmospheric transport is involved in all three, but with different time scales: minutes for local, hours to days for regional, and many weeks or more for global distribution. General oceanic circulation may be significant for global transport, but only over a time scale of years (Wollast *et al.*, 1974). Transport by water (rivers) can be important for regional distribution, while surface run-off and subsurface water must sometimes be considered for local pollutant distribution (transport of dust by air, or of sediment by water, is considered to be included). It should be specifically mentioned that although a general mixing and dilution occurs in most cases, there can also be situations involving surprisingly little general dispersion or dilution. Atmospheric transport may be confined to a relatively narrow plume immediately downwind of a source (e.g. Kao, 1974); even in oceans or large lakes, pollutants may remain in the top few metres of water for some time, taking many years before mixing is truly complete (Robinson, 1973). Another complicating factor is that each of these processes may occur several times, such as when dust is resuspended by wind or when sediment is moved downstream in a river by successive spring floods.

It is true that each particular mode of movement is a part of the overall global circulation of a pollutant. However, movement on scales of different time and distance may often be thought of as essentially decoupled. The study of transport of a chemical in a river, for example, will typically require that exchange to and from the atmosphere be considered; this does not, however, require that global atmospheric movement be considered simultaneously, for a particular river will contribute negligible amounts to the atmosphere, and the latter can be considered a fixed reservoir of large size for the purpose of studying the river (Bolin, 1976).

Concerning actual transport rates, approaches differ for each class of problem. Predicting airborne transport at local distances depends very much on having information on vertical structure of the atmosphere as well as current wind conditions. The techniques of Pasquill (1974) and others are quite well developed for this sort of problem, and one variant or another of this approach is in wide use today.

For regional atmospheric transport, these methods fail since the weather pattern itself will typically change in a time comparable to that taken by the pollutant to travel a few hundred kilometres, and knowledge of wind conditions at the source provides an inadequate basis. Here, one must use either the as-yet-underdeveloped general circulation models, or trust to the simpler but justly criticized approach of

Trajectory Analysis, which also requires more detailed wind and weather data than are easily accessible (Nordo et al., 1974). Global transport may be even more difficult, and only general large-scale mixing and transfer rates are normally used (Bolin, 1976).

Transfer from atmosphere to ground level is by absorption into, or adsorption onto, ground-based substances, or by wet or dry deposition. Rates at which these take place are moderately well understood when airborne pollutants are attached to large particles which settle out rapidly (dry deposition), in which case one can speak of deposition velocities estimated on the basis of laboratory and field experiments. For removal from the atmosphere by rain or snow, less is known, but some guidelines are available; one needs to estimate the fraction of the pollutant contained in the air column up to the height at which the precipitation forms, and to use a fractional clearance factor based on the degree of precipitation (Slade, 1968). Some substances, such as radioactive noble gases, may be assumed to be uniformly mixed so that concentrations may be calculated directly. (In all these estimates, many further details must be considered, such as the fact that pollutants do not normally rise above the tropopause in local- or regional-scale problems; see Slade, 1968; Nordo et al., 1974; Pasquill, 1974.)

Movements in the reverse direction, i.e. movements into the atmosphere by resuspension, volatilization, etc., are much less well understood. At this time, few generalizations are available and the phenomena must be investigated separately for each pollutant examined.

Transport by water (normally fresh water) over local and regional distances, has been extensively studied. Generally, polluting materials will dissolve (phosphates, oxides of sulphur) or be attached to particulates (heavy metals, water-insoluble hydrocarbons) (Goodman, 1974). In the latter case, transport is determined by the behaviour of the particulates in the water itself, and the characteristics of the river will determine whether the pollutants move only during movement of bed sediment, or whether a significant amount is bound to particles small enough to remain in suspension. Typically, most of the pollutant is at any time bound to larger bed sediment particles, but the suspended materials account for most of the transport (De Groot and Allersma, 1975).

When a pollutant comes into contact with ground or water, most of it adsorbs or binds chemically to soil or sediment particles. Here, complicated chemical and biological reactions may take place which again have been the subject of large amounts of research but permit few generalizations (Wood, 1974). Typically, the soil or sediment constitutes a large reservoir of relatively unavailable polluting material, while chemical equilibria or biologically mediated transformations keep small but significant amounts available for transport or uptake by biota; particularly for heavy metals, these amounts may be in much more toxic chemical forms (Krenkel, 1973).

In general, then, pathways are many and varied. Each of the mechanisms mentioned is under active study by many scientists, whose work we clearly cannot hope to summarize in the space available. However, once a particular pollutant release is identified for consideration, it becomes clear that only a few of the pathways are important. We hope to illustrate this with the examples to follow.

4.3. TROPHIC-LEVEL AND SPECIES DISTRIBUTION

There is no end to the discussion about subdividing an ecosystem into compartments. Many biologists point out (correctly!) that inter-species, inter-age, and inter-sex differences are great and that there is no possibility of discussing how 'an invertebrate', for example, processes or reacts to any given substance. Nonetheless, the concept of energy, mass, and nutrient transfer between trophic levels in an ecosystem has proved extremely useful in many studies, and can certainly be applied to pollutant transfer problems (Rigler, 1975). Furthermore, the alternative of studying all species in sufficient detail to be able to predict their responses to a given pollutant is simply not practicable and we are forced to imagine the ecosystem as subdivided into rather large and internally heterogeneous compartments, and to describe each in overall terms only.

One saving feature of this approach is that the individual scientist can frequently add to the information about a compartment by using his own more detailed understanding of it. Thus, for example, if in an aquatic ecosystem it were known that the average concentration of a particular heavy metal in fish was of a certain value, the biologist would typically expect this concentration to be higher in larger and in older fish, and certainly higher in piscivorous species (Bligh, 1971).

The greatest advantage of the trophic-level approach is that pollutants typically move from one species to another, either through the environment or along the food chain, and the latter is precisely what the energy and nutrient cycling studies are concerned with. If one knows the food intake and gut absorption, as well as the body retention function for that particular combination of animal and pollutant, then the resulting body levels can be calculated (Fagerstrom and Asell, 1973).

It is here that we encounter again the importance of the chemical form of the pollutant since absorption, retention, and toxicity (not to mention solubility and other physical variables) change as a pollutant takes different chemical forms (Goodman, 1974). This happens, for example, when a heavy metal is changed from inorganic to organic form or the reverse.

Because of these features of the problem, it may be necessary either to restrict one's consideration to the form of the substance that is most toxic, or to consider simultaneously several 'superimposed' diagrams of the system, one for each chemical form, with interconnections at (and only at) those points where chemical conversion may take place (see the example of mercury, below). The rate of transfer due to chemical change, especially in respect of biologically mediated transformations but to a lesser extent in those governed by chemical equilibria, remains one of the areas of greatest ignorance. Nonetheless, such transfer must be quantified since we need to know the total toxicity of the forms present, not just the amount of the substance, if ecotoxicity is to be made a subject of prediction and precise analysis (Truhaut, 1974).

4.4. MODEL CONSTRUCTION: GENERAL

To the word 'model' different writers attach quite different meanings, sometimes with unfavourable connotations. Let us begin, therefore, by saying that we do not use the word to describe something different from or unrelated to the actual ecosystem under study. To us, a 'model' is simply a quantitative summary statement of what is known about the important processes going on in the system. It can be no more elaborate than knowledge of the various real processes permits — failure to accept this limitation has caused difficulties for many modellers — but it can provide quantitative information that intuition could not (Holcomb, 1976).

There are still many types of models, and for the present purposes we choose to identify three (Smith, 1974). To begin with, one can talk about simple calculations designed to elucidate, for example, the relation between biomasses and energy transfers or perhaps to point out how simple the bioaccumulation of a pollutant really is. It is this type of mathematics that the field biologist or other scientist understands most easily, and the kind that is generally taken most seriously.

A second type of model involves overall balances of inputs and outputs of a given substance to and from a system (Odum, 1971). While not designed to predict future history or to address mechanisms by which pollutants move, this type is critical for an understanding of the relative importance of different pathways, and the exercise of producing the appropriate numerical estimates sometimes leads to the realization that what had previously been regarded as a dominant pathway is not the most important at all. This sort of formulation has the added advantage of helping to identify mechanisms for which understanding is lacking, and focuses our attention on pathways that merit closer attention and further research (Fagerstrom and Asell, 1973).

A third and quite different kind of 'model' is one that might conceivably be used where no idea of overall performance is available, and is designed to simulate future history in a descriptive and predictive way, rather than to expose the simplicity of its structure or its overall balance. This type of 'model' has many good and bad (mainly bad) characteristics: it is not accessible to people most competent to criticize its reality; a cumbersome process of computer programming is required before its predictions can begin to be examined; and its verification and validation require techniques for which the basic methodology, not to mention general confidence in them, has not been developed (Mar, 1974). Nonetheless, it is the only type that can provide information totally inaccessible by intuitive approaches (Robinson, 1973).

So far, in using the word 'model' we have been referring to the third type. This is deliberate; highly simplified calculations will never have sufficient predictive value, reliability or subtlety to support the complex decisions involved in setting environmental standards. However, it might help if an example or two of the simpler concepts were given before the more complicated one is attacked.

Consider a substance that passes unchanged through several predator–prey interactions, and for which retention in the body of the predator is rather long. Specifically, assume that each day a predator, with body weight b_1, eats a_1 g of prey which contains a concentration x_0 of pollutant. Assume further that the predator absorbs a fraction f_1 of the pollutant taken in, and clears it at a fractional rate per day of k_1. Then the equilibrium concentration of pollutant in the body of the predator will be

$$x_1 = \frac{a_1 f_1}{b_1 k_1} x_0$$

If this predator is a prey for some higher organism, then the equilibrium level for pollutant in its flesh will be in turn

$$x_2 = \frac{a_2 f_2}{b_2 k_2} \frac{a_1 f_1}{b_1 k_1} x_0$$

and this may be repeated several more times. Since the k_i are typically small ($k = 0.01$ for a biological half-life of 70 days), a many-fold magnification can take place at each level.

It is perhaps worth saying explicitly that this is the end of the example; the calculations are meant to clarify the mechanism rather than to predict future history.

Sometimes mechanisms can be greatly clarified on the basis of simple calculations. As an example of this, consider the situation with the highly toxic monomethylmercury, as opposed to less toxic inorganic mercury compounds, and the observation that the fraction of mercury in the organic form increases sharply as it moves up the food chain, independent of the total mercury concentration. Typical values for mercury in the organic form (Miller, 1977) are the following:

invertebrates	25–30%
fish	65–75%
fish (piscivorous)	85–95%

The suggestion has often been made, based largely on these numbers, that fish must be able to convert inorganic mercury to the organic form.

However, organic mercury is much more efficiently absorbed through the intestine; experiments demonstrate that perhaps 95% of organic mercury is absorbed, compared to perhaps 15% for the inorganic form (de Freitas, 1977). Suppose that the ratio of organic to inorganic mercury in invertebrates is 3:7 (i.e. 30% organic, 70% inorganic), then the corresponding ratio for the mercury actually absorbed by their predator is

$$\frac{3}{7} \times \frac{0.95}{0.15} = 2.7$$

which is equivalent to 73% organic. A second trophic level increases this to a ratio of 17 : 1, or a percentage of 94% organic. These are close enough to observed values to make it clear not only that we do not need to postulate a methylating mechanism, but that in fact if such a mechanism were present, we would be hard pressed to explain why levels of organic mercury are not even higher. (In fact, the fraction in organic form is higher than this, since inorganic mercury undergoes faster clearance.)

Many other examples could be given, each describing a single mechanism, generally a simple one, by which one can explain most or all of some observations on a system. Such examples are satisfying; however this situation is not what one encounters in environmental questions. The ecotoxicologist is forced to consider the problem of describing many interacting processes. Naturally, we lose the ready appeal and clarity of simple models, and must proceed cautiously; however we may not avoid a problem or a solution simply because it is complex.

The next section deals with the overall cycling or movement of a pollutant, using mercury as an example.

4.5. COMPARTMENT MODELS

One begins by dividing the ecosystem into a collection of separate (mutually exclusive) compartments, generally numbering around 10 or fewer (Smith, 1974). This leads, admittedly, to a very rough description, but the larger the number of compartments the more interactions must be investigated; in fact the number of interactions increases roughly with the square of the number of compartments (Miller, 1977). One must make some sort of trade-off between a model so simple that it can provide no answers, and one so complex that more interactions are involved than one can possibly quantify.* Probably this figure of 10 is fairly close to the optimum at present, although more complex models are feasible where research resources are plentiful (Patten, 1971).

For each compartment one identifies a small number of key quantities adequate to describe it. Examples are total amount of pollutant it contains, and total mass or biomass (or energy or carbon equivalent). Normally these are regarded as varying in

---

*There does exist a school of thought which holds that these constraints can never be satisfied together, and any model simple enough to calibrate will be too simple to be meaningful. The author does not subscribe to this point of view but cannot disprove it.

time, with steady-state values considered as a special (simplified) case. Procedures for working with a time-dependent model are important.

One must always bear in mind that the describing variables (often called State Variables) chosen at this point represent everything that the model will ever be able to tell the user about the system; it is at this stage, not later, that one must decide on the level of detail of the questions that the model will be expected to answer. Thus, for example, if only total biomass is considered for some compartment, it is not meaningful to ask later whether shifts in age or species distribution might or might not be predicted by the mathematical formulation.

Since each of these variables changes over time, one must describe how their values are to be calculated, normally as a time-dependent differential equation. If the variables represent quantities that are conserved, e.g. persistent chemicals, or mass or energy, then any change in a compartmental level must be explained by input from or output to interconnected compartments. In these other compartments, one also knows the level of the same substance, and it is only a matter of choosing an appropriate form for the 'rate equation' and numerical value for the 'rate constant' (as well as deciding how the rate constant is to be assumed to vary with time, weather, temperature, and a variety of other physical and chemical factors).

In many cases, it is reasonable to assume that the rate of transfer of material out of some compartment is proportional to the instantaneous amount present (a constant death rate, for example, or a radioactive decay). Then, one can write the governing equations in the form often referred to as 'first-order kinetics' or, more simply, as a linear equation. In the (very unlikely) case in which it was reasonable to approximate *all* such interactions with linear terms, one would have a fully linear system; the mathematical properties of such systems have been extensively investigated. Indeed, some workers deliberately choose a fully linear formulation, not because they are convinced that it properly describes all interactions, but because the added analytical power of a linear model may make up for some inaccuracies (Waide and Webster, 1976).

In general, there are compelling reasons for believing that not all interactions can be described by linear terms, and some non-linear expressions will generally be incorporated in the system model (Burns, 1975). In principle, this does not alter the ability of a reasonably sized computer to predict the future from the model. It does, however, make life difficult for the analyst or theoretician who wishes to make generalizations about model behaviour and its response to disturbances.

We should insert a comment about the relative difficulty of the various steps in model formulation, namely the comment that the mathematics is not the problem. Complex formulations involving sophisticated mathematical notation abound; what is missing in most cases is a rational consideration of what forms the various interactions may take and, even more important, estimates of the numerical value of the rate constants and how they vary with environmental conditions. Since it is immediately clear that these questions can occupy years of effort for each interaction, it follows that any formulation requiring the determination of many such coefficients is unlikely to be implemented or to yield useful results (Holcomb Research Institute, 1976).

4.6. AN EXAMPLE: MERCURY IN WATER

The basic qualitative information about mercury behaviour in water has been known for some time (Goldwater, 1972). Generally, mercury enters a river in a moderately harmful chemical form, sinks, and resides in the sediments for lengthy periods. For the most part, what geographical transport does take place is desorption into the water column coupled with some movement of the sediment itself (Krenkel, 1973).

The hazard resulting from sedimentary mercury became evident when it was observed that fish were contaminated with low, but toxicologically significant, levels of monomethylmercury (Fimreite, 1970), a form causing a variety of neurological symptoms through poorly understood mechanisms (Bidstrup, 1964). Subsequent experiments showed that this methylmercury can be synthesized *in situ* by bacterial action, probably in the sediment (Jensen and Jernelov, 1969), or perhaps photochemically in the water column itself (Akagi and Takabatake, 1973).

In accordance with what was said earlier, we must be careful to subdivide the system into a minimum number of compartments consistent with the known transport and transfer processes. At the very least, it seems clear that water and sediment (suspended and deposited) must be separated; in general we know that the nature and amount of suspended material in the water greatly affects its binding capacity for heavy metals. For overall transport of the bulk of the mercury, probably nothing else needs to be considered (Miller, 1977).

From the standpoint of ecotoxicology, however, biota must be considered, especially fish, since fish serve as a bioaccumulator and the vector to man, and since the vanishing of fish is an observed ecological effect of mercury contamination (Fimreite, 1970). In order to describe the pathways of mercury to fish, it would seem that compartments for fish, higher plants, and invertebrates are needed.

Further subdivisions could be considered. Piscivorous fish are observed to contain higher mercury levels than others (Bligh, 1971), and could be modelled separately; also, it might well be worthwhile to subdivide further both suspended materials and sediments into organic and inorganic fractions.

Finally, as was emphasized earlier, the chemical form of the mercury must be considered. In the present case monomethylmercury must obviously be singled out for special consideration because of its much greater toxicity (Bidstrup, 1964), and for simplicity we may hope to include all other forms in some compartment such as 'inorganic mercury', since our interest is only to describe the transfer through fish to man.

In most presentations, one sees a diagram much like that in Figure 4.1, in which are shown the compartments and the important interactions (Miller, 1977).

Models for Total Transport

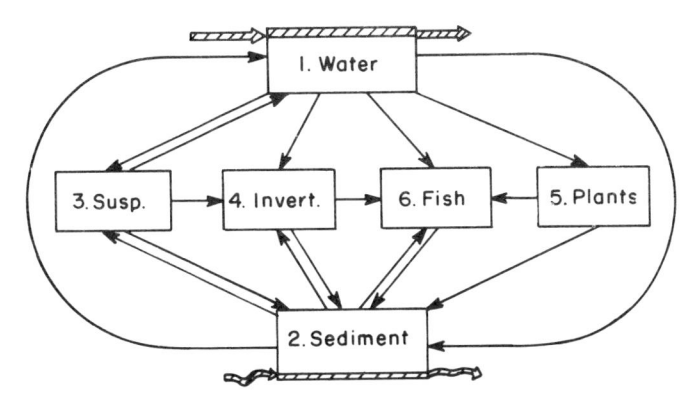

Figure 4.1   Pollutant transport model: mercury in an aquatic ecosystem

However, such a diagram is misleading. It does not make clear whether each box represents amount of pollutant, amount of biomass, or some combination; nor is it clear whether arrows represent transfer of mass but not necessarily pollutant (as from fish to sediment) or pollutant but not biomass (water to fish), etc. It must be remembered that pollutant moves in several ways; it is carried along by biomass transfer but also moves by adsorption, desorption, etc., which have no associated biomass movement. Thus such transport phenomena must be considered separately.

To be consistent with what was said earlier, we might model the system at several levels simultaneously, presumably having one set of variables for biomasses and another for each of the critical chemical species. In these terms, it would seem that the 'minimum resolution' model would involve six compartments, within each of which we keep track of biomass and two kinds of mercury, and would require 18 descriptive variables and interactions which, if not further restricted, would number approximately 80.

However, further consideration shows that some of the compartments and many of the interactions may be omitted. For example, we have agreed that biota play little part in the movement of total mercury, and are little affected by it, so three compartments may be ignored in the second level. Similarly the various arrows often indicate transfers that are clearly not real (such as movement of biomass from water to fish) or have been demonstrated to be negligible in studies undertaken for the specific purpose (e.g. uptake of mercury by plants directly from sediment). Detailed considerations of this type produce the much-simplified version shown in Figure 4.2 (in which no arrows have been omitted). Now, the number of compartments is reduced to 15 and the number of interactions to around 40. These numbers are still high; and we are forced to resist the temptation to subdivide further in such ways as mentioned above.

82   *Principles of Ecotoxicology*

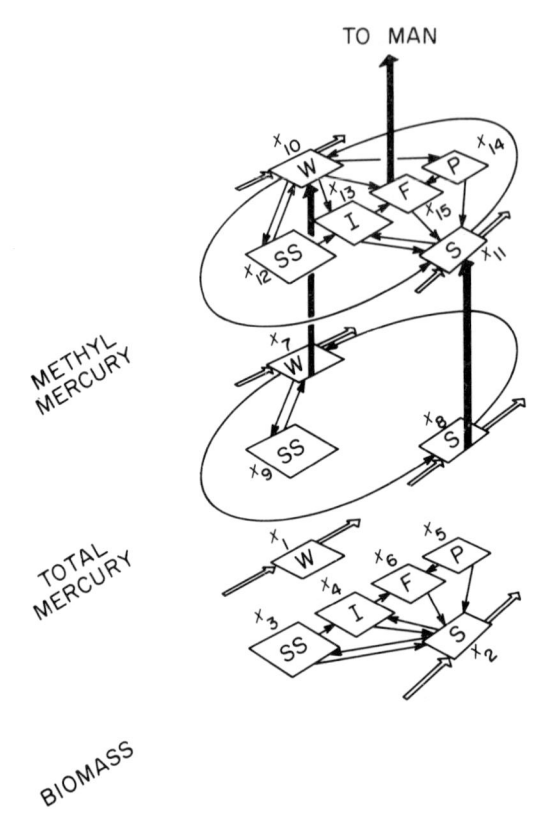

Figure 4.2   The dynamics of methylmercury production

We now assign symbols x_1 to x_{15} shown in Figure 4.2 to represent the masses of each compartment. Notice that these are all in units of mass, but represent masses of different substances, and will be greatly different in magnitude, ranging from millions of kilograms (x_1, x_2) to one gram or less (x_{12} to x_{15}). (This may also lead to numerical difficulties but such computational problems will not be discussed here.)

Each mass must next be made the subject of a rate-of-change equation of the form

$$\frac{dx_i}{dt} = \text{Sum of several terms, each representing one interaction}$$

and the remaining task is to decide on the form of each term. The symbol $k_{i,j}$ conventionally represents the rate constant for the transfer *from x_j to x_i*. Thus, for example, $k_{7,9}$ would be involved in the term describing movement of total mercury in the suspended-solid compartment (x_9) to mercury associated with water (x_7), i.e. a process of desorption and solubilization (Miller, 1977).

Finally, we can address the actual terms. For obvious reasons, we do not include an exhaustive listing of them all, but rather select just one as an example. Somewhat arbitrarily, let us consider x_{12}.

For x_{12}, there are three mechanisms: uptake of methylmercury from water, clearance into water, and browsing by invertebrates. Remembering that x_{12} represents the methylmercury content of the suspended solids, not their mass or any such quantity, we can conjure up an expression for each providing that we know the dynamics of the mass itself, which is the subject of the equation for x_3, not discussed here.

Table 4.1 Coefficients for the Mercury Transport Model (Partial List)

Coefficient	Numerical value	Units
$k_{2,3}$	0 Summer 0.077 (May only)	fraction/day
$k_{2,4}$	0.001–0.003	fraction/day
$k_{2,5}$	0.0 Winter–summer 0.006 (Aug. to Nov.)	fraction/day
$k_{2,6}$	0.002 Average	fraction/day
$k_{3,3}$	0.05	fraction/day
$k_{4,2}$	0.100	kg sed/kg invertebrate/day
$k_{4,3}$	0.05	kg/kg inv/day
$k_{5,5}$	0.005–0.03 (May to July)	fraction/day
$k_{6,4}$	0.056–0.063	kg/kg fish/day
$k_{6,5}$	0.02 (June to mid-Sept.)	kg/kg fish/day
$k_{6,6}$	2.0×10^{-7}	fractional desorption/day
$k_{8,7}$	5.545	fractional sorption/day
$k_{7,9}$	0.00016	fractional desorption/day
$k_{9,7}$	16.6	fractional sorption/day
$k_{11,12}$	0.224	fractional desorption/day
$k_{12,11}$	5.545	fractional sorption/day
$k_{11,13}$	0.00049	fractional desorption/day
$k_{13,11}$	16.6	fractional sorption/day
$k_{12,8}$	4.0×10^{-5}	kg MeHg/kg Hg/day
$k_{12,14}$	0.009	fractional clearance/day
$k_{12,15}$	1.0	excretion efficiency
$k_{12,16}$	0.014	fractional clearance/day
$k_{14,11}$	52.04	kg water cleared/kg invertebrate/day
$k_{14,13}$	0.85	digestion efficiency (MeHg)
$k_{15,11}$	81.0	kg water cleared/kg plant/day
$k_{16,11}$	96.0	kg water cleared/kg fish/day
$k_{16,14}$	0.85	digestion efficiency of MeHg
$k_{16,15}$	0.85	digestion efficiency of MeHg

Uptake rate of methylmercury from water will be proportional to the concentration (not the mass) of methylmercury in water, which is x_{10}/x_1. It will also be proportional to the mass of the solids, i.e. to x_{12}, but will be limited by the concentration present. Thus when we include a rate constant which is to be determined, we have a form like

$$\frac{dx_{12}}{dt} = + k_{12,10} x_{10} x_{12} (C_{12,\max} - x_{12}/x_3) x_1$$

(a non-linear term). For clearance to water, a term of the form

$$-k_{10,12}(x_{12}/x_3)(C_{10,\max} - x_{10}/x_1)$$

may be appropriate. Removal by invertebrates is equal to the browsing rate of the equation for x_3 multiplied by the concentration of methylmercury; the term might have a form like

$$-k_{13,12}(k_{4,3} x_4 x_3/x_{3,\text{avg}})(x_{12}/x_3)$$

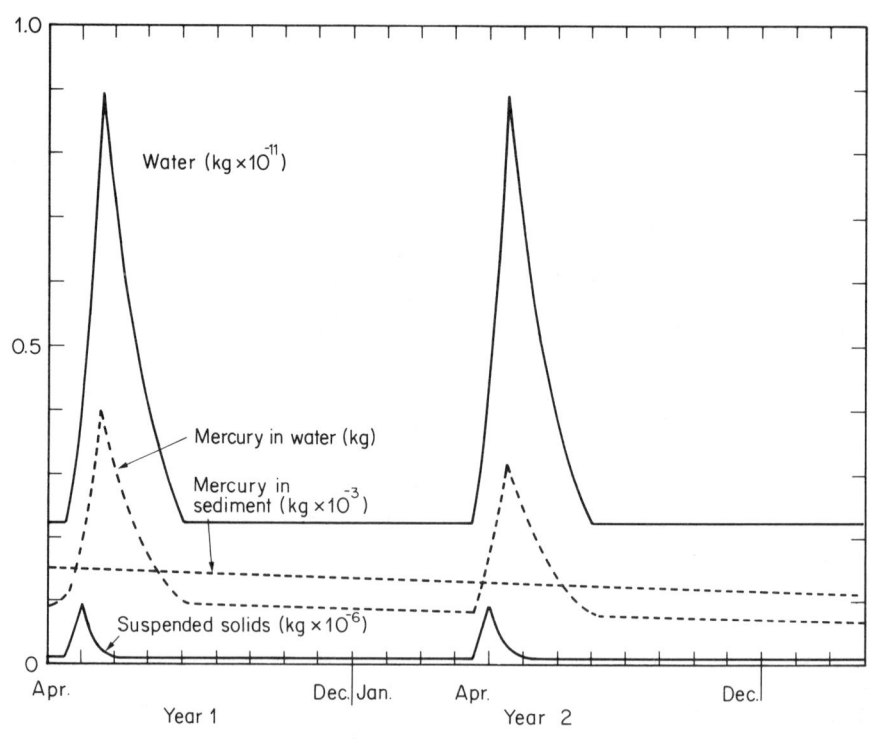

Figure 4.3 Behaviour of mercury in an aquatic system. W = total mass of water; SS = mass of suspended material; TMW = total mercury associated with water; TMS = total mercury associated with sediment. Spring floods cause violent fluctuations in all but the final quantity, which decreases smoothly with time

with the constant $k_{13,12}$ now representing an efficiency of ingestion of methylmercury when contained in food particles, i.e. a gut uptake coefficient determined in the laboratory. These terms would be combined with an overall transport term, also determined by measurement.

We need go no further to observe how terms are constructed (or how much uncertainty there is in the actual mechanisms). The next step is to summarize numerical values for the various $k_{i,j}$ (leaving it to the reader to surmise the functional form of the term). This is done in Table 4.1.

Finally, we display the results of a two-year run in Figure 4.3. It can be seen that overall mercury levels in sediment are predicted to decrease substantially each year and concentration of mercury in fish lags noticeably behind. We will comment further on this model in section 4.7.

4.7. USES OF MODELS

There are many uses for models formulated in this fashion. Roughly in temporal sequence rather than in order of importance, the main ones would seem to be the following (Naylor *et al.*, 1968):

1. to force participants to agree on clear definitions of compartments and pathways;
2. to clarify the extent to which each mechanism of transport or transfer can be described in quantitative terms;
3. to make numerical predictions of future behaviour;
4. to estimate the uncertainties in those predictions;
5. to assist in research planning and resource allocation by identifying critical and poorly understood mechanisms;
6. to allow the scientist to test the response of the simulated system to specified disturbances; and ultimately
7. to predict the response of the system to various management scenarios.

There are other purposes, of course, including such things as suggesting validation experiments and procedures. However, those listed above seem to be the most important to the ultimate user or decision-maker as opposed to the analyst. The main point is that many uses exist for such a formulation; it is unfortunate that so much attention has been devoted to the third and so little to those that follow it (Frenkiel and Goodall, 1977).

The first two probably do not need to be discussed further since they have been illustrated by this chapter. Of course, it is always worthwhile emphasizing that, in the actual research process, these stages may involve substantial personal and professional conflict. The resolution of this conflict early in the course of the work is essential for acceptance, if not success, of the steps to follow (Holcomb Research Institute, 1976).

The third, prediction of the future state of the system, is the one most of us are most aware of. It is necessary to repeat that what is predicted is only the value of those describing variables that have been defined explicitly in the model formulation. Generally, this constitutes *less* information than the ecologist could gain by the most cursory of on-site inspections (although it may be more quantitative) and in most cases the biologist can and should interpret and extend the model predictions in the light of his more detailed understanding of the system.

Estimation of uncertainty, or probable error level, in model prediction, is a relatively simple procedure which has only recently begun to attract as much attention as it deserves (Burns, 1975). Essentially, error arises because of our lack of knowledge about both the functional form of the interactions and the numerical values (and variations) of the parameters, or rate constants, involved. The latter are more important over the short term, the former over the long.

For parameter values, the procedure for error estimation depends on the linearity or non-linearity of the model. One asserts that the output variables depend on the input parameters in some way,

$$x_j(t) = f(p_1, p_2, \ldots, p_m)$$

and identifies the 'best estimate', say $x_j^0(t)$, as that sequence of values produced by the 'best' parameter values (say p_i^0) so that

$$x_j^0(t) = f(p_1^0, p_2^0, \ldots, p_m^0).$$

If this dependence is linear, the changes in the x_j will be proportional to the changes in the p_i, so that changing any number of the parameters p_i will change the output in a way that may be calculated as

$$x_j - x_j^0 = \Sigma_i \frac{\partial f}{\partial p_i}(p_i - p_i^0).$$

(In practice, these coefficients $\partial f/\partial p_i$ are simple to calculate numerically.) Next, one estimates how uncertain the actual values used for p_i^0 really are, and uses this formula to produce an estimate of the uncertainty in the x_j.

If the behaviour of the system is non-linear, the same answers can be obtained without using a linear formula. To do this, one again estimates the probable range within which each 'true' parameter value is thought to lie, and solves the model many times (i.e. uses the computer to generate many future behaviours), each time using for the parameters values chosen at random from somewhere within the appropriate range (the Monte Carlo technique). Finally, one observes the magnitudes of run-to-run differences in output variables and uses these as estimates of real uncertainties.*

---

*This description of a Monte Carlo process has been kept very general largely because of space limitations and in the hope that this work may appeal to a wide audience. The technical details have been discussed elsewhere (Miller, 1974; Burns, 1975).

The technique of Sensitivity Analysis is closely allied to these ideas. In that approach, one tries to identify some measure of disturbance, or some other effect on the ecosystem, which depends in a simple way on the describing variables. For example, one might use the sum of squares of the differences between the disturbed output variables and those for the 'best' run, integrated over time.

$$D = \int_0^T \Sigma_j [x_j(t) - x_j^0(t)]^2 \, dt$$

Next, usually by solving the computer model again with a single parameter altered, one calculates the change produced in D by a unit change in that parameter. This is the Sensitivity Coefficient for that parameter:

$$S_i = \frac{\partial D}{\partial p_i}$$

In most practical cases, these coefficients differ from one parameter to another by orders of magnitude. By observing their relative sizes one can then consider allocating greater research effort to the investigation of those mechanisms to which model behaviour is particularly sensitive. The approach has been rather widely used to guide data gathering efforts in, for example, geophysical exploration (Meyer, 1971); its use in assignment of research priorities is more recent.

In some cases it may be observed that all the sensitivity coefficients are small, so that model performance and predictions are relatively stable over a range of parameter variations and other disturbances. Such a model is referred to as 'robust', and one has some confidence that its predictions represent the real performance of the system. That is to say, one can argue that it is likely to be a valid representation of reality.

The concept of validity is, naturally, of great concern to modellers; any use of a model depends ultimately on the assumption that its performance does correspond to the actual behaviour of the system, and the process of demonstrating this, the so-called validation process, is a vital part of model development (Miller, 1977). Unfortunately, there is no general procedure for carrying it out, any more than there is a general procedure for verifying any scientific theory. The best that one can do is to demonstrate that the behaviour of the universe is consistent with the model over a certain range of conditions, subject to the limited predictive abilities of the model and a recognized level of uncertainty. The validation process is often thought of as the demonstration that the model is able successfully to predict the state of the actual system. Actually, validation should be regarded as a continuing effort or at least as an iterative process.

4.8. FINAL THOUGHTS

Ultimately, the purpose of this or any model is to provide information on which judgments of ecotoxicology can be based. For this purpose, the final product is the time course of exposure of some target organism to a polluting substance. For the case of effects of methylmercury on man, for example, the critical pathway is through consumption of contaminated fish. The level of contamination at any time is the quantity $x_{15}(t)$ in the model described. Total exposure will then be given by

$$E = \int_0^\infty C(t) x_{15}(t) \, dt$$

where $C(t)$ describes rate of fish consumption, and $x_{15}(t)$ may result from any specified combination of environmental inputs. Once this function and the resulting integral are specified (along with the likely errors), the environmental transport model has done its work; the rest is a matter of assessing responses and effects, as described in the chapters to follow.

4.9. CONCLUSIONS

(i) No single type of model can be useful for all pollutant transport problems; particular situations must be characterized in terms of time, distance, and the nature of the ecosystem.

(ii) The fundamental issue in specific model formulation is identification of the different chemical forms of the pollutant and their interconversions, together with the persistence and toxicology of each.

(iii) For specific chemicals, the geographical transport is typically less well understood than sources, bioaccumulation, or toxicology.

(iv) In given cases, models can be formulated which will contribute materially to the prediction of pollutant movement and the making of policy decisions. However, more attention needs to be paid to credibility of models, and more work devoted to verification and validation procedures. In particular, model predictions should not be quoted unless some measure or estimate of the uncertainty in those predictions is included.

4.10. REFERENCES

Akagi, H. and Takabatake, E., 1973. Photochemical formation of methylmercuric compounds from mercuric acetate. *Chemosphere*, 2, 131–3.

Argentisi, F., DiCola, G., and Verkeyden, N., 1973. *Biosystems Modeling: A Preliminary Bibliographic Survey*, Commission of the European Communities EUR 4966 e, Luxembourg.

Bidstrup, P. L., 1964. *Toxicity of Mercury and its Compounds*, Elsevier, New York, 112 pp.

Bligh, E. G., 1971. Mercury levels in Canadian fish. In *Proceedings of the Symposium on Mercury in Man's Environment*, Feb. 15–16, 1971, Royal Society of Canada, Ottawa, pp. 73–90.

Bolin, B., 1976. Transfer processes and time scales in biogeochemical cycles. In *Nitrogen, Phosphorus, and Sulphur-Global Cycles* (Eds. B. H. Svensson and R. Söderlund), *SCOPE Report 7, Ecol. Bull. (Stockholm)*, **22**, 17–22.

Burns, J., 1975. Error analysis of nonlinear simulations. *IEEE Trans. Sys. Man. Cyb.*, **SMC-5**, 331–40.

de Freitas, A. S. W., 1977. Uptake and retention of mercury by fish. In Miller, D. R., *op. cit.*

De Groot, A. J. and Allersma, E., 1975. Field observations in the transport of heavy metals in sediments. In *Heavy Metals in the Aquatic Environment* (Ed. P. A. Krenkel), Pergamon Press, pp. 85–96.

Fagerstrom, T. and Asell, B., 1973. Methyl mercury accumulation in an aquatic food chain. Model and some implications for research planning. *Ambio*, **2**, 164–71.

Fimreite, N. 1970. Mercury uses in Canada and their possible hazards as sources of mercury contamination. *Environ. Pollut.*, **1**, 1.

Frenkiel, F. and Goodall, D. W. (Eds.), 1977. *Simulation Modelling, Report of SCOPE Project 5*, in press.

Goldwater, L. J., 1972. *Mercury: A History of Quicksilver*, York Press, Baltimore.

Goodall, D. W., 1974. Problems of scale and detail in ecological modelling. *J. Environ. Management*, **2**, 149–57.

Goodman, G. T., 1974. How do chemical substances affect the environment? *Proc. Roy. Soc. Lond.*, B**185**, 127–48.

Holcomb Research Institute, 1976. *Environmental Modelling and Decision Making, Report for the Scientific Committee on Problems of the Environment*, Praeger Publishers, 111 Fourth Avenue, New York.

Jeffers, J. N. R. (Ed.), 1972. *Mathematical Models in Ecology*, The 12th Symposium of the British Ecological Society, Grange-over-Sands, Lancashire, 23–26 March 1971, Blackwell, Oxford.

Jensen, S. and Jernelov, A., 1969. Biological methylation of mercury in aquatic organisms. *Nature*, **223**, 753–4.

Kao, S. K., 1974. Basic characteristics of global scale diffusion in the troposphere. *Adv. Geophys.*, **18B**, 15–32.

Krenkel, P. A., 1973. Mercury: environmental consideration. *CRC Critical Reviews in Environmental Control*, May 1973, pp. 303–73 and August 1974, pp. 251–339.

Mar, B. W., 1974. Problems encountered in multidisciplinary resources and environmental simulation models development. *J. Environ. Management*, **2**, 83–100.

Meyer, C. F., 1971. Using experimental models to guide data gathering. *J. Hydraul. Div., Amer. Soc. Civil Eng.*, **97**, 1681–97.

Miller, D. R., 1974. Sensitivity analysis and validation of computer simulation models. *J. Theor. Biol.*, **48**, 345–60.

Miller, D. R. (Ed.), 1977. *Distribution and Transport of Pollutants in Flowing Water Ecosystems*, Final Report of the Ottawa River Project, National Research Council, Ottawa, Canada, May 1977.

Miller, D. R., Butler, G. and Bramall, L., 1976. Validation of ecological system models. *J. Environ. Management*, **4**, 383–401.

Naylor, T. H., Balintfy, J. L., Burdick, D. S., and Chu, K., 1968. *Computer Simulation Techniques*, Wiley, New York.

Nordo, J., Elaissen, A., and Saltbones, J., 1974. Large-scale transport of air pollutants. *Adv. Geophys.*, **18B**, 137–50.

Odum, E. P., 1971. *Fundamentals of Ecology*, 3rd edn., Saunders, Philadelphia.

Pasquill, F., 1974. *Atmospheric Diffusion*, 2nd edn., Wiley, New York.

Patten, B. C., 1971. A primer for ecological modelling. In *Systems Analysis and Simulation in Ecology* (Ed. B. C. Patten), Vol. I, Academic Press, New York, pp. 1–122.

Rapoport, A., 1972. Explanatory power and explanatory appeal of theories. *Synthese*, 24, 321–42.

Rigler, F. H., 1975. The concept of energy flow and nutrient flow between trophic levels. In *Unifying Concepts in Ecology* (Eds. W. H. van Dobben and R. H. Lowe-McConnell), Report of the Plenary Sessions, First International Congress of Ecology, The Hague, the Netherlands, Sept. 8–14, 1974. B. V. Junk Publishers, The Hague, pp. 15–26.

Robinson, J., 1973. Dynamics of pesticide residues in the environment. In *Environmental Pollution by Pesticides* (Ed. C. A. Edwards), Plenum Press, London.

Slade, D. H. (Ed.), 1968. *Meteorology and Atomic Energy*, U.S. Atomic Energy Commission TID-24190, National Technical Information Service, U.S. Dept. of Commerce, Springfield, Virginia.

Smith, J. M., 1974. *Models in Ecology*, The University Press, Cambridge.

Truhaut, R., 1974. Ecotoxicology – A New Branch of Toxicology: A General Survey of its Aims, Methods and Prospects. In *Ecological Toxicology Research: Proceedings of the NATO Science Committee Conference on Ecotoxicology held at Mt. Gabriel, Quebec, Canada, May 6–10, 1974*. Plenum Press, New York, 1975.

Waide, J. B. and Webster, J. R., 1976. Theory in Ecosystem Analysis. In *Systems Analysis and Simulation in Ecology*, Vol. 4, Ed. B. C. Patten. Academic Press.

Wollast, R., Billen, G. and Mackenzie, F. T., 1974. Behaviour of Mercury in Natural Systems and its Global Cycle. In *Ecological Toxicology Research: Proceedings of the NATO Science Committee Conference on Ecotoxicology held at Mt. Gabriel, Quebec, Canada, May 6–10, 1974*. Plenum Press, New York, 1975.

Wood, J. M., 1974. Biological cycles for toxic elements in the environment. *Science*, 183, 1049–52.

CHAPTER 5

Estimation of Doses and Integrated Doses

G. C. BUTLER

*Division of Biological Sciences, National Research Council of Canada,
Ottawa, Canada, K1A 0R6*

5.1 INTRODUCTION	91
5.2 UPTAKE	93
5.3 RETENTION	96
5.4 EXCRETION	99
5.5 PATTERNS OF UPTAKE	100
(i) Single isolated uptake	101
(ii) Chronic uptake	102
(iii) Declining uptake following a single exposure	102
(iv) Several uptakes in a limited period	103
5.6 DOSE COMMITMENT	104
5.7 SAMPLE CALCULATIONS OF DOSE, DOSE COMMITMENT AND HARM COMMITMENT	107
(i) Radioactive iodine	108
(ii) Methylmercury	109
5.8 CONCLUSIONS	110
5.9 REFERENCES	110

5.1. INTRODUCTION

Assessment of the environmental impact of pollutants requires extensive knowledge of doses and effects for a variety of receptors and for a variety of effects in those receptors. Most discussions (verbal and written) of the results of environmental pollution concentrate on effects and neglect the problem of estimating doses. Without accurate knowledge of the dose there can be no quantitative assessment of the effects. This problem of estimating dose is one of the things that distinguishes ecotoxicology from classical toxicology. In the latter the dose and the route of intake are known because they are determined by the experimenter, but in the former they can often be estimated only after much investigation. It was perhaps in radiation protection and radiobiology that the science of dosimetry was first given its proper recognition.

The treatment of the subject in this chapter is strongly coloured by the experience of the author in radiation dosimetry for radionuclides that enter the human body. The concepts and knowledge for this situation are much better developed than they are for the behaviour of chemical pollutants and in species other than man. Nevertheless the 'radiation' approach is offered without apology because it has proved to be practical and because it indicates the kind of information, thinking, and procedures that could be useful for other pollution problems.

Classically 'dose' was used to mean the amount of substance inhaled, ingested, or absorbed through the skin by receptors, or the amount injected or administered. This was called 'exposure' by the Stockholm Conference (U.N., 1972) which gave the definition: *'exposure*: the amount of a particular physical or chemical agent that reaches the target.' A more recent definition of dose was given for use in establishing dose—response relations for heavy metals (Nordberg, 1976), viz. 'the amount or concentration of a given chemical at the site of effect.' To permit the calculation of integrated dose as well as dose rate it is recommended that this definition should have added to it, '. . . at all times following a single uptake or the beginning of a chronic uptake.'

It is important at this point to begin distinguishing between intake and uptake. Intake is the entry of a substance into the lungs, the gastrointestinal tract, or subcutaneous tissues of animals; the fate of this material will be governed by processes of absorption. Uptake is the absorption of the substance into extracellular fluid. The fate of the material taken up will be governed by metabolic processes. This distinction will be observed throughout the report.

Since it is rarely possible to measure the dose directly by non-destructive means (except in the case of radionuclides emitting electromagnetic radiations) the dose may be estimated by indirect means, such as:

calculation from the measured or calculated uptake and the retention equation;

measurement of the concentration in tissues or excreta and the calculation of the amount in the body from excretion equations and 'standard' tissue distributions. The commonest media for measurement are: blood, urine, faeces, exhaled air, hair.

Some clues to methods of calculating doses may be obtained by considering the units or dimensions in which the dose is expressed*. Dose rate is usually a function of concentration and total dose is a function of concentration multiplied by time. In the case of air pollutants such as oxides of sulphur or nitrogen the dose may be expressed as the concentration inhaled x time of exposure. For systemic poisons the dose will be concentration in the body (or organ) x the residence time. If the concentration is constant, dose = concentration x time. If the concentration is not constant, dose = mean concentration x time, or

$$\text{Dose} = \int_0^t C(u) du \tag{5.1}$$

*Recently there have been proposals (Bridges 1973, 1974; Crow, 1973; Latarjet, 1976) to describe doses of chemical mutagens in terms of the amount of X- or Y-rays producing the same effect. The units 'radiquiv' or roentgen equivalent dose (RED) have been proposed. Although some unifying concept is needed this particular one has been much critized in the light of experimental results (Auerbach, 1975; Hahn, 1975).

$$C(t) = \frac{q(t)}{m} \qquad (5.2)$$

where

$C(u)$ = concentration at time u
$C(t)$ = concentration at time t
$q(t)$ = amount of pollutant in mass m at time t
m = mass of body or organ

Standard values of m for man can be found in ICRP (1975) and for other animals in Spector (1956) and Altmann and Dittmer (1972).

The body or organ content as a function of the time following the beginning of intake may be calculated from the equation (Butler, 1972)

$$q(t) = \int_0^t I(\zeta) R_s(t - \zeta) d\zeta \qquad (5.3)$$

where

$q(t)$ = the amount of pollutant in the body or organ at time t
$I(\zeta)$ = the rate of uptake at time ζ
$R_s(t - \zeta)$ = the fraction of a single uptake remaining after time $(t - \zeta)$

5.2. UPTAKE

The information required, as well as the amount available, about intake and uptake is different for various biota and will be presented under four headings: mammals, fish, plants (terrestrial), and plants (aquatic).

The mammal of greatest interest is man and for this species we have the largest amount of information. For average values the Reference Man data of ICRP (ICRP, 1975) are most useful.

Skin absorption of several potential environmental pollutants has been demonstrated experimentally in man. A small proportion of the radioactive iodide applied to human skin appeared in the thyroid gland (Harrison, 1963). Osborne (1966) found that in workmen one-third of the total absorption of tritiated water (HTO) vapour took place by way of the skin compared with two-thirds by way of the lungs. Wahlberg (Dukes and Friberg, 1971) found that when compounds of several metals in aqueous solutions were applied to skin the greatest absorption took place from a tenth-molar solution for mercuric and chromate ions. All these findings may be important for occupational health but it is difficult to conceive that percutaneous absorption is an important route of exposure to environmental pollutants for humans. More, needs to be known, however, about the magnitude of this mode of uptake for aquatic animals.

The behaviour of ingested materials can be predicted on the basis of the physiological model of Eve (1966), which was developed for radiation dosimetry

(Dolphin and Eve, 1966). Accordingly, one of its main purposes was to give average residence times and concentrations of ingested radionuclides in each section of the gastrointestinal tract. This is of less important for non-radioactive pollutants; for these it is essential to have values for f_1, the fraction of ingested pollutant which is absorbed into extracellular fluids. Values of f_1 for many elements, including the heavy metals, are given in ICRP (1977a). For most organic compounds such as DDT and for metal-organic compounds such as methylmercury the value of f_1 could be taken to be near unity. This crude estimate of the total fraction taken up during passage through the intestines is adequate for most calculations of dose in terms of organ concentration x time (long times). More detailed knowledge about the circumstances of the ingestion may make it possible to modify the average value of f_1 (Forth, 1971). For calculations which require a knowledge of the maximum concentration reached in a tissue at some time (short time) after a single ingestion a more refined approach is needed (Goldstein and Elwood, 1971).

Terrestrial animals may be exposed to airborne pollutants of various kinds and particle sizes (Fennelly, 1975). The fraction of inhaled particulate material that is deposited in the parts of the respiratory tract varies with particle size and the fraction absorbed from the tract into the body varies with the chemical nature of the aerosol. The ICRP Task Force on Lung Dynamics produced the 'Lung Model' for Reference Man (Bates *et al.*, 1966). In this model inhaled substances were divided into three classes according to the time they remained in the lungs: class D (days), W (weeks), Y (years). The most recent version (ICRP, 1977a) of that model is displayed in Figure 5.1 which represents the behaviour of particles with a median aerodynamic diameter of 1 micron. The information in this scheme can be

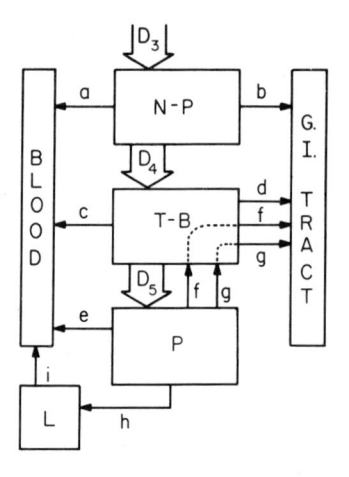

Compartment		Class					
		D		W		Y	
		T	f	T	f	T	f
N - P (D_3 = 0.30)	a	0.01	0.5	0.01	0.1	0.01	0.01
	b	0.01	0.5	0.4	0.9	0.4	0.99
T - B (D_4 = 0.08)	c	0.01	0.95	0.01	0.5	0.01	0.01
	d	0.2	0.05	0.2	0.5	0.2	0.99
P (D_5 = 0.25)	e	0.5	0.8	50	0.15	500	0.05
	f	n.a.	n.a.	1.0	0.4	1.0	0.4
	g	n.a.	n.a.	50	0.4	500	0.4
	h	0.5	0.2	50	0.05	500	0.15
L	i	0.5	1.0	50	1.0	1000	0.9

Figure 5.1 ICRP lung clearance model. N-P = nasopharyngeal compartment, T-B = techoebronchial compartment, P = alveolar compartment, L = lymphatic compartment. (Reproduced by permission of Pergamon Press, Ltd., from ICRP, 1977a)

Table 5.1 Fraction of Inhaled Material Absorbed

Inhalation class	
D	$f_D = 0.48 + 0.15 f_1$
W	$f_W = 0.12 + 0.5\ f_1$
Y	$f_Y = 0.05 + 0.6\ f_1$

displayed in another form (Table 5.1) more readily available for application. This shows the amount absorbed into the body, quickly and slowly, for classes D, W, and Y. The classification of substances into D, W, and Y behaviour is given in Bates et al. (1966). Revisions to this classification are given for many inorganic materials in ICRP (1977a). Most organic pollutants could be assumed to be of class D.

This model was meant to be applicable to human beings but in the absence of more pertinent information it might be used for other mammals. It should always be remembered that this model applied to Reference Man as defined in ICRP (1975) under normal conditions of light work in a temperate environment. There can be significant departures from the quantitative model due to species and the attendant differences in morphology. Within any one species there can be variations due to age, sex, state of health, and environmental conditions. For all these reasons the numbers in the model should be used only when there is no information about the individual(s) being considered.

The rate of uptake of pollutant by freshwater fish by the ingestion of contaminated food has been shown to depend on both the maintenance metabolic rate and on the rate of growth (Norstrom et al., 1975). At 20°

$$I(t)_{\text{ing}} = C_f \left(0.25\, m^{0.8} + 2\, \frac{dm}{dt} \right) f_1 \text{ g/day} \tag{5.4}$$

where

$I(t)_{\text{ing}}$ = rate of uptake from the intestinal tract (g/day)

C_f = concentration of pollutant in food (g/g)
m = body mass (g)
$\frac{dm}{dt}$ = rate of increase of body mass (g/day)
f_1 = fractional absorption from the gastrointestinal tract

The respiratory uptake of pollutant (methylmercury) through the gills of freshwater fish was shown experimentally to be dependent on the metabolic rate (de Freitas and Hart, 1975b). At 20°

$$I(t)_{\text{resp}} = 1000 \times m^{0.8} \times C_w \times f_r \text{ g/day} \tag{5.5}$$

where

$I(t)_{resp}$ = rate of uptake through the gills (g/day)

m = body mass (g)
C_w = concentration of pollutant in water (g/g)
f_r = fractional absorption through the gills

The uptake of pollutants by terrestrial plants has been studied because it is a first step in the transport to human beings through food chains. The approach used for ^{90}Sr was (Burton et al., 1960) to develop the equation

$$C = p_d F_d + p_r F_r \qquad (5.6)$$

where

C = yearly average ^{90}Sr/Ca ratio in cows' milk (pCi/g)
p_d = 'soil' factor
F_d = total accumulated deposit of ^{90}Sr in soil (mCi/km^2)
p_r = 'rate' factor
F_r = yearly fall-out rate of ^{90}Sr(mCi/km^2)

Calculations from field observations in 1958 gave values of P_d = 0.15–0.2 and for p_r = unity. From these values it is clear that foliar deposition is much more important than uptake by the roots as the mechanism for contamination of the forage plants. This approach has been much used by the United Nations Scientific Committee on the Effects of Atomic Radiation (UNSCEAR) (U.N., 1964). It is probable that for heavy metals also, foliar deposition is a more important route of contamination than is root absorption (see NRC, 1973, for a summary). The importance of deposition on the aerial part of mosses is so well recognized that analysing mosses for pollutants is a useful method of assessing past and present air pollution (Goodman and Roberts, 1971).

In higher aquatic plants the uptake of water-borne pollutants by stems and leaves is much more important than the absorption from sediments by the roots (Eriksson and Mortimer, 1975). The rate of uptake from the water of both inorganic and methylmercury by growing aquatic plants (*Elodea densa*) can be expressed as

$$\frac{[Hg] \text{ plants}}{[Hg] \text{ water}} = 3000t$$

where t is the duration of growth in days. This relation was found to hold for water concentrations ranging from 1–10,000 ng/l (Mortimer, 1976).

5.3. RETENTION

Equations describing body or organ content as a function of time are needed for $R_s(t)$ in equation (5.3) and in the excretion equations to be presented later. The most direct way to determine retention equations is to introduce the substance into the man, animal, or plant in a short period of time and measure the amount remaining at intervals thereafter until the amount retained has declined by several orders of magnitude. For experimental animals and plants the measurements can be made by analysis of the tissues after excision; for man and some experimental organisms it can be done by external counting of the γ-rays emitted by a radioactive tracer. When the amount retained on day τ, $R_s(\tau)$, is plotted against time a die-away type of curve is obtained. Further analysis may show either of the following:

(a) A plot of $\log R_s(\tau)$ against time may give a straight line or two or more straight lines joined by curved portions. The general equation describing this result is

$$R_s(t) = q_0 \sum_i K_i e^{-\beta_i t} \tag{5.7}$$

where

q_0 = the amount administered (retention at time 0)
β_i = the slope of each segment
K_i = the coefficient of each term such that $\sum_i K_i = 1$

The number of terms in the equation and the values of K_i and β_i for each term are determined by the standard procedures of curve fitting. It will be found that the number of terms varies from one to five depending on the metabolic behaviour of the substance in the organism under investigation.

(b) A plot of $\log R_s(\tau)$ against the log of time may yield a straight line over some intermediate range of times beginning at a few hours or days. The equation for such a straight line is of the form

$$R_s(t) = q_0 \epsilon^b (t + \epsilon)^{-b} \tag{5.8}$$

where

ϵ = a constant with a value between 0.1 and 4
b = a positive number <1

With man and experimental animals, if the amount of substance remaining in the body cannot be measured a retention equation may be obtained by measuring the total amount excreted per unit time. The results, plotted against time, may show the same two possible types as above

(a) $E_s(\tau) = q_0 \sum_i k_i e^{-\beta_i \tau}$ \hfill (5.9)

where

$k_i = K_i \beta_i$

(b) $E_s(\tau) = q_0 \times a(\tau + \epsilon)^{-(b+1)}$ \hfill (5.10)

where

$$a = be^{-b}.$$

The retention equation can then be obtained by integrating the excretion equation which gives

(a) equation (5.7) and
(b) equation (5.8).

Examples of retention equations obtained in this way are:
tritiated water in man,

$$R_s(t) = e^{-\frac{0.693}{10}t} \qquad \text{(ICRP, 1977a)}$$

lead in man,

$$R_s(t) = 0.7\, e^{-\frac{0.693}{12}t} + 0.17\, e^{-\frac{0.693}{180}t} + 0.13\, e^{-\frac{0.693}{12000}t} \qquad \text{(ICRP, 1977a)}$$

cobalt in man,

$$R_s(t) = 0.5\, e^{-\frac{0.693}{0.5}t} + 0.3\, e^{-\frac{0.693}{6}t} + 0.1\, e^{-\frac{0.693}{60}t} + 0.1\, e^{-\frac{0.693}{800}t}$$

$$\text{(ICRP, 1977a)}$$

strontium in man, a combination of (a) and (b) types,

$$R_s(t) = 0.4\, e^{-0.25t} + 0.45\, (t+0.2)^{-0.18}\, (0.55\, e^{-6.6 \times 10^{-5} t}$$

$$+ 0.45\, e^{-2.65 \times 10^{-4} t}) \qquad \text{(Marshall et al., 1973)}$$

methylmercury in fish,

$$R_s(t) = e^{-0.029 m^{-0.58} t} \qquad \text{(de Freitas et al., 1975a)}$$
m = body mass, in grams

methylmercury in aquatic plants (*Utricularia* and *Elodea*),

$$R_s(t) = 0.4\, e^{-\frac{0.693}{140}t} + 0.6\, e^{-\frac{0.693}{700}t} \qquad \text{(Mortimer and Kudo, 1975)}$$

There have been studies of the variation of retention time with mammalian species (see, for example, Stara *et al.*, 1970). It has been found that the longest half-life for the retention of caesium, i.e. that for the third exponential term, T_3, increases with body weight according to the equation

$$T_3 = 2m^{0.35} \qquad \text{(Stara et al., 1970; Guillot, 1972)}$$

Estimation of Doses and Integrated Doses

Two comments can be made about these retention equations:

The type of equation shown for strontium in man (a product of a power function and an exponential) is difficult to integrate and therefore it is preferable to use a retention equation which is a sum of exponentials for substitution in equation (5.3).

In carrying out the research to develop a retention equation it is important to make a special effort to determine the last term or two (with the slowest rate of loss) accurately. When the equation is integrated to calculate dose these 'slow' terms make the greatest contribution to the integral.

One organ or tissue of the body may be affected most by the uptake of a pollutant, in which case it may be called the critical organ. There are a number of possible reasons for such a designation in that the organ may:

(a) receive the highest dose;
(b) be the most easily damaged;
(c) be the most important to the body.

It is important to have the same kind of retention function for a critical organ as those shown above for whole organisms. If this information is not available one should know what fraction of the body content is in the critical organ. If this fraction is F_c then the retention function for the critical organ is assumed to be $F_c \times R_s(t)$. It may happen that one of the exponential terms in a retention equation can be identified with retention in a particular organ. For example, for iodide,

$$R_s(t) = 0.7\, e^{-\frac{0.693}{0.35}t} + 0.3\, e^{-\frac{0.693}{100}t} \qquad \text{(ICRP, 1968)}$$

and the amount in the thyroid gland at any time t is represented by the second term of the equation.

5.4. EXCRETION

From the (above) description of how the equations for $R_s(t)$ are determined it will have been realized that the relation between retention and excretion equations for a single uptake is

$$Y_s(t) = -\frac{d}{dt}R_s(t) = -R'_s(t) \qquad (5.11)$$

where

$Y_s(t)$ = the fraction of the original uptake excreted per day on day t if t is in days

$$E_s(t) = q_0 \times Y_s(t) \qquad (5.12)$$

where

$E_s(t)$ = the amount excreted per day on day t

For a prolonged uptake the equation developed by Butler (1972) for the rate of excretion is

$$E(t) = -\int_0^t I(\xi)R_s'(t-\xi)d\xi \qquad (5.13)$$

This comes from equation 5 of Butler (1972) and has the term for radioactive decay omitted.

The body content may be calculated from the amount excreted with the help of equation (5.12) or (5.13). $E_s(t)$ is measured and $R_s(t)$, $R_s'(t)$, and $Y_s(t)$ are known; therefore q_0 or $I(t)$ may be calculated. If q_0 is substituted into equation (5.16), or $I(t)$ into equation (5.20), the dose may be calculated. In the case of monitoring contaminated workers, this may be the only method possible. Since only one route of excretion, usually urinary, is monitored at one time it is necessary to know the fraction of the total excretion that takes place by each route. This information is given for Reference Man for about thirty elements in ICRP (1975), chiefly the fractions excreted in urine and faeces. These fractions may not be constant with time as was found in the case of technetium (ICRP, 1977b) and methylmercury (Miettinen et al., 1971). Sometimes the concentration of a heavy metal in hair can be measured to calculate the concentration in the body, for example, the concentration of methylmercury in hair is 300 times that in blood (Expert Group, 1971).

5.5. PATTERNS OF UPTAKE

In real life four main patterns of uptake, depending on the times and routes of intake, are encountered. They are illustrated in Figures 5.2, 5.3, 5.4 and 5.5. The equations for retention and excretion are slightly different for each case so they will be presented in turn.

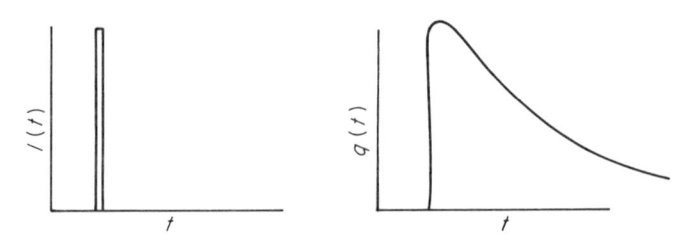

Figure 5.2 Single uptake. (Reproduced by permission of Pergamon Press, Ltd., from ICRP, 1971)

Estimation of Doses and Integrated Doses 101

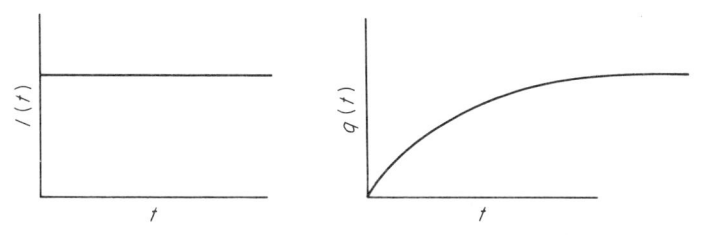

Figure 5.3 Chronic uptake. (Reproduced by permission of Pergamon Press, Ltd., from ICRP, 1971)

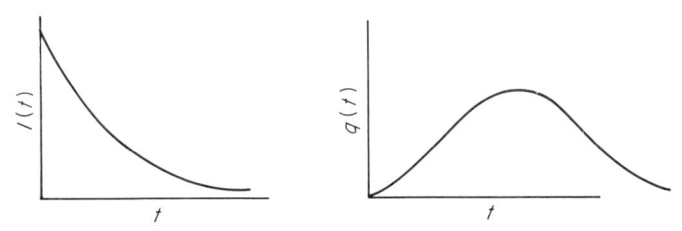

Figure 5.4 Declining uptake resulting from initial contamination of the lungs or a wound. (Reproduced by permission of Pergamon Press, Ltd., from ICRP, 1971)

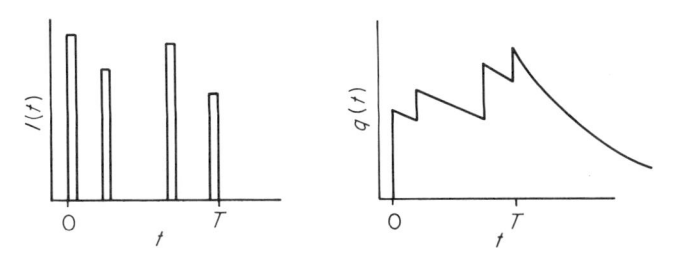

Figure 5.5 Several uptakes in a limited period. (Reproduced by permission of Pergamon Press, Ltd., from ICRP, 1971)

(i) Single Isolated Uptake

If there is a short-term uptake of q_0 at time 0 and if, thereafter, $I(t) = 0$, then

$$q(t) = q_0 R_s(t) \tag{5.14}$$

$$E(t) = q_0 \times Y_s(t) = -q_0 R_s'(t) \tag{5.15}$$

The dose during time v

$$= q_0 \int_0^v R_s(t) \times H \, dt \tag{5.16}$$

where H is a dose coefficient.

(ii) Chronic Uptake

If the intake continues at a steady rate

$I(t) = $ a constant I

$$q(t) = I \int_0^t R_s(u)du \tag{5.17}$$

$$E(t) = I \int_0^t Y_s(u)du \tag{5.18}$$

If the intake continues for a long time the rate of uptake is balanced by the rate of loss and the body content becomes constant at

$$q = I \int_0^\infty R_s(t)dt \tag{5.19}$$

In this situation the annual dose

$$= q \times 365 \times H = 365 \, I \int_0^\infty R_s(t)dt \times H \tag{5.20}$$

and the total future dose (Dose Commitment — see below) from a year of uptake

$$= 365 \, I \int_0^\infty R_s(t)dt \times H$$

which is identical with equation (5.20). This is the basis of the theorem, important in assessing environmental pollutants: 'Under conditions of constant intake and constant body content, the Dose Commitment from a year's uptake is numerically equivalent to the annual dose.'

(iii) Declining Uptake Following a Single Exposure

There may be a short-term intake of insoluble contaminant to the body resulting in an initial deposition (e.g. in lungs or in subcutaneous tissue) followed by a slow uptake to extracellular fluid. The case of greatest practical interest is the inhalation of a class W or Y substance and the subsequent movement from the lungs into the systemic circulation. According to the lung clearance model there will be a rapid uptake occurring within a day or two followed by a slow uptake, directly from the lungs and indirectly from the gastrointestinal tract. According to the lung clearance model (Figure 5.1), the rapid uptake can be formulated as

$q_0 = 0.3a + 0.08c + (0.3b + 0.08d + 0.25f)f_1$

and the slow uptake can be formulated as (approximately)

$I(t) = 0.25L'(t)(e + h + gf_1)$

where

$$L'(t) = \frac{d}{dt} \times L(t) = \lambda e^{-\lambda t}$$

$$L(t) = e^{-\lambda t}$$

$$\lambda = \frac{0.693}{50} \text{ for class W and } \frac{0.693}{500} \text{ for class Y}$$

Since the parameters for class W and Y are different they are given separately.

Class W

$$q_0 = 0.7 + 0.41 f_1 \text{ of that inhaled}$$

$$I(t) = 3.5 \times 10^{-3} (0.2 + 0.4 f_1) e^{-\frac{0.693}{50}t}$$

Class Y

$$q_0 = 0.004 + 0.48 f_1$$

$$I(t) = 3.5 \times 10^{-4} (0.2 + 0.4 f_1) e^{-\frac{0.693}{500}t}$$

Then the total amount in the systemic metabolism of the body at any time t will be given by substituting q_0 in equation (5.14) and $I(t)$ in equation (5.3) and adding.

This approach could also be used to calculate the effective toxic concentration of DDT or PCB when it moves out of the adipose tissue reservoir during starvation or exercise (Findlay and de Freitas, 1971; de Freitas and Norstrom, 1974). But more information is needed about the rates of migration and metabolism of these compounds.

(iv) Several Uptakes in a Limited Period

In practice it is not uncommon to encounter repeated and unequal exposures in some period of interest such as a year. If these are well separated in time (by 3 or 4 half-lives) each one can be considered as a single uptake, evaluated according to equation (5.14) and summed for the period of interest. If, however, the uptakes overlap a different approach is needed. It is simplest to consider the total uptake A in the period T days giving an average daily rate of uptake of A/T. The equations describing the body content at various times are given in Butler (1972). The one for the end of the period T is

$$q(T) = \frac{A}{T} \int_0^T R_s(u) du \tag{5.21}$$

The corresponding equations for excretion and integral dose are also to be found in Butler (1972).

5.6. DOSE COMMITMENT

Since doses of environmental contaminants and their concomitant harm or detriment to a population are to be used to measure the cost of the practice causing the contamination, they must be estimated quantitatively as a function of the operation. Most results of human activities are audited annually; doses and detriment are no exception but in some cases the yearly audit will not provide an adequate measure.

In an environmental situation there may be many possible sources each emitting a number of possible pollutants. From the point of view of calculating doses and harm the types of releases and pollutants may be considered as follows:

Release	Contaminant
A. Single	1. Short-lived
	2. Long-lived
B. Continued	1. Short-lived
	2. Long-lived

For case A.1 the annual dose will encompass the total and will be adequate to assess the total detriment. For the other three cases the estimates must extend beyond one year and for these cases UNSCEAR developed the concept of 'Dose Commitment'. Although that Committee used the concept for assessing the total effects of detonating a nuclear weapon, the concept has many other potential uses in evaluating the results of some activity that contaminates the environment. The assessment of effects beyond one year may be necessary because:

the amount of the practice may increase in the future;

the pollutant may persist in the environment long after its release;

the pollutant may be produced in the environment long after the release of its parent substance, e.g. methylmercury;

the pollutant may remain in the body of the receptor long after its intake;

the biological effects may be expressed long after the initial intake.

The foregoing summarized the reasons for estimating 'Dose Commitment'; the essential steps in the process will now be reviewed. Details of the individual steps have already been presented in the chapter on Environmental Transport, and in the earlier part of this chapter on calculations of doses to Reference Man. The information required for calculating 'Dose Commitment' usually consists of monitoring data, varying in time and space, on sources, pollutants released, receptors and information from research on the organs and processes of the receptor affected by the pollutant. When it is realized that each of these elements has multiple possibilities, some appreciation of the complexity of the assessment emerges. For simplicity of presentation the sequel will consider only a single possibility for each element and will treat the four cases displayed above, viz. A.1 and 2, B.1 and 2.

Earlier in this chapter it was shown how in principle the content of pollutant in the receptor could be calculated from the rate of intake and the retention function (equation 5.3). As modified for the present purpose the equation would be

$$q(t) = \int_0^t I \times C(\tau) \times A \times R_s(t - \tau) d\tau \tag{5.22}$$

where

$q(t)$ = the amount in the body or organ
I = the amount of air, water, or food taken in per unit time
$C(\tau)$ = the concentration of pollutant in the medium taken in at time τ
A = the fractional absorption to the blood of the pollutant ingested or inhaled
$R_s(t)$ = the fraction retained in the body at time t after a single uptake

Some pollutants can be metabolically detoxified (by conversion or excretion) at an appreciable rate. In these cases, if there is no accumulation of subclinical effects, no effect will result until the concentration of pollutant in the receptor, $q(t)/m$, exceeds its detoxifying capacity, i.e. exceeds a critical concentration (threshold). There would therefore be a threshold dose and the effective dose would be the tissue concentration minus the critical concentration or threshold. Equation (5.3) would be appropriate for calculating the tissue contents relevant to an acute effect resulting from this situation.

For other pollutants such as radioactive substances, and possibly some heavy metals, the relevant dose is the time integral of tissue concentration, i.e.

$$D(v) = \int_0^v \frac{q(t)}{m} dt \tag{5.23}$$

where

$D(v)$ = the dose delivered in time v

$q(t)$ is defined in equation (5.22)

m = mass of tissue

The calculation and integration of $q(t)$ for the four cases A.1 and 2, B.1 and 2, will now be discussed.

Case A.1. The total intake from a single release of short-lived material should take place in a time much shorter than a year. Thus the total dose can be calculated from equations (5.3) or (5.14) by integrating the latter over any time from one year to infinity, all giving the same result. In other words the annual dose is the same as the 'Dose Commitment'.

Case A.2. Even from a single release, an appreciable fraction of the pollutant may persist in the environment beyond one year; it is usually assumed, however, that it does not persist beyond three decades. To calculate the total effect (dose) it is necessary to carry out the integration of equation (5.23) beyond one year; it is usually integrated to infinity.

Case B.1. With continuous release of a short-lived pollutant $C(\tau)$, the rate of uptake and the body or organ content become constant after a time equivalent to three or four effective half-lives. When this condition is reached the yearly dose from the constant content becomes equivalent to the total dose, integrated to infinity from a year's uptake. Earlier in the chapter this proposition was demonstrated algebraically; it makes it possible to calculate annual doses from a knowledge of either the constant body content or the constant annual uptake. This state of affairs is the one most frequently assumed by UNSCEAR in its calculations of annual doses and 'Dose Commitments'. For a constant practice the total dose for the duration of the practice is calculated by multiplying the constant annual dose by the number of years the practice is expected to continue.

Case B.2. In this case the annual dose is not all-inclusive nor does it become constant. Because the pollutant, constantly produced, persists in the environment for a very long time, $C(\tau)$, the rate of uptake and the total dose continue to increase without limit. It is important yet difficult to decide over what period the integration of equation (5.23) should be carried out. The only possibility is to guess how long the practice will last and use this time as the limit of integration.

The foregoing has explained how to calculate annual or total expected doses for an individual from a knowledge of his mass, intake, uptake, and retention. It is not practicable to know these things about all individuals being exposed but fortunately it is not necessary since the assessment of harm or detriment is concerned with the total response. It is calculated by multiplying the average dose received by an individual by the number of individuals receiving it. The product so obtained is called the collective dose; if it is an annual dose the product is called the annual collective dose; if it is the total expected dose the product is called the collective dose commitment.

To calculate these collective doses the procedure is the same as for an individual (equation 5.23) except that the values or formulae used for I, A, m, and $R(t)$ are averages for the population at risk. Since $C(\tau)$ may not be reported as a continuous function but sometimes as an integral to infinite time it is preferable to rewrite the dosimetric equation as

$$D(\infty) = \int_0^\infty \frac{q(t)}{m} dt = \frac{I \times A}{m} \int_0^\infty dt \int_0^\infty C(\tau) R(t - \tau) d\tau$$

$$= \frac{I \times A}{m} \int_0^\infty \left(C(\zeta) d\zeta \right) \left(\int_0^\infty R(\eta) d\eta \right) \quad (5.24)$$

where

$\zeta = t - \tau$

$\eta = \tau$

The average total dose calculated from this equation is multiplied by the number of individuals in the population to give the collective dose commitment, the total of all doses to the whole population for all time.

Such a summation is possible in this straightforward way only if there is a constant proportionality between effects and doses and if there is no dose without an effect (threshold dose). The ICRP assumed these conditions for making recommendations about lifetime radiation doses to workers. They further assumed that each unit of dose had the same effect regardless of the rate at which it was received. For calculating collective dose commitments UNSCEAR made the same assumptions.

This concept of collective dose commitment was developed for, and is applicable to, the stochastic* effects of irradiation. It is probably also applicable to single releases of persistent heavy metals such as lead and mercury. It may also be used for the release of shorter-lived pollutants by a process that goes on for a very long time without any foreseeable termination. It would be interesting and probably profitable to search, by epidemiological research, for other pollution problems that lend themselves to this kind of assessment.

5.7. SAMPLE CALCULATIONS OF DOSE, DOSE COMMITMENT AND HARM COMMITMENT

It will be helpful to the reader to illustrate the foregoing principles by using real data to calculate doses of interest in protecting human populations. Two cases will be discussed: the aerial release of radioactive iodine and the aquatic release of mercury.

---

*ICRP (1977c) defined stochastic effects as those for which the probability of the occurrence rather than the severity varies with dose and for which there is no dose without an effect (threshold).

(i) Radioactive Iodine

Isotopes of iodine are among the most abundant products of nuclear fission in uranium and plutonium. Being gaseous the iodines may escape and contaminate the air around nuclear reactors or fuel reprocessing plants. It will be sufficient to base the calculations on one isotope of iodine, viz. ^{131}I with a radioactive half-life of 8 days. Metabolic data for a child 1–4 years old (MRC, 1975).

mass of thyroid gland, 1.8 g
f_1, 1
biological half-life of iodine in thyroid, 23 days
fraction of iodine in blood deposited in thyroid, 0.35
milk intake, 0.7 l/day

Chronic release rate of 1 Ci/yr at ground level (ICRP, 1977d):

air concentration at ground zero, 0.27 pCi/m³
above surface food crop at 1 km, 320 pCi/m²
pasture grass at 1 km, 310 pCi/m²
milk concentration at 1 km, 120 pCi/l
Rate of uptake to thyroid = 0.7 × 0.35 × 120 pCi/days = 29.4 pCi/day

$$R_s(t) = e^{-0.693\left(\frac{1}{23}+\frac{1}{8}\right)t} = e^{-0.117t}$$

Constant thyroid content = $29.4 \int_0^\infty e^{-0.117t} dt = \frac{29.4}{0.117} = 251$ pCi

Concentration × time (1 year) = $\frac{251}{1.8} \times 365$

Radiation dose = $\frac{251 \times 365}{1.8} \times 51 \times 0.23 \times 10^{-6}$ (ICRP, 1977d) = 0.6 rads/yr

Single release of 1 µCi at ground level (ICRP, 1977d):

milk concentration at 1 km, 1.3 µCi-days/l

integrated thyroid concentration = $\frac{1.3 \times 0.7}{1.8} \times 0.35$ µCi-days

radiation dose = 0.18 rads

Lifetime risk of thyroid cancer from irradiation in childhood ≃ 200 cases per rad per million people exposed including about 7 fatal cases/M/rad (U.N., 1977).

(ii) Methylmercury

The situation for which calculations will be made results from the release of inorganic mercury (Hg^0, Hg^+ or Hg^{2+}) into water, its concentration in sediments, bioconversion to monomethylmercury (CH_3-Hg^+), release of CH_3-Hg^+ to the water, followed by its uptake and bioconcentration in fish. When man eats the fish the CH_3-Hg^+ is further concentrated in the human body.

The concentration in fish is calculated for the Ottawa River at Ottawa (Ottawa, 1976).

velocity of flow = 17 Km/day
area of sediment in 17 Km = 17×10^6 m^2
volume of flow = 1.5×10^{11} l/day
Rate of CH_3-Hg^+ production by sediment = 2 µg/m^2/day (Langley, 1973)

$$\therefore \text{Concentration of } CH_3-Hg^+ \text{ in river} = \frac{17 \times 10^6 \times 2 \times 10^{-6}}{1.5 \times 10^{14}}$$

$$= 2.3 \times 10^{-13} \text{ g/g}$$

Uptake rate = $2 \times 1000 \times m^{0.8} \times C_{pw}$ (5.25)
 = $2000 \times 100^{0.8} \times 2.3 \times 10^{-13}$
 = 20 ng/day for a 100 g fish

$R_s(t)$ for a 100 g fish = $e^{-0.002t}$ (de Freitas et al., 1975a)

∴ at equilibrium

$$q(t) = 20 \int_0^\infty e^{-0.002t} dt = \frac{20}{0.002} = 10{,}000 \text{ ng}$$

Concentration of CH_3-Hg^+ in a 100 g fish = 100 ng/g = 0.1 ppm

From studies of Japanese fish-eaters some of whom suffered from poisoning with CH_3-Hg^+, it was found that there were no symptoms with blood levels below 0.1 µg/g of whole blood or 30 µg/g of hair (Skerfving, 1972). It was also found that 10% of the CH_3-Hg^+ in the body was in the blood (Miettinen et al., 1971) (volume = 5,500 ml).

∴ 0.1 µg/g blood corresponds to a body content of $0.1 \times 5{,}500 \times 10$ µg = 5.5 mg

$$R_s(t) = e^{-0.01t} \text{ (SCOPE, 1977)}$$

$$q(t) \text{ at equilibrium} = I \int_0^\infty e^{-0.01t} dt = \frac{I}{0.01}$$

$$\text{If } \frac{I}{0.01} = 5.5 \text{ mg}, I = 55 \text{ µg/day}$$

If the subject eats x g of fish per day the concentration of CH_3-Hg^+ in fish to give an equilibrium concentration of 0.1 µg/g blood = $55/x$ ppm.

5.8. CONCLUSIONS

1. With a knowledge of the rate of uptake and the retention equation of a pollutant, it is possible to calculate the body content and its time integral (dose) for the four patterns of exposure actually encountered.
2. There are metabolic models for the uptake by man resulting from ingestion and inhalation. Similar models are needed for other receptors.
3. The application of the dose commitment concept requires that the dose–effect relation be of the 'linear non-threshold' type. This concept becomes more difficult to apply as the period of release, the time of disappearance, and the latent period of effects of the pollutant become longer.
4. From two examples it can be seen that, when adequate data are available, it is simple to calculate the dose, dose commitment, and harm commitment resulting from the release of a contaminant to the environment.

5.9. REFERENCES

Altmann, P. L. and Dittmer, D. S. (Eds.), 1972. *Biology Data Book*, Federation of American Biological Societies.

Auerbach, C., 1975. The effects of six years of mutagen testing on our attitude to the problems posed by it. *Mutation Research*, 33, 3–9.

Bates, D. V., Fish, B. R., Hatch, T. F., Mercer, T. T., and Morrow, P. E., 1966. Deposition and retention models for internal dosimetry of the human respiratory tract. Report of the Task Group of Lung Dynamics, International Commission on Radiological Protection, *Health Phys.*, 12, 173.

Bridges, B. A., 1973. Some general principles of mutagenicity screening and a possible framework for testing procedures. *Environ. Health Persp.*, 6, 221–7.

Bridges, B. A., 1974. The three-tier approach to mutagenicity screening and the concept of radiation — equivalent dose. *Mutation Research*, 26, 335–40.

Burton, I. D., Milbourn, G. M., and Russell, R. S., 1960. Relationship between the rate of fall-out and the concentration of strontium-90 in human diet in the United Kingdom. *Nature*, 185, 498.

Butler, G. C., 1972. Retention and excretion equations for different patterns of uptake. In *Assessment of Radioactive Contamination in Man*, IAEA, Vienna, STI/PUB/290, p. 495.

Crow, J. F., 1973. Impact of various types of genetic damage and risk assessment. *Environ. Health Persp.*, 6, 1–5.

de Freitas, A. S. W., Gidney, M. A. J., McKinnon, A. E., and Norstrom, R. J., 1975a. Factors affecting whole body retention of methylmercury in fish. *Proc. 15th Hanford Life Science Symposium on the Biological Implications of Metals in the Environment*, Richland, Wash., Publ. in the *Energy Research and Development Administration Symposium Series*.

de Freitas, A. S. W., and Hart, J. S., 1975b. Effect of body weight on uptake of methylmercury in fish. *Water Quality Parameters*, ASTM STP 573, Amer. Soc. Testing Materials, p. 356.

de Freitas, A. S. W., and Norstrom, R. J., 1974. Turnover and metabolism of polychlorinated biphenyls in relation to their chemical structure and the movement of lipids in the pigeon. *Can. J. Physiol. Pharmacol.*, 52, 1080.

Dolphin, G. W. and Eve, I. S., 1966. Dosimetry of the gastrointestinal tract. *Health Phys.*, 12, 163.

Dukes, K. and Friberg, L. (Eds.), 1971. Absorption and excretion of toxic metals. *Nordisk Hygienisk Tidskrift*, 53, 70.

Eriksson, C. and Mortimer, D. C., 1975. Mercury uptake in rooted higher plants; laboratory studies. *Verh. Internat. Verein Limnol.*, 19, 2087.

Eve, I. S., 1966. A review of the physiology of the gastrointestinal tract in relation to radiation doses from radioactive materials. *Health Phys.*, 12, 131.

Report from an Expert Group, 1971. Methyl mercury in fish. *Nordisk Hygienisk Tidskrift*, **Suppl. 4**.

Fennelly, P. F., 1975. Primary and secondary particulates as pollutants. *J. Air Polln. Control Assn.*, 25, 697.

Findlay, G. M. and de Freitas, A. S. W., 1971. DDT movement from adipocyte to muscle cell during lipid utilization. *Nature*, 229, 63.

Forth, W., 1971. Factors influencing the absorption of heavy metals. *Acta Pharmacol. Toxicol.*, 29 (**Suppl. 4**), 78.

Goldstein, R. A. and Elwood, J. W., 1971. A two-compartment, three-parameter model for the absorption and retention of ingested elements by animals. *Ecology*, 52, 935.

Goodman, G. T. and Roberts, T. M., 1971. Plants and soils as indicators of metals in the air. *Nature*, 231, 287.

Guillot, P., 1972. Comments on 'Comparative metabolism of radionuclides in mammals'. *Health Phys.*, 22, 107.

Hahn, G. M., 1975. Radiation and chemically induced potentially lethal lesions in noncycling mammalian cells: recovery analysis in terms of X-ray and ultraviolet-like-systems. *Radiation Research*, 64, 533–45.

Harrison, J., 1963. The fate of radioiodine applied to human skin. *Health Physics*, 9, 993.

Recommendations of the International Commission on Radiological Protection, 1968. *Report of Committee 4 on Evaluation of Radiation Doses to Body Tissues from Internal Contamination Due to Occupational Exposure, ICRP Publication 10*, Pergamon Press.

Recommendations of the International Commission on Radiological Protection, 1971. *Report of Committee 4 on the Assessment of Internal Contamination Resulting from Recurrent or Prolonged Uptakes, ICRP Publication 10A*, Pergamon Press.

International Commission on Radiological Protection, 1975. *Report of the Task Group on Reference Man, ICRP Publication 23*, Pergamon Press, Oxford.

ICRP, 1977a. *Limits for Intakes of Radionuclides by Workers*, Pergamon Press, Oxford, in press.

ICRP, 1977b. *Monitoring for Internal Contamination due to Occupational Exposure*, Pergamon Press, Oxford, in press.

ICRP, 1977c. *Publication 29, Recommendations of the International Commission on Radiological Protection*, Pergamon Press, Oxford.

ICRP, 1977d. *Principles and Methods for Use in Radiation Protection Assessments*

Related to Planned and Unplanned Releases of Radioactive Materials into the Environment, Pergamon Press, in press.

Langley, D. G., 1973. Mercury methylation in an aquatic environment. *J. Water Polln. Control Fed.*, **45**, 44.

Latarjet, R., 1976. First European Symposium on Rad-equivalence. Euratom.

Marshall, J. H., Lloyd, E. L., Rundo, J., Liniecki, J., Marotti, G., Mays, C. W., Sissons, H. A., and Snyder, W. S., 1973. Alkaline earth metabolism in adult man. *Health Phys.*, **24**, 125.

Medical Research Council, 1975. *Criteria for Controlling Radiation Doses to the Public after Accidental Escape of Radioactive Material*, Her Majesty's Stationery Office.

Miettinen, J. K., Rahola, T., Hattula, T., Rissanen, K., and Tillander, M., 1971. Elimination of ^{203}Hg-methyl mercury in man. *Ann. Clin. Res.*, 3, 116.

Mortimer, D. C., 1976. Private communication.

Mortimer, D. C. and Kudo, A., 1975. Interaction between aquatic plants and bed sediments in mercury uptake from flowing water. *J. Environ. Qual.*, **4**, 491.

National Research Council of Canada, 1973. Associate Committee on Scientific Criteria for Environmental Quality. *Lead in the Canadian Environment*, NRCC No. 13682.

Nordberg, G. F., 1976. *Effects and Dose–Response Relationships of Toxic Metals*, Elsevier, Amsterdam.

Norstrom, R. J., McKinnon, A. E., and de Freitas, A. S. W., 1975. A bioenergetics-band model for pollutant accumulation by fish. Simulation of PCB and methylmercury levels in Ottawa River perch (*Perca flavescens*). *J. Fish. Res. Bd. Canada*, **33**, 248.

Osborne, R. V., 1966. Absorption of tritiated water vapour by people. *Health Physics*, **12**, 1527.

Ottawa River Project, 1976. University of Ottawa — National Research Council of Canada, *Distribution and Transport of Pollutants in Flowing Water Ecosystems*, Report Number 3, January 1976.

SCOPE, 1977. Environmental Issues (Eds. M. W. Holdgate and G. F. White), *SCOPE Report 10*, John Wiley and Sons, London.

Skerfving, S., 1972. Mercury in fish – some toxicological considerations. *Fd. Cosmet. Toxicol.*, **10**, 545.

Spector, W. S., 1956. *Handbook of Biological Data*, Saunders, Philadelphia.

Stara, J. F., Nelson, N. S., Della Rose, R. J., and Bustad, L. K., 1970. Comparative metabolism of radionuclides in mammals: a review. *Health Phys.*, **20**, 113.

United Nations, 1964. General Assembly, *Report of the United Nations Scientific Committee on the Effects of Atomic Radiation*, A/5814: G. A. Official records, 21st sess., Suppl. No. 14, p. 35.

United Nations, 1972. *Identification and Control of Pollutants of Broad International Significance*, Document A/CONF. 48/8 prepared for the United Nations Conference on the Human Environment, Stockholm, 1972. Para. 124.

United Nations, 1977. *United Nations Scientific Committee on the Effects of Atomic Radiation, Report to the General Assembly*, United Nations, New York, in press.

SECTION II
THE STATISTICAL ANALYSIS OF DOSE–EFFECT RELATIONSHIPS

CHAPTER 6

The Statistical Analysis of Dose-Effect Relationships

C. C. BROWN

Biometry Branch, National Cancer Institute, National Institute of Health, Bethesda, Maryland 20014, U.S.A.

6.1 INTRODUCTION 115
6.2 EXPERIMENTAL DESIGN 117
6.3 QUANTAL RESPONSES 120
6.4 MATHEMATICAL MODELS OF TOLERANCE DISTRIBUTIONS 122
6.5 ESTIMATION OF A QUANTAL DOSE–RESPONSE CURVE 126
6.6 QUANTITATIVE RESPONSES 136
6.7 TIME-TO-OCCURRENCE MODELS 136
6.8 APPLICATION TO ECOLOGICAL RISK ASSESSMENT 140
6.9 CONCLUSIONS 141
6.10 REFERENCES 142
6.11 APPENDIX: THE MAXIMUM-LIKELIHOOD METHOD OF FITTING
     DOSE–RESPONSE MODELS TO QUANTAL DATA 145

6.1. INTRODUCTION

The relation of responses to doses of environmental toxicants is an important element in the control and prevention of ecological problems. In general terms, dose–response is the relation between any measurable stimulus, physical, chemical, or biological, and the response of living matter in terms of the reactions the stimulus produces over some range of the amount of stimulus. In toxicological situations, there is normally, though not always, a monotone relation between the intensity of the stimulus and the particular response it elicits.

The reactions to any one stimulus may be multiple in nature, e.g. loss of weight, decrease in blood sugar, central nervous system disorders, decrease in organ function, or even death. Each reaction will have its own unique relation with the degree of the stimulus. In addition, the measurement of any specific reaction can be made in terms either of the magnitude of the effect produced, including whether the effect is produced or not, or of the time required for the appearance of a specific effect. These responses may be acute reactions, sometimes occurring within minutes of the stimulus, or they may be long-delayed effects such as cancer, which may not appear clinically until most of the animal's normal lifespan has elapsed.

Other responses may not even appear in the exposed subject, but may become manifest in some later generation.

The degree of stimulus, or in general terms, the dose level, may be measured in different ways. For example, consider some animal that is exposed to a chemical toxicant in the environment, either through the air breathed, the food eaten, or through some other external exposure. The magnitude of the stimulus, or exposure level, may be quantified as parts per million in the air or food, or may be measured as the quantity of the substance actually reaching the target receptor, some internal organ or other tissue. The former quantity can be thought of as the 'actual' exposure, or dose level, while the latter may be termed the 'effective' exposure level. The actual level may be modified by absorption, distribution, metabolism, detoxification, and excretion of the chemical substance. Therefore, the effective level may well be some complex function of the actual level along with the biochemical and physiological dimensions of the host resulting in a potentially quite different relation between the levels of the stimulus and the magnitude of response, depending upon the manner in which the stimulus is measured. Swartz and Spear (1975) recommend relating the time-integrated internal exposure to the observed response in carcinogenesis studies since mechanistic conclusions, which are dependent upon this relation, may be obscured by the relation of external to internal exposure.

The relation between the dose level of a toxicant and the resulting reaction is often impossible to estimate by direct measurements of the stimulus itself unless its mechanism of action is known. A toxic element in different chemical constitutions may produce quite different biological effects in the same host due to different modes of interaction with the organism and its tissues. Therefore, lacking complete mechanistic knowledge, the relation between the toxicant and its effects on a biological system must be either observed in nature or tested in the laboratory or field. Experimentation under laboratory conditions is ideal for the control of extraneous variation or possible bias. The experimental conditions may be rigidly controlled and the results are subject to considerations of reproducibility. However, because of potential physical limitations of the laboratory, the experimental conditions may well be unrealistic images of the natural environment and hence, extrapolation to the real world may be difficult and sometimes unwarranted. Experiments on plants or animals may also be conducted in the field. Here, the idea is to reproduce the natural environment as closely as possible, but with this type of experiment one cannot easily control the many factors which may have an influence on the results of the test.

This section is concerned with the statistical techniques used for the estimation of dose—response relations. It will be assumed that the particular agent in question is known to be generally toxic and that the purpose of the following statistical analyses is to obtain an indication of the relation between dose level and the toxic response. The general methods are applicable to any experimental situation, be it in a highly controlled laboratory environment, in the field, or gradations in between.

The Statistical Analysis of Dose-Effect Relationships 117

One feature common to all experiments in any field, biological or other, is the variability in the measured effects from a given stimulus. In experiments with living matter this variability will usually be much greater than in the common chemical or physical measurements. In addition to the simple variation inherent in the measuring device, such as a scale to measure weight or a more complex assay of the amount of a certain chemical in the blood, the response of the experimental subjects, be they plants or animals, may also be influenced by biological and physiological factors such as sex, age, or some other physical conditions. The test subjects themselves will not be a completely homogeneous group with respect to all important factors which may affect the stimulus–response reaction. The susceptibilities of the subjects to the stimulus may well be dependent upon genetic differences which, even if known, cannot be completely controlled. In any experiments used to estimate a dose–response relation, the results of the experiment and its analysis must include some measures of the variability of the results, for such results to be properly interpretable.

6.2. EXPERIMENTAL DESIGN

The assessment of any stimulus–response relation will depend upon the data on which it is based. Inadequate data, whether experimental or observational, will not permit estimation of this relation. As an extreme example, when the effect is measured in terms of the per cent incidence of a particular response, any experiment in which the dose levels are either too low to show any response, or so high that they all show 100% response, can obviously give little information on the dose–effect relation. Any study, properly designed to elicit adequate information on the relation between dose and response, will entail the consideration of many factors (Emmens, 1948; Jerne and Wood, 1949; Finney, 1964).

The initial step is selection of the species of test subjects. Ideally, the subjects should be that species to which the results of the experiment are to be applied, or a closely comparable species. If one wishes to estimate the toxicological effect of some chemical pollutant in the environment, then representative species in the environment, either terrestrial or aquatic or both, should be tested. However, this approach is often not feasible for a variety of reasons. The species to be tested should be chosen on the basis of its susceptibility to the induction of the response of interest and on the basis of its biochemical and physiological similarity to the species to which the experimental results are to be applied. Similar dose–effect relations for many species will produce invaluable information on the general applicability of the dose–effect relation. Dissimilarities, which can be explained by known physiological differences between the species, may also be useful for extrapolating from one species to another.

The next step in the experimental design is the selection of the route and duration of the exposure which should be comparable to those occurring in nature. An inhalation study cannot be extrapolated to oral exposure without making a number of assumptions concerning the fate of the toxic agent between initial exposure and its reaching the target tissue. Metabolic pathways or physiological barriers may vary with different exposure routes. The results of single exposure studies cannot always be applied to chronic exposures in the environment; a single dose of 100 units of some toxicant may be either more or less effective than 100 fractionated one-unit doses of the same agent distributed over a period of time. For example, Matsumura (1972) showed that the ratio of the chronic to acute oral dosage for mallards required to produce 50% mortality is 50 for DDT and 1 for dieldrin and 'Zectran', with other insecticides falling within this range. In addition, single exposure situations may depend on physiological factors such as the age of the subject. Huggins *et al.* (1961) have shown that the rate of induction of mammary cancer by 3-methycholanthrene in rats is strongly affected by the age of the animal at exposure.

The selection of dose levels for an experimental bioassay is a critical step in the design of the study and will depend upon the purpose of the study. If the primary purpose is to show that a certain effect can be produced by the test agent, then the ideal design would have one treated group at the highest dose level that can be tolerated by the test subjects, i.e. a dose that will not produce other toxic effects that may obscure the response of interest. To guard against the possibility of incorrectly using too high a level of the stimulus, a second, lower level is also commonly incorporated in the design. If the response under consideration may have a spontaneous occurrence, then an untreated control group must also be included. This type of design will not, however, produce much information about the shape of the dose—response relationship. An efficient design to ascertain this shape should consist of a number of dose levels selected to produce a range of response rates between 10% and 90%. If little or no prior information is available on the expected response rates at various dose levels, then some type of pre-experiment should be performed. Dixon and Mood (1948) proposed what has been termed an up-and-down method, using single-test subjects, to estimate the dose level required to produce a specific response rate of some acute effect. Their technique has been generalized and extended by Robbins and Munro (1951) and Hsi (1969) to estimate the dose levels that produce 10%, 50%, and 90% responses; the final dose—response bioassay study can then be designed to use dose levels within this range. Bartlett (1946) suggested the use of an 'inverse sampling rule' for this problem. Ideally, the more dose levels used, the better the dose—response relationship can be estimated, but the cost of the study will normally determine the total number of subjects in the experiment. Therefore there is some trade-off between having many dose levels and having adequate numbers of test subjects at each dose.

The optimal spacing between the chosen number of dose levels is unknown, but a common approach is to choose equal spacing on the dose—level scale that is to be used in the analysis. For example, if it is planned to use a probit or log-logistic model for the analysis, then the doses should be selected to have equal spacing on a logarithmic scale. Hoel and Levine (1964) have given an optimal spacing solution to the general polynomial regression problem when the purpose is to estimate a value outside the observation range. Wetherill (1963) gives an optimal spacing for use in a logistic model.

Once the total number of test subjects has been decided, a decision must be made as to their allocation over the dose levels; this will depend upon the purpose of the study. If the design calls for k dose-level groups, each to be compared to a control group simply to find if an effect of treatment exists at some dose, then an optimal design is to allocate an equal number, say N_t, to each treated group, and place $\sqrt{k}N_t$ subjects in the control group. However, if the purpose is to estimate the response at each dose level equally well, then the number of subjects allocated to the ith level should be proportional to $p_i(1 - p_i)$, where p_i is the expected response rate. This means that one should allocate progressively less subjects to the extreme dose levels, both low and high. However, this does not imply that the number of control subjects should be zero unless the unexposed response rate is known to be zero. If no prior information exists on the expected response rates, then equal allocation should prove a reasonable strategy.

More complex bioassays involving combinations of toxic agents or the examination of different factors that modify the effects of a single toxicant may also be designed. The combined exposure to different toxicants may produce independent, additive, or synergistic effects, or they may be antagonistic to one another. Definitions of these actions, and the construction of theoretical models to explain them have been proposed by many authors; Plackett and Hewlett (1967), Hewlett and Plackett (1959, 1964), and Ashford and Smith (1964, 1965) are among them. Street *et al.* (1970) discuss the ecological significance of such interactive effects and the complexities inherent in their measurement. The subject is also discussed in Chapter 9.

A single experimental bioassay can measure the effect of a stimulus under only one fixed set of experimental conditions. However, nature presents more than a single face, and the ability to extrapolate experimental results to the natural environment will depend upon the generality of the bioassay design. Muirhead-Thompson (1971) discusses the influence of physical factors, such as temperature, water hardness, and pH, on the pesticide impact on fresh water. Matsumura (1972) showed that the mortality rate of brine shrimp exposed to various chlorinated hydrocarbon insecticides was dependent upon the salt concentration, the effect being greatest at either extremely low or high concentrations. Therefore, since we do not know whether these potential modifying factors exert independent or interactive effects, multifactorial studies should be designed in such a manner as to allow for the measurement of all possible joint effects. A complete multifactorial experiment will normally require a sizeable study. For example, a study with 3 factors having 4, 3, and 2 levels respectively along with 5 dose levels would require $120 = (4 \times 3 \times 2 \times 5)$ groups of test subjects. At twenty subjects per group, we would be faced with an experiment containing 2,400 subjects. However, using experimental design techniques as described by Kempthorne (1952) and Cochran and Cox (1957) for general statistical problems and Finney (1964) and Das and Kulkarni (1966) for bioassays, the size of these studies may be reduced if one is willing to assume that certain interactions between the factors, the dose and the response are negligible. However, this reduction in experimental units produces greater complexity in the analysis of such studies, so great care should be exercised in their design.

6.3. QUANTAL RESPONSES

One type of response commonly measured in toxicological studies is the quantal, or all-or-nothing, response, e.g. death. Measurements of degree of effect will normally provide a more refined measure of the response to a stimulus, but quantification is often very difficult or, for some responses, impossible.

When the response is quantal, its occurrence, for any particular subject, will normally depend upon the degree of the stimulus. For this subject, under constant environmental conditions, there will usually be some level of the stimulus below which the response will not occur and above which it will. This level is referred to as the subject's tolerance. Because of the biological variability among the population of individuals, their tolerance levels will also vary, sometimes within quite wide limits.

For quantal responses it is therefore natural to consider the distribution of tolerances over the population in question. If D represents the dose level of a particular stimulus, then the distribution of tolerances may be mathematically expressed as $f(D)dD$ which represents the proportion of individuals having tolerances between D and $D + dD$, where dD is small. If we are willing to assume that all members of the population will respond to a sufficiently high level of the stimulus, then the sum of these proportions should equal unity, or

$$\int_0^\infty f(D)dD = 1. \tag{6.1}$$

If a population is exposed to a dose of D_0, then all members with tolerances less than or equal to D_0 will respond, and the proportion they represent of the total population can be calculated as,

$$P(D_0) = \int_0^{D_0} f(D)dD. \tag{6.2}$$

Figure 6.1 shows a hypothetical tolerance distribution, $f(D)dD$, along with its corresponding cumulative distribution, $P(D)$, as defined in equation (6.2). The function $P(D)$ represents the dose—response relationship for the population when the response is quantal in nature. For any individual, the dose—response curve would be a step function, zero less than its tolerance and unity greater than its tolerance. However, for a population of individuals, the response can be measured

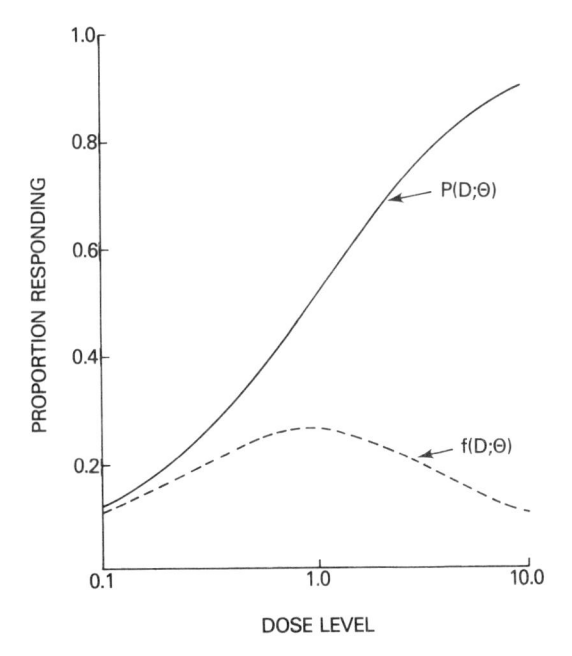

The Statistical Analysis of Dose-Effect Relationships

Figure 6.1 Example of relation between threshold distribution $f(D)$ and dose–reponse curve $P(D)$

as the proportion responding. The curve defined by $P(D)$ can also be considered as the probability that one individual, selected at random from the population, will respond to a dose level D. This particular distribution in Figure 6.1 assumes that $P(0) = 0$ (no responders for a zero dose) and $P(\infty) = 1$ (all will respond to some high dose). Either or both of these two assumptions may be untrue. The first assumes no spontaneous occurrence of the particular response which is false for a number of responses, while the second may be false if either an immune group exists within the population or the particular response in question becomes overwhelmed at high dose levels by a different response and does not have a chance to become manifest.

If a group of N test subjects were chosen at random from some population having a tolerance distribution given by $f(D)dD$, and each subject was exposed to the same dose level D, then the number of subjects showing the response would be a random variable having a binomial probability distribution. In mathematical terms, the probability of R responders out of N subjects each given a dose of D, is given by,

$$\text{Prob}(R) = \binom{N}{R} P(D)^R Q(D)^{N-R}, \tag{6.3}$$

where $Q(D) = 1 - P(D)$ and $P(D)$ is the probability of response for one randomly chosen test subject as defined in Equation 6.2. The observed proportion of responders in the test sample, $p(D) = R/N$, is then an estimate of the true proportion in the population, $P(D)$. This observed proportion can be either larger or smaller than the population value because of sampling variation, but the variation is centered around the value $P(D)$ and decreases as the size of the test sample increases.

An estimate of the dose–response relation can be obtained by testing various groups of subjects at different dose levels. Each value of the observed proportion of responders, $p(D)$, is an estimate of its corresponding $P(D)$, and from these quantities, the population cumulative tolerance distribution can be estimated. In general, $P(D)$ will increase with the dose D, but if the number of test subjects at each dose level is small, then sampling variation may interfere with the regularity of trend in the observed proportions $p(D)$.

The preceding discussion has assumed that the subjects chosen to be tested at each dose level have been randomly selected from some larger group of subjects and that the experimental conditions are the same for each dose. Deviations from these assumptions, either by choice or by chance, may result in the binomial model equation (6.3) being inapplicable. For example, non-random selection of test subjects, such as different age groups at the different dose levels or the selection of a group of subjects from the same parent to be tested at the same dose, would cause the ratio of the variation among dose groups to the variation within groups to be larger than expected under the binomial model. In addition, differences between the dose groups with respect to factors known to be associated with the response, such as age, sex, weight, or some conditions of the experiment itself, could produce an undesired bias in the observed proportions. In any experimental situation, care must be given to select the groups at random and to control other variables of the experiment.

6.4. MATHEMATICAL MODELS OF TOLERANCE DISTRIBUTIONS

A toxicological dose–response experiment, or bioassay, will result in a series of dose levels, D_i, $i = 1, \ldots, k$, along with their corresponding observed proportions responding, $p_i = p(D_i)$. These pairs of values (D_i, p_i) provide an estimate of the dose–response relation for only a limited number, k, of dose levels. An estimate of the entire dose–response curve can be made only by assuming some general functional form relating dose to response, i.e. $P(D) = g(D; \theta)$, where the function g represents some particular class, the member of the class being defined by the unknown constant, or constants, θ. For example, g may be a simple linear function, $g(D; \theta) = \theta_0 + \theta_1 D$, or a more complex function such as $g(D; \theta) = 1/2 + (1/\pi)\tan^{-1}(\theta_0 + \theta_1 \log(D))$.

The results of toxicity tests often show that the observed proportion of responders monotonically increases with dose and shows a sigmoid relation with some function of the exposure level. This led to the development of the normal, or probit, model of the dose–tolerance distribution. This model assumes that the population distribution of tolerances is given by the normal probability model,

$$f(D;\theta) = (2\pi\theta_1^2)^{-1/2} \exp\left[-\frac{1}{2}\left[\frac{x(D)-\theta_0}{\theta_1}\right]^2\right], \theta_1 > 0 \tag{6.4}$$

or

$$P(D;\theta) = \int_{-\infty}^{\frac{x(D)-\theta_0}{\theta_1}} (2\pi)^{-1/2} \exp(-\tfrac{1}{2}t^2)\,dt,$$

where $x(D) = x$ is some transformed value of the dose level D. Some transformations commonly used in practice are

$$x = \log_{10}(D),$$

and, more generally

$$x = D^a,$$

where $a \leq 1$. The validity of any transformation will depend upon the mechanism of the response to the stimulus in question and, as such, is beyond the scope of this discussion. We, as Finney (1949), propose to use any transformation that appears to fit the observational data, but any additional mechanistic knowledge for a particular problem should be incorporated into the model. The normal model in equation (6.4) has commonly been used with the logarithmic dose transformation. A history of the development of this model is given by Finney (1952).

Other mathematical models of tolerance distributions which lead to the sigmoid appearance of their corresponding dose–response curves are:

1. The logistic curve,

$$P(D;\theta) = \{1 + \exp[\log(\theta_0) - \theta_1 \log(D)]\}^{-1}, \theta_0, \theta_1 > 0$$

$$= \frac{D^{\theta_1}}{\theta_0 + D^{\theta_1}} \tag{6.5}$$

which is derived from chemical kinetic theory and was proposed by Wilson and Worcester (1943) and Berkson (1944) for bioassay analyses, and

2. The sine curve,

$$P(D;\theta) = \tfrac{1}{2}\{1 + \sin[\theta_0 + \theta_1 \log(D)]\} \tag{6.6}$$

which is applicable only over a limited range of doses, $-\pi/2 \leq \theta_0 + \theta_1 \log(D) \leq \pi/2$, and has no theoretical justification other than computational simplicity (Knudsen and Curtis, 1947). Other dose–response models have been proposed on the basis of what has been called 'hit theory'. These models do not start with an assumed dose–tolerance distribution to produce a dose–response curve, but are derived on general mechanistic dose–response assumptions. Turner (1975) summarizes all these models. The more commonly used models are the single hit model,

$$P(D;\theta) = 1 - \exp(-\theta D), \quad \theta > 0 \tag{6.7}$$

and the multi-hit model,

$$P(D;\theta) = 1 - \sum_{h=0}^{m-1} \frac{1}{h!} (\theta D)^h \exp(-\theta D), \quad \theta > 0 \tag{6.8}$$

where m is the minimum of 'hits' on a receptor required to obtain a response. Another model, which has been proposed as a mechanism in carcinogenesis, has been termed the multi-stage model. One derivation of this general process, due to Armitage and Doll (1961) and extended by Peto (1975), leads to the mathematical relation,

$$P(D;\theta) = 1 - \exp\left(-\sum_{h=1}^{m} \theta_h D^h\right), \quad \theta_h \geq 0, \quad h = 1, \ldots, m \tag{6.9}$$

where m is the number of stages in the process affected by the agent. A dose–tolerance distribution may be obtained from these models by inverting equation (6.2), i.e.

$$f(D;\theta) = \frac{dP(D;\theta)}{dD} \tag{6.10}$$

Therefore, the tolerance distribution corresponding to the multi-hit model (6.8) is the gamma distribution

$$f(D;\theta) = \frac{\theta^m}{\Gamma(m)} D^{m-1} \exp(-\theta D) \tag{6.11}$$

which looks similar to both the log-normal and log-logistic tolerance distributions.

Table 6.1 compares the dose–response relations of three models. It might be thought that the selection of one particular model over the others would be obvious from inspection of the calculated responses but this table shows that three of the most commonly used models, log-normal, log-logistic, and single-hit give results that differ by little over a 256-fold dosage range (FDA Advisory Committee on Protocols for Safety Evaluation, 1971). It would take an inordinately large experiment to be able to conclude which of the three models best described the observational data.

If the calculated dose–response curve is to be used to estimate the response rate that would be expected from an exposure level within this range of observable responses, then all three models will give comparable results. Interpolation between observed data points within the range of approximately 5%–95% response rates will not be greatly affected by the mathematical model selected. However, extrapola-

Table 6.1 Expected Per Cent Responding for Various Models over a Range of Dose Levels

Dose level	Log-normal model	Log-logistic model	Single-hit model
16	98	96	100
8	93	92	99
4	84	84	94
2	69	70	75
1	50	50	50
$\frac{1}{2}$	31	30	29
$\frac{1}{4}$	16	16	16
$\frac{1}{8}$	7	8	8
$\frac{1}{16}$	2	4	4

tion to exposure levels expected to give very low response rates is highly dependent upon the choice of mathematical models, which is shown in Table 6.2 extending the previous table to much smaller doses. It can be seen that the further one extrapolates from the observed response range, the more divergent the various models become. At a dose level that is 1/1000 of the 50% response dose, the single-hit model gives an estimated response rate 200 times as large as the log-normal model. The fact that a moderate-sized experiment conducted at dose levels high enough to give observable response rates cannot discriminate among these various models, and the fact that these same models show a substantial divergence at small dose levels present major difficulties for low dose extrapolation problems. Brown (1976a) has suggested the use of a multi-stage model (equation 6.9) along with an estimate of both sampling and model variability for this problem, since the multi-stage model has the extrapolation characteristics of most other models, depending upon the number of stages used.

It should be noted that all the mathematical dose—response models presented have the feature that for any dose $D > 0$, $P(D) > 0$, i.e. there is no absolute threshold level below which the probability of response is zero. A common

Table 6.2 Expected Per Cent Responding at Low Doses for Models Describing Observed Responses in the 5%–95% Range Equally Well

Dose level	Log-normal model	Log-logistic model	Single-hit model
0.01	0.05	0.4	0.7
0.001	0.00035	0.026	0.07
0.0001	0.0000001	0.0016	0.007

toxicological problem is whether or not a threshold level actually exists. Thresholds should, however, be considered from two different viewpoints: a 'theoretical' level below which a toxic response is impossible; and a 'practical' level below which the chance of response is highly unlikely or unobservable. Theoretical thresholds could vary substantially in an outbred population due to genetic differences, and their existence cannot be proven by statistical arguments; this proof would have to come from complete knowledge of the mechanism of toxicological action.

Experimental or observational evidence for the existence of a threshold is commonly presented in the form of a dose—response graph in which the response rate is plotted against dose level. Either the existence of doses not showing an increase in response over controls or the extrapolation of such curves to low doses which apparently would result in no increased response are cited as indications of the existence of some threshold below which no response is possible. This type of evidence is of little value. In the first situation, the observation of no responders does not guarantee that the probability of response is actually zero. From a statistical viewpoint, zero responders out of N at risk is consistent at the 5% significance level, with an actual response rate between zero and approximately $3/N$.

In the second case, when a graph of observed dose against responses is extrapolated downward to a no-effect level, the observed dose—response relation, often linear, is assumed to persist throughout the entire range of dose levels. This assumption can easily lead to an erroneous conclusion when the true dose—response curve has a rising slope, i.e. is convex. Brown (1976b) discusses this problem in detail when the response is carcinogenic in nature. He shows that statistical analyses of bioassay results cannot discriminate between mathematical models which assume the existence or non-existence of an actual threshold. Therefore, without a knowledge of the mechanism producing the response, when extrapolating below the observable response range, it would be prudent to assume that no threshold exists.

6.5. ESTIMATION OF A QUANTAL DOSE—RESPONSE CURVE

In this section we shall assume that we have concluded a typical quantal dose—response experiment, or bioassay, and, having observed the proportions responding at various dose levels, we wish to estimate the population dose—response relation.

First we have to select one of the mathematical models with which we shall perform the analysis. For the sake of simplicity, we shall use the log-logistic model (equation 6.5), though the general technique to be described is applicable to any model. The logistic model can be written as,

$$P(x; a, b) = \{1 + \exp[-(a + bx)]\}^{-1}, \; b \geqslant 0 \qquad (6.12)$$

where P is the probability of a randomly selected individual from the population responding to an exposure of $x = \log(\text{dose})$. The series of log dose levels are denoted as $x_1, x_2, \ldots, x_k$ and their corresponding observed proportions responding as $p_i = p(x_i) = r_i/n_i$, $i = 1, 2, \ldots, k$, where r_i is the number of test subjects responding out of n_i exposed at the ith dose level. The estimation of the dose—response curve consists of estimating the two unknown parameters a and b, which can be accomplished in a variety of ways, ranging from simple graphical techniques (DeBeer, 1945; Litchfield and Wilcoxon, 1949), to more sophisticated methods such as maximum likelihood or minimum chi-square. The maximum-likelihood method is an extremely general, fully efficient estimation technique but is computationally difficult (Cornfield and Mantel, 1950). Bliss (1935) and Finney (1952) give its application to the probit model and the details of this method are given for any general quantal response model in the Appendix to this chapter. A simpler computational approach has been given by Grizzle et al. (1969) for the general logistic model and by Berkson (1949) for bioassay data.

The logistic model in (6.12) can be rewritten as,

$$\log\left[\frac{P(x;a,b)}{Q(x;a,b)}\right] = a + bx \qquad (6.13)$$

where $Q(x;a,b) = 1 - P(x;a,b)$. The transformation $\log(P/1 - P)$ has been termed the logit transformation by Berkson (1944). This has reduced the problem from a non-linear to a simpler linear problem which can be solved by the method of least squares. The linear model in equation (6.13) relates the response to the dose for the ith experimental group as,

$$z_i = a + bx_i$$

where z_i is the logit of the response rate, and x_i is logarithm of the dose level.

Since the variances of the observed logits are not necessarily equal, we should properly use weighted, as opposed to unweighted, least squares. The variance of the ith response rate is $P_i Q_i/n_i$, where $Q_i = 1 - P_i$ and P_i is the population response rate. From asymptotic statistical theory, the variance of the logit of the response rate is $1/n_i P_i Q_i$, which is an approximate variance for finite samples. Therefore, the weighted least squares method uses $W_i = n_i P_i Q_i$ as the weights, which can be approximated by the observed response rates $w_i = n_i p_i q_i$. In the event that $p_i = 0$ or 1, which would result in the weighting factor being zero, Berkson (1953) suggests using $p_i = 1/2n_i$ in place of zero, and $p_i = 1 - 1/2n_i$ in place of unity. The weighted least squares technique produces the following estimates for a and b,

$$\hat{a} = \frac{\Sigma w_i x_i^2 \Sigma w_i z_i - \Sigma w_i x_i \Sigma w_i x_i z_i}{\Sigma w_i \Sigma w_i x_i^2 - (\Sigma w_i x_i)^2} = \bar{z} - \hat{b}\bar{x},$$

and (6.14)

$$\hat{b} = \frac{\Sigma w_i \Sigma w_i x_i z_i - \Sigma w_i z_i \Sigma w_i x_i}{\Sigma w_i \Sigma w_i x_i^2 - (\Sigma w_i x_i)^2} = \frac{\Sigma w_i (x_i - \bar{x})(z_i - \bar{z})}{\Sigma w_i (x_i - \bar{x})^2},$$

where $\bar{x} = \Sigma w_i x_i / \Sigma w_i$ and $\bar{z} = \Sigma w_i z_i / \Sigma w_i$. Once a and b are estimated, the population response rate can then be estimated for any dose level D in the following manner. First estimate the logit of the response rate using equation (6.13),

$$\hat{z} = \hat{a} + \hat{b}\log(D),$$

and then transform the logit to the response probability using equation (6.12),

$$\hat{P}(D) = (1 + e^{-\hat{z}})^{-1}$$

The estimated variances of these parameter estimates are given by,

$$S_a^2 = \frac{\Sigma w_i x_i^2}{\Sigma w_i \Sigma w_i x_i^2 - (\Sigma w_i x_i)^2} = \frac{\Sigma w_i x_i^2 / \Sigma w_i}{\Sigma w_i (x_i - \bar{x})^2},$$

and (6.15)

$$S_b^2 = \frac{\Sigma w_i}{\Sigma w_i \Sigma w_i x_i^2 - (\Sigma w_i x_i)^2} = \frac{1}{\Sigma w_i (x_i - \bar{x})^2}$$

The variance of the estimated logit z of the response rate for any value of the log dose, $X = \log(D)$, can be obtained from the relation,

$$\text{Var}(z) = \text{Var}(a + bX) = S_a^2 + X^2 S_b^2 + 2X S_{ab}^2,$$

where $S_{ab}^2 = -\bar{x}/\Sigma w_i (x_i - \bar{x})^2$ is the covariance between the two parameter estimates, $\hat{a}$ and $\hat{b}$. This can be simplified to become,

$$\text{Var}(z) = S_z^2 = \frac{1}{\Sigma w_i} + \frac{(X - \bar{x})^2}{\Sigma w_i (x_i - \bar{x})^2} \tag{6.16}$$

This variance can be used to place statistical confidence limits on the estimated logit,

$$z \pm Z_{\alpha/2} S_z \tag{6.17}$$

where $Z_{\alpha/2}$ is a standard normal deviation corresponding to a total tail area of α. For 95% confidence limits, $\alpha = 0.05$ and $Z_{\alpha/2} = 1.96$. Once these limits are placed on the estimated logit, they can be transformed by equation (6.12) to give confidence limits on the estimated population response,

$$\{1 + \exp[-(z - Z_{\alpha/2} S_z)]\}^{-1} \leqslant P(D) \leqslant \{1 + \exp[-(z + Z_{\alpha/2} S_z)]\}^{-1} \tag{6.18}$$

The regression equation (6.13) can also be used in reverse to find that dose level which is expected to produce a certain population response rate. A common method of characterizing the toxicity of a stimulus is by means of the median effective dose. This value is defined as the dose which will produce a response in half the population, and thus is sometimes called the median tolerance. This median effective dose is commonly denoted by ED_{50}. When the response in question is death, then the term LD_{50}, the median lethal dose is used. Analogous symbols may be denoted dose levels which affect other proportions of the population, such as the ED_{10}, the dose which causes 10% of the population to respond.

In general, an estimate of the ED_p, the dose resulting in a proportion p of responders, can be obtained from equation (6.13) as,

$$\hat{X}_p = \log(D_p) = (\log(p/q) - \hat{a})/\hat{b} \qquad (6.19)$$

The approximate variance of this log dose is given by,

$$S^2_{\hat{X}_p} = \left(\frac{1}{\hat{b}}\right)^2 \left(\frac{1}{\Sigma w_i} + \frac{(\hat{X}_p - \bar{x})^2}{\Sigma w_i (x_i - \bar{x})^2}\right). \qquad (6.20)$$

As can be seen from this expression, the variance of this log dose increases as the estimated log dose moves away from the average log dose of the experiment. As before, confidence limits for the dose D_p giving the desired response rate p can be obtained from equations (6.19) and (6.20) as,

$$\exp(\hat{X}_p - Z_{\alpha/2} S_{\hat{X}_p}) \leq D_p \leq \exp(\hat{X}_p + Z_{\alpha/2} S_{\hat{X}_p}) \qquad (6.21)$$

The following is an example of the concepts and calculations of these statistical procedures. The observed data come from an experiment designed to measure the lethal effect of rotenone sprayed on the chrysanthemum aphis. The data are given in Table 6.3, reproduced from Finney (1952). The proportions responding to treatment, (response here is death) are shown in column 5 for each of the five concentrations used in the experiment. The next five columns are quantities used in the calculation of the parameters a and b and their variances in the log-logistic model. The calculation of these parameters is shown below the table. The log dose associated with 50% mortality is estimated from equation (6.19) where $\log(P/Q) = \log(0.5/0.5) = 0$ resulting in the estimate being simply, $-\hat{a}/\hat{b}$. The last two columns of the table give the estimated proportion of responders at each dose level, calculated from equation (6.12), and the contributions to a chi-square test for comparing the fitted logistic model with the observed data. Figure 6.2 shows the observed proportions responding at each dose level and a graph of the log-logistic model fitted to these data along with 95% confidence limits on the fitted curve. The limits were obtained from equation (6.18). The figure shows the sigmoid nature of the log-logistic model and that it appears to describe adequately the observed dose–response relationship. One should, however, use a statistical test to evaluate the fit of the model to the data.

The Pearson chi-square statistic, denoted X^2, is commonly used to test whether or not the assumed model, here a log-logistic, fits the observed data. The statistic X^2 is given by,

$$X^2 = \Sigma \frac{(\text{observed} - \text{expected})^2}{\text{expected}}$$

Table 6.3 Example of Log-logistic Model Applied to Quantal Response Data. Toxicity of Rotenone. (Reproduced by permission of Cambridge Univ. Press from Finney, 1952)

Concentration (mg/l) D	Log dose x	Responders r	Number at risk n	Proportion responding p	$w = npq$	wx	wx^2	wz	wxz	Estimated proportion responding	Chi square
2.6	0.956	6	50	0.120	5.380	5.048	4.826	−10.518	−10.055	0.129	0.036
3.8	1.335	16	48	0.333	10.661	14.232	19.0	−7.388	−9.863	0.322	0.027
5.1	1.629	24	46	0.522	11.477	18.696	30.456	1.010	1.645	0.539	0.054
7.7	2.041	42	49	0.857	6.007	12.260	25.023	10.759	21.959	0.805	0.844
10.2	2.322	44	50	0.880	5.280	12.260	28.468	10.518	24.423	0.907	0.432
					38.705	62.496	107.773	4.381	28.109		$\chi^2_{(3)} = 1.393$

$$\bar{x} = \frac{\Sigma wx}{\Sigma w} = 1.615$$

$$\bar{z} = \frac{\Sigma wz}{\Sigma w} = 0.113$$

$$\hat{a} = \bar{z} - \hat{b}\bar{x} = -4.838$$

$$\hat{b} = \frac{(\Sigma w \Sigma wxz - \Sigma wz \Sigma wx)}{(\Sigma w \Sigma wx^2 - (\Sigma wx)^2)} = 3.065$$

$$\hat{x}_{50} = \frac{-\hat{a}}{\hat{b}} = 1.578$$

$$\hat{LD}_{50} = 4.85$$

$$S_a^2 = \frac{\Sigma wx^2}{\Sigma w \Sigma wx^2 - (\Sigma wx)^2} = 0.406$$

$$S_b^2 = \frac{\Sigma w}{\Sigma w \Sigma wx^2 - (\Sigma wx)^2} = 0.146$$

95% confidence limits on LD_{50}: $4.37 \leq LD_{50} \leq 5.37$

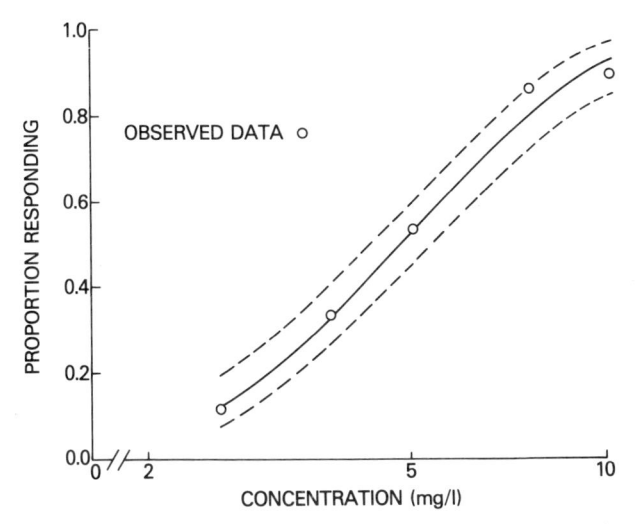

Figure 6.2 Estimated dose–response curve with 95% confidence limits (rotenone toxicity example). (Reproduced by permission of Cambridge Univ. Press from Finney, 1952)

which, for this type of bioassay experiment, is equal to,

$$X^2 = \Sigma_i \frac{n_i}{\hat{p}_i \hat{q}_i} (p_i - \hat{p}_i)^2 \tag{6.22}$$

where $\hat{p}_i$ is the estimated response probability for the ith dose level, and $\hat{q}_i = 1 - \hat{p}_i$. The degrees of freedom for this statistic are given by the number of dose levels minus 2, the number of parameters in the model that were estimated from the data. For this example, the tables show that $X^2 = 1.39$ with three degrees of freedom, which is not statistically significant. Therefore, we may conclude that the model adequately fits the observed data.

This is not always the case, however, as shown in the following example. The data are from an experiment by Sinnhuber et al. (1968) in which Rainbow trout, held in 200-gallon tanks, were fed with different commercial diets plus additives. The amount of aflatoxin in the diet and additives was measured and the response of interest was the incidence of hepatomas. The fish were sampled at 6, 9, and 12 months. Table 6.4 shows the results of the 12 month sample.

The last column of the table, the chi-square calculation to test the adequacy of the log-logistic model, yields a chi square of 21.9 with 8 degrees of freedom which is statistically significant, $P < 0.01$. An examination of the standardized differences between the observed and expected proportions responding, given by

$$R_i = \frac{\hat{p}_i - p_i}{\sqrt{\hat{p}_i \hat{q}_i / n_i}} \tag{6.23}$$

Table 6.4 Example of Log-logistic Model Applied to Heterogeneous Quantal Data. Incidence of Hepatomas in Rainbow Trout Fed Diets Containing Aflatoxin for 12 Months. (Reproduced by permission of the authors from Sinnhuber et al., 1968)

Concentration (ppb) D	Log dose x	Responders r	Number at risk n	Proportion responding p	$w = npq$	wx	wz	wx^2	wxz	Estimated proportion responding	Chi square
0.8[a]	−0.223	0	12	0.0	0.444	−.099	−1.429	.022	0.319	0.104	1.393
3.7	1.308	5	10	0.5	2.5	3.27	0.0	4.277	0.0	0.380	0.611
4.0	1.386	2	13	0.154	1.907	2.643	−2.910	3.663	−4.033	0.401	3.302
5.0	1.609	9	18	0.5	4.5	7.241	0.0	11.651	0.0	0.460	0.116
7.9	2.067	5	12	0.417	2.929	6.054	−0.908	12.514	−1.877	0.584	1.378
8.0[a]	2.079	20	30	0.667	6.720	13.970	4.495	29.044	9.346	0.587	0.792
15.3	2.728	62	82	0.756	15.251	41.604	17.005	113.496	46.388	0.742	0.084
19.0	2.944	34	39	0.872	4.625	13.616	8.492	40.086	25.0	0.785	1.749
36.5	3.597	31	40	0.775	7.121	25.613	8.538	92.130	30.710	0.881	4.287
42.0	3.738	89	90	0.988	1.459	5.454	5.966	20.387	22.301	0.896	8.175
					47.456	119.366	39.249	327.270	128.154		

$\chi^2_{(8)} = 21.887$

[a] Fed for 13 months.

$$\bar{x} = \frac{\Sigma wx}{\Sigma w} = 2.515 \qquad \bar{z} = \frac{\Sigma wz}{\Sigma w} = 0.827$$

$$a = \frac{\Sigma wx^2 \Sigma wz - \Sigma wx \Sigma wxz}{(\Sigma w \Sigma wx^2 - (\Sigma wx)^2)} = -1.912$$

$$b = \frac{(\Sigma w \Sigma wxz - \Sigma wz \Sigma wx)}{(\Sigma w \Sigma wx^2 - (\Sigma wx)^2)} = 1.089$$

$$S_a^2 = \frac{\Sigma wx^2}{(\Sigma w \Sigma wx^2 - (\Sigma wx)^2)} = 0.255$$

$$S_b^2 = \frac{\Sigma w}{(\Sigma w \Sigma wx^2 - (\Sigma wx)^2)} = 0.037$$

may sometimes give a clue as to where the assumed model fails. Draper and Smith (1966) give an excellent discussion of residual analysis. In this situation, there appear to be no systematic deviations related to dose level nor to response rate, so we would have to conclude that the assumption of binomial sampling may be in error. The variation in the data is larger than would be expected under the binomial assumption. The original experiment consisted of 15 groups being fed various diets containing different amounts of aflatoxin. This variation in diets and additives may well have created the extra sampling variation. Kleinman (1973) and Cochran (1943) discuss this problem of proportions with extraneous variation from the viewpoint of statistical tests in the analysis of variance. They consider the extra variation to be additive to the binomial sampling variation. Finney (1952) suggests assuming that the true variation is simply a multiple of the binomial variation, and he proposes multiplying each of the variance estimates in equations (6.15), (6.16), and (6.20) by a heterogeneity factor, X^2/k, where k is the number of degrees of freedom for the chi-square statistic X^2. In addition, in order to allow for the uncertaintly in estimating this heterogeneity factor, the normal deviates, $Z_{\alpha/2}$, used in calculating confidence limits, equations (6.17) and (6.21), should be replaced by t values, $t_{k,\alpha/2}$ where t has k degrees of freedom. Since these t values are larger than the corresponding normal deviations, this has the effect of producing wider confidence intervals. In the previous example with heterogeneous data, the 95% confidence limits on the probability of response at a concentration of 5.0 p.p.b. of aflatoxin in the diet would be, using no heterogeneity factor and using normal deviations,

$$0.35 \leqslant P(5 \text{ ppb}) \leqslant 0.571$$

However, after multiplying the variance in equation (6.16) by the heterogeneity factor 21.887/8 = 2.736, and using a t value with 8 degrees of freedom, 2.306, in equation (6.18), the proper 95% confidence limits become,

$$0.264 \leqslant P(5 \text{ ppb}) \leqslant 0.669$$

which are substantially wider than those calculated assuming no heterogeneity. It should be noted, however, that this correction procedure is only an approximation and should not be used if the Pearson X^2 statistic is not statistically significant.

The preceding results have been obtained by tacitly assuming that no response was possible at a zero dose, i.e. $P(0) = 0$. This is seen to be true for the log-logistic model defined in equation (6.12). When the dose is zero, the log dose, x, is minus infinity, and since the parameter b is restricted to be non-negative, the probability of response becomes zero. This assumption of no response for no stimulus will not necessarily be true. In the two previous examples, both experimental designs contained a control, non-treated, group which produced no responders, so there was no apparent reason to consider the possibility of spontaneous, non-dose-related, response. However, if the response under consideration is one which either appears among a group of untreated subjects, or if such controls are not available, it may be expected to have a spontaneous occurrence, then any mathematical dose–response model should properly allow for this possibility.

Two methods have been proposed to incorporate the possibility of response due to factors other than the stimulus in question. The first is known as 'Abbott's correction', attributed to W. S. Abbott (1925), and is based on the assumption of an independent action between the stimulus and the background, or spontaneous response. If a proportion C of the subjects would respond in the absence of any stimulus, then the total response rate at a dose level D, assuming independent actions, would be,

$$P'(D) = C + (1 - C)P(D) \tag{6.24}$$

From this equation it follows that the proportion responding due to the stimulus alone is,

$$P(D) = \frac{(P'(D) - C)}{(1 - C)} \tag{6.25}$$

If the spontaneous response rate C were known exactly, then the previous log-logistic model could be used with only minor modifications. The observed response rates p_i' should be transformed to their corresponding dose-related response rates p_i by use of equation (6.25). These p_i would be transformed to logits, $\log(p/1 - p)$. The weights w_i used for the calculation of the parameters a and b, and their variances, become,

$$w_i' = \frac{n_i p_i (1 - p_i)}{1 + \dfrac{C}{p_i (1 - C)}} \tag{6.26}$$

which are equal to w_i when $C = 0$. These modifications should not add undue complexities to the estimation of a dose–response curve. However, the spontaneous response rate is usually unknown, and even the knowledge of the proportion responding in some control group will only provide a range of plausible spontaneous rates. In this common situation of uncertainty, more complex methods of estimation, such as maximum likelihood, will be required and the general technique described in the Appendix to this chapter can be applied.

The second method of allowing for the spontaneous occurrence of response was proposed by Albert and Altshuler (1973) in which they suggested that the natural environment contains some additive background level of the stimulus, D_0, so that the response rate for a dose D would be,

$$P'(D) = P(D + D_0) \tag{6.27}$$

Abbott's correction factor would then be given as $C = P'(0) = P(D_0)$. Therefore, if C and the true dose–response curve were known, the background level D_0 could also be computed. This method, which will give results similar to the previous approach, can only be employed using more complex estimation procedures. The maximum-likelihood method, assuming that D_0 is an unknown parameter, can be used.

In general, when the response under consideration may have a spontaneous occurrence independent of the stimulus, one should take care when using a model that assumes no such possibility. Figure 6.3 shows the effect of a spontaneous occurrence on the log-logistic dose–response function. The probability of response is given by equation (6.24) where $P(D)$ is the log-logistic model. The figure shows that, on a logit scale, whereas the log-logistic model is a linear function of the log dose (equation 6.13), the addition of a spontaneous response (here $C = 0.05$) induces a curvature at the low response rates. This could produce a bias in estimating the parameters of the log-logistic model assuming no spontaneous occurrence. The model from which the 'true' curve was produced has parameters $a = -5$ and $b = 3$, but the straight line fitted by eye through this 'true' curve would estimate them as $\hat{a} = -3.2$ and $\hat{b} = 2$. In general, when the 'true' dose–response curve contains a spontaneous element, then any model which assumes no such possibility will lead to an underestimate of the slope and an overestimate of the intercept. Looking at the difference between the estimated and 'true' curves in Figure 6.3 shows this difference to be negative for low dose levels, positive for moderate dose levels, and negative for high dose levels. This pattern provides a useful clue when examining the standardized residuals from the fit of a model to experimental data.

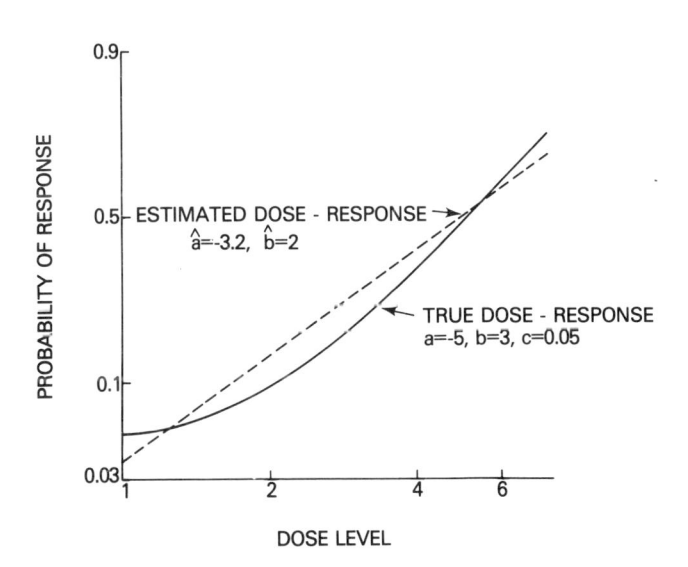

Figure 6.3 Comparison of true dose–response curve having spontaneous occurrence with estimated curve assuming no spontaneous occurrence

6.6. QUANTITATIVE RESPONSES

The preceding discussion has been concerned only with quantal responses. In many biological investigations, however, it is possible to quantify the magnitude of the response. Such data could be reduced to the quantal form by a simple dichotomy of the measurements into those greater than or less than some selected value. Since this procedure is wasteful of information, it would be better to use the actual measurements to construct a dose–response relation.

Dose–response data obtained over a limited range of dose levels may be analysed by general statistical regression techniques. The estimated regression equation will however, not be valid outside this range; the magnitude of the response is often limited to be non-negative and to have a maximum possible response at high doses. The general techniques of the previous sections can be applied to these quantitative data by transforming the responses to lie between zero and one which can be done by dividing each one by the maximum response (if unknown it is treated as an unknown parameter to be estimated from the data). Therefore, quantitative responses can be thought of as a special case of quantal responses. The statistical treatment of such data is similar to, but somewhat different from, that for quantal responses. Finney (1952, 1964) provides a discussion of these analyses.

6.7. TIME-TO-OCCURRENCE MODELS

The estimation and analysis of the dose–response models discussed in the previous section are based on observational data that are quantal in nature, i.e. the subject either responds or not. However, for many experimental situations, there is an additional piece of information that has been infrequently used in the past, viz. the time since initiation of treatment at which the subject responded, or in the case of no response, the total length of time the subject was observed without a response. This additional information may be of benefit in two ways. It should add data to define more clearly the dose–response relation, especially in those situations for which the response rates at the high dose levels are close to 100%. Those experiments for which most dose levels produce 100% response will not provide much information on the quantal dose–response relation, but upon examination of the times to response, a monotone relationship between dose levels and the mean, or median, time to occurrence may be revealed, Brown (1973). The further advantage of using these occurrence times is that it allows for the construction of mathematical models which relate the dose level to the expected time of response. These models can then be used to estimate the number of responders in some population at risk at various points in time. A third potential advantage will be important in some studies but not in others. This relates to the problem of competing risks. If an agent produces two or more toxic responses, then relating dose to the incidence of a late-occurring response may be obscured by a different, earlier toxic response such as death.

The construction of a mathematical time-to-occurrence model for dose and response is made up of two parts. The first is the mathematical form for the probability distribution of the response-time random variable t,

$$\text{Probability } (t \leq T) = F(T; \theta_1, \ldots, \theta_k) \tag{6.28}$$

where $F(\cdot)$ is some cumulative distribution function having probability function $f(\cdot) = dF(\cdot)/dt$, and $\theta_1, \ldots, \theta_k$ are unknown parameters. It is assumed that the mathematical form of F is the same for each dose level, and that one, or more, of the parameters θ_i are functions of dose. It is also assumed that all subjects will eventually respond, i.e. $F(\infty) = 1$, but because of competing risks such as death without evidence of the desired response, the subject may be removed from observation before the response can occur. This assumption is probably more valid for chronic exposure than acute exposure situations.

The second assumption concerns the relation between the dose level D and the parameters in equation (6.28). A general empirical relationship that has been proposed by Busvine (1938) and others is,

$$\theta = \alpha D^\beta$$

or

$$\log(\theta) = \log(\alpha) + \beta \log(D) \tag{6.29}$$

There is no presumed biological basis for this relationship, but when θ is the median time to response, a linear relation between the logarithm of θ and the logarithm of dose has often been observed.

Many mathematical time-to-occurrence models have been proposed and all have a direct correspondence with quantal dose–response models (Chand and Hoel, 1974). One of the first models was proposed by Druckrey (1967) in which he assumed that the probability distribution of the occurrence times was log normal,

$$f(t; \theta_1, \theta_2) = (\sqrt{2\pi}\theta_2 t)^{-1} \exp\left[-\frac{1}{2}\left(\frac{\log(t) - \log(\theta_1)}{\theta_2}\right)^2\right] \tag{6.30}$$

where θ_1 is the median time to occurrence and θ_2 is the standard deviation of the log response times. The value of θ_1 is assumed to be related to dose through equation (6.29) while θ_2 is assumed to be independent of the dose. The probability of a response at or before time T can be written as,

$$\begin{aligned} P(D) = \int_0^T f(t; \theta_1, \theta_2) dt &= \Phi[(\log(T) - \log(\theta_1))/\theta_2] \\ &= \Phi[(\log(T) - \log(\alpha) - \beta\log(D))/\theta^2] \\ &= \Phi(a + b \log(D)) \end{aligned}$$

where $\Phi(\cdot)$ is the cumulative normal distribution function, $a = (\log T - \log \alpha)/\theta_2$, and $b = \beta/\theta_2$. This shows that the log normal time-to-occurrence model produces the normal, or probit, quantal response model.

Other time-to-occurrence models include the Weibull model proposed by Pike (1966) and Peto et al. (1972),

$$f(t; \theta_1, \theta_2) = \theta_1 \theta_2 t^{\theta_2 - 1} \exp(-\theta_1 t^{\theta_2}) \quad (6.31)$$

where θ_1 is assumed to be related to dose and θ_2 is independent of dose to give,

$$P(D) = \int_0^T f(t; \theta_1, \theta_2) dt$$
$$= 1 - \exp(-\alpha D^\beta T^{\theta_2})$$
$$= 1 - \exp\{-\exp[a + \beta \log(D)]\}$$

which corresponds to the one-hit model when $\beta = 1$. The compound Weibull, or generalized Pareto model,

$$f(t; \theta_1, \theta_2, \theta_3) = \frac{\theta_1 \theta_2 t^{\theta_2 - 1}}{(1 + \theta_1 t^{\theta_2}/\theta_3)^{\theta_3 + 1}} \quad (6.32)$$

leads to the log-logistic quantal response model if θ_1 is assumed to be related to dose through equation (6.29) and θ_3 is assumed to be unity,

$$P(D) = \int_0^T f(t; \theta_1, \theta_2, \theta_3 = 1) dt$$
$$= (1 + \alpha D^\beta T^{\theta_2})^{-1}$$
$$= \{1 + \exp[a + \beta \log(D)]\}^{-1}$$

These and other models have been studied by Shortley (1965) and Gart (1965). The estimation of the parameters for any of these models from observed data cannot be done in a simple straightforward manner. Methods such as the iterative maximum-likelihood procedure previously discussed will have to be used. Figure 6.4 shows the results of fitting a Weibull time-to-response model (equation 6.31) to the results of an experiment in which benzopyrene was chronically painted on the backs of mice and the time until appearance of a skin tumour was noted. Lee and O'Neill (1971) found that, using equation (6.29) to describe the relation between induction time and dose, the parameter β was approximately equal to 2 indicating a relationship of time to response to the square of the dose level. This is an excellent example of the degree to which time-to-response models can describe observed data.

When such models are used to estimate risks from pollution of the natural environment, it should be noted that the estimated median response time, from equation (6.29), may often be greater than the natural lifespan of the species under consideration. This does not mean that the pollutant is without observable hazard. Due to the variation in response times around this median, some proportion of the subjects at risk may respond within their lifespan.

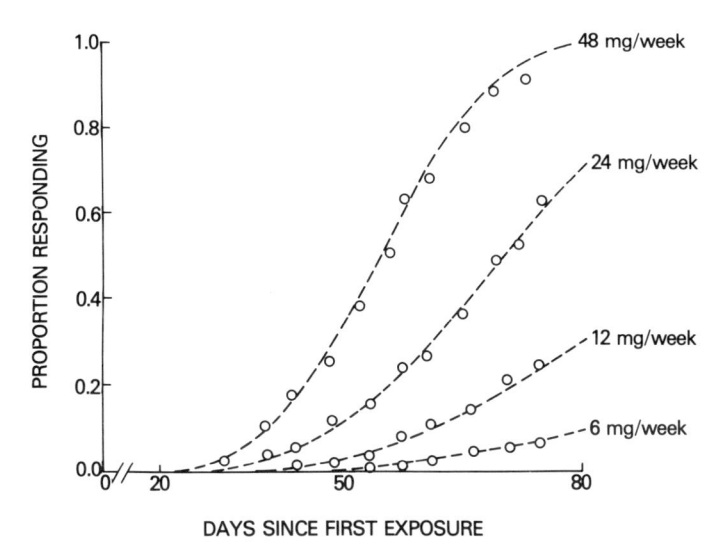

Figure 6.4 Comparison of estimated (– – –) and observed (○) proportions of animals with skin tumours at four dose levels. (Reproduced by permission of H. K. Lewis and Co. Ltd., from Lee and O'Neill, 1971)

Another aspect of time-related mathematical models which has received little attention is the effect of chronic exposure upon later generations of the organism in question. Brown (1958) discusses two possible changes in the dose–response relation from one generation to the next:

1. A reduced slope of the dose–response curve with a corresponding increase in the LD_{50} as the more susceptible phenotypes are eliminated from the population. This can be followed by a steepening of the slope of the dose–response curve as the population becomes more homogeneous for resistance.

2. Increases in the LD_{50} may also occur without changes in the slope due to an increased vigour of the strain rather than elimination of specific phenotypes.

A number of experiments have submitted colonies of insects to selection pressure from a specific insecticide, and when the LD_{50} levels are plotted against the generation number, they are often found to constitute a sigmoid curve. In the first few generations there is little increase in the LD_{50}, then a sharp increase followed by a flattening out, implying that a maximum resistance has been reached. Such increased resistance may be expected to develop faster when the selection pressure is high rather than when only a small proportion is killed in each generation. Once resistance has been increased and the selection pressure removed, it has often been found that the strain shows a reversion toward the original susceptibility.

This process of adaptation to a toxicant has also been found with phytoplankton. The period of adaptation before resumption of normal exponential growth will depend upon the level of exposure. Stockner and Antia (1976) found that this adaptation period for the marine diatom *Skeletonema costatum* in various concentrations (10%–30%) of kraft pulpmill effluent ranged from 2 to 12 days. No mathematical models have been proposed to explain either of these phenomena.

6.8. APPLICATION TO ECOLOGICAL RISK ASSESSMENT

Once the proper experimental or observational data have been obtained in the laboratory and one can estimate the relation between exposure level and response, the impact of an exposure upon the total ecosystem or its individual components can in theory be evaluated. The changes to be produced within the entire ecosystem will entail knowledge of the complex interactions among the component parts, which is beyond the scope of this chapter. We shall consider only the assessment of risk to the one individual component which relates to the data in hand.

This risk assessment can be based on a variety of measures and responses: increased incidence of some undesirable response in the exposed population, e.g. death, decreased reproductive capacity, or other response; life-shortening due to one or more toxicological responses; or a manifestation of these responses passed through genetic mutations from the exposed population to their offspring. The selection of particular responses and their measurement may depend upon practical considerations but a variety of such measurements should be made to obtain a more complete assessment of the total risk.

A common problem in the ecological assessment of risk is that of extrapolating from the results of experiments on laboratory animals which are generally conducted under highly controlled conditions with genetically homogeneous animals, to animals of different species not genetically homogeneous, living under diverse environmental conditions, and exposed to a variety of other toxicological agents. Nutritional differences and the physical environment can affect the response to many stimuli. In addition, the natural chemical environment with its great variety of substances provides the possibility for either synergistic or antagonistic activity. Genetic differences can affect many aspects of toxicological susceptibility. Therefore, the environment and genetic variability of the target population should, whenever possible, be considered in the extrapolation process. (In general, this heterogeneity should reduce the slope of any experimentally derived dose–response curve.)

The animal species used in the laboratory experiment is often different from the species in the ecosystem to which the experimental results are to be extrapolated. Many simplistic methods for extrapolating from one species to another have been proposed. A useful first approximation is provided by the surface area rule. The basic assumption is that the locus of action of any chemical is on some surface area; which particular surface may be unknown. If we assume an essential similarity, except size, between different mammalian species and assume about equal densities, then any surface area in an organism will be approximately proportional to the $\frac{2}{3}$ power of its weight. This surface area extrapolation rule is, however, only an approximation. After an animal is exposed to a chemical toxicant, a number of events may occur which can influence the observed toxic effect. These events include: absorption, distribution and storage, metabolism, excretion and re-absorption. Comparisons of the similarity of these events should be made among various animal species to improve the extrapolation of toxic effects between the different species. Absorption rates of chemicals through the gastrointestinal tract, the lung, and the skin are measurable in animals. The surface area-to-volume ratio in the gastrointestinal tract often differs among species. The presence of bacteria in the gastrointestinal tract may indirectly affect absorption. Once a chemical is absorbed, it is distributed through, and stored in, various parts of the body. The toxicant must pass through a variety of barriers before reaching its site of action. Intra-species variation in this distribution system should be taken into account when extrapolating from one species to another. Since a metabolite of the chemical to which an animal is exposed may be the toxic agent, rather than the original compound, metabolic differences among species will also enter the extrapolation equation. Differences in excretory rates of compounds through the kidneys, liver, and intestines may also be important.

These intra-species comparisons should provide for an 'equivalent dose' rule among species. For example, suppose we have two species, one of which is the experimental test animal denoted species T, and the other is the species in the ecosystem, denoted species E, for which we wish to make a risk assessment. Then we can relate the ecological exposure to species E, d_E, to the equivalent dose for the test species, $d_T = H(d_E)$, where the function H will depend upon differences in the biochemical and physiological properties between the two species. However, since this extrapolation rule between species will not be known with certainty, some measure of the error in estimating the species' properties should be incorporated into the final extrapolation. Therefore, the estimation of risk to any population in the ecosystem will have some measure of uncertainty which is a combination of sampling errors in the data, choice of a particular mathematical model among the many possible models when extrapolating outside the observable range, and the uncertainties inherent in extrapolating between different animal species. All these sources of variation can lead to large potential errors in the final assessment of risk.

6.9. CONCLUSIONS

In the past, research on the analysis of dose—response relations has been limited to describing the results of an experiment conducted on a single species under controlled conditions. Due to the magnitude of interactions inherent in any ecosystem, these techniques will not be generally applicable for predicting the effect of pollutants upon an entire system or even subsystems. Future work on doses and responses for estimating risks to ecosystems should examine the following problems.

1. Time-dependent models of ecosystems, with as many interactions as feasible, should be designed and tested. This will entail, as a first step, definitions of ecosystem response variables. These responses can range from a simple measure of the population size for a particular species, to a measure of both population size and mix of a number of species. The time-dependent nature of the model would allow for the estimation of these response variables over an extended time frame.
2. With respect to estimation of the effects on single species, more work should be done on modelling the changes over time in resistance or susceptibility to chronic exposure to toxicants.
3. Increased effort should be devoted to study both the interactive effects of combined toxicants and the modifying effects of other non-toxic environmental conditions since extrapolation of experimental results to a natural ecosystem will depend upon the generality of the bioassay.

In general, the weakness of current techniques used to measure the relations of doses and responses is that they are aimed at single species under controlled conditions. We must begin to consider an ecosystem as a single entity, albeit a complex one, and formulate new methods of estimating dose response of the system as a whole.

6.10. REFERENCES

Abbott, W. S., 1925. A method of computing the effectiveness of an insecticide. *J. Econ. Entomol.*, 18, 265–7.

Albert, R. E. and Altshuler, B., 1973. Considerations relating to the formulation of limits for unavoidable population exposures to environmental carcinogens. In *Radionuclide Carcinogenesis* (Ed. J. E. Ballou), AEC Symposium Series CONF-72050, Springfield, Virginia NTIS.

Armitage, P. and Doll, R., 1961. Stochastic models for carcinogenesis. *Fourth Berkeley Symposium on Mathematical Statistics and Probability*, University of California Press, pp. 19–38.

Ashford, J. R. and Smith, C. S., 1964. General models for quantal response to the joint action of a mixture of drugs. *Biometrika*, 51, 413–28.

Ashford, J. R. and Smith, C. S., 1965. An alternative system for the classification of mathematical models for quantal responses to mixtures of drugs in biological assay. *Biometrics*, 21, 181–8.

Bartlett, M. S., 1946. A modified probit technique for small probabilities. *J. Roy. Stat. Soc. Suppl.*, 8, 113–7.

Berkson, J., 1944. Application of the logistic function to bioassay. *J. Am. Stat. Assoc.*, **39**, 357–65.

Berkson, J., 1949. Minimum chi-square and maximum likelihood solution in terms of a linear transform, with particular reference to bioassay. *J. Am. Stat. Assoc.*, **44**, 273–8.

Berkson, J., 1953. A statistically precise and relatively simple method of estimating the bioassay with quantal response, based on the logistic function. *J. Am. Stat. Assoc.*, **48**, 565–99.

Bliss, C. I., 1935. The calculation of the dosage–mortality curve. *Ann. Appl. Biol.*, **22**, 134–67.

Brown, A. W. A., 1958. *Insecticide Resistance in Arthropods*, World Health Organization Monograph Series No. 38, Geneva.

Brown, C. C., 1976a. Variability of risk estimation in dose–response experiments. Presented at NIEHS conference on problems of extrapolating the results of laboratory animal data to man.

Brown, C. C., 1976b. Mathematical aspects of dose–response studies in carcinogenesis–the concept of thresholds. *Oncology*, **33**, 62–5.

Brown, V. M., 1973. Concepts and outlook in testing the toxicity of substances to fish. In *Bioassay Techniques and Environmental Chemistry* (Ed. G. E. Glass) Ann Arbor Science, Ann Arbor.

Busvine, J. R., 1938. The toxicity of ethylene oxide to *Calandra oryzae* C. *granaria*, *Tribolium castaneum*, and *Cimex lectularis*. *Ann. Appl. Biol.*, **25**, 605–32.

Chand, N. and Hoel, D. G., 1974. A comparison of models for determining safe levels of environmental agents. In *Reliability and Biometry SIAM*, Philadelphia, pp. 681–700.

Cochran, W. G., 1943. Analysis of variance for percentages based on unequal numbers. *J. Am. Stat. Assoc.*, **38**, 287–301.

Cochran, W. G. and Cox, G. M., 1957. *Experimental Designs*, 2nd ed., Wiley and Sons, New York.

Cornfield, J. and Mantel, N., 1950. Some new aspects of the application of maximum likelihood to the calculation of the dosage response curve. *J. Am. Stat. Assoc.*, **45**, 181–210.

Das, M. N. and Kulkarni, G. A., 1966. Incomplete block designs for bioassays. *Biometrics*, **22**, 706–29.

DeBeer, E. J., 1945. The calculation of biological assay results by graphic methods. The all-or-none type of response. *J. Pharmacol. Exp. Ther.*, **85**, 1–13.

Dixon, W. J. and Mood, A. M., 1948. A method for obtaining and analyzing sensitive data. *J. Am. Stat. Assoc.*, **43**, 109–26.

Draper, N. R. and Smith, H., 1966. *Applied Regression Analysis*, Wiley and Sons, New York.

Druckrey, H., 1967. Quantitative aspects of chemical carcinogenesis. In *Potential Carcinogenic Hazards from Drugs (Evaluation of Risks)* (Ed. R. Truhaut), Springer-Verlag, New York.

Emmens, C. W., 1948. *Principles of Biological Assay*, Chapman and Hall, London.

Finney, D. J., 1949. The choice of a response metameter in bioassay. *Biometrics*, **5**, 261–72.

Finney, D. J., 1952. *Probit Analysis*, 2nd ed., University Press, Cambridge.

Finney, D. J., 1964. *Statistical Methods in Biological Assays*, 2nd ed., Griffin, London.

Food and Drug Administration Advisory Committee on Protocols for Safety Evaluation, 1971. Panel on carcinogenesis report on cancer testing in the safety

evaluation of food additives and pesticides. *Toxicol. Appl. Pharmacol.*, **20**, 419–38.
Gart, J. J., 1965. Some stochastic models relating time and dosage in response curves. *Biometrics*, **21**, 583–99.
Grizzle, J. E., Starmer, C. F., and Koch, G. G., 1969. Analysis of categorical data by linear models. *Biometrics*, **25**, 489–504.
Hewlett, P. S. and Plackett, R. L., 1959. A unified theory for quantal responses to mixtures of drugs: non-interactive action. *Biometrics*, **15**, 591–610.
Hewlett, P. S. and Plackett, R.L., 1964. A unified theory for quantal responses to mixtures of drugs: competitive action. *Biometrics*, **20**, 566–75.
Hoel, P. G. and Levine, A., 1964. Optimal spacing and weighting in polynomial prediction. *Ann. Math. Stat.*, **35**, 1553–60.
Hsi, B. P., 1969. The multiple sample up-and-down method in bioassay. *J. Am. Stat. Assoc.*, **64**, 147–62.
Huggins, C., Grand, L. C., and Brillantes, F. P., 1961. Mammary cancer induced by a single feeding of polynuclear hydrocarbons, and its suppression. *Nature*, **189**, 204–7.
Jerne, N. K. and Wood, E. C., 1949. The validity and meaning of the results of biological assays. *Biometrics*, **5**, 273–99.
Kempthorne, O., 1952. *The Design and Analysis of Experiments*, Wiley and Sons, New York.
Kleinman, J. C., 1973. Proportions with extraneous variance: single and independent samples. *J. Am. Stat. Assoc.*, **68**, 46–54.
Knudsen, L. F. and Curtis, J. M., 1947. The use of the angular transformation in biological assays. *J. Am. Stat. Assoc.*, **42**, 282–96.
Lee, P. N. and O'Neill, J. A., 1971. The effect of both time and dose applied on tumour incidence rate in benzopyrene skin painting experiments. *Br. J. Cancer*, **25**, 759–70.
Litchfield, J. T. and Wilcoxon, F., 1949. A simplified method of evaluating dose effect experiments. *J. Pharmacol. Exp. Ther.*, **96**, 99–113.
Matsumara, F., 1972. Biological effects of toxic pesticidal contaminants and terminal residues. In *Environmental Toxicology of Pesticides* (Eds. F. Matsumara, G. M. Boush, and T. Misato), Academic Press, New York and London.
Muirhead-Thompson, R. C., 1971. *Pesticides and Freshwater Fauna*, Academic Press, London and New York.
Peto, R., 1975. Presentation to the NIEHS conference on extrapolation of risks to man from environmental toxicants on the basis of animal experiments.
Peto, R., Lee, P. N., and Paige, W. S., 1972. Statistical analysis of the bioassay of continuous carcinogens. *Br. J. Cancer*, **26**, 258–61.
Pike, M. C., 1966. A method of analysis of a certain class of experiments in carcinogenesis. *Biometrics*, **22**, 142–61.
Plackett, R. L. and Hewlett, P. S., 1967. A comparison of two approaches to the construction of models for quantal responses to mixtures of drugs. *Biometrics*, **23**, 27–44.
Robbins, H. and Munro, S., 1951. A stochastic approximation method. *Ann. Math. Stat.*, **22**, 400–7.
Shortley, G., 1965. A stochastic model for distributions of biological response times. *Biometrics*, **21**, 562–82.
Sinnhuber, R. O., Wales, J. H., Ayres, J. L., Engebrecht, R. H., and Amend, D. L., 1968. Dietary factors and hepatoma in rainbow trout (*Salmo gairdneri*). I. Aflatoxins in vegetable protein foodstuffs. *J. Natl. Cancer Inst.*, **41**, 711–8.

Stockner, J. G. and Antia, N. J., 1976. Phytoplankton adaptation to environmental stresses from toxicants, nutrients, and pollutants – a warning. *J. Fish. Res. Board Can.*, 33, 2089–96.

Street, J. C., Mayer, F. L., and Wagstaff, D. J., 1970. Ecological significance of pesticide interactions. In *Pesticides Symposia* (Ed. W. B. Deichman), Halos, Miami.

Swartz, J. and Spear, R. C., 1975. Dynamic model for studying the relationship between dose and exposure in carcinogenesis. *Math. Biosc.*, 26, 19–39.

Turner, M. E., 1975. Some classes of hit-theory models. *Math. Biosc.*, 23, 219–35.

Wetherill, G. B., 1963. Sequential estimation of quantal response curves. *J. Roy. Stat. Soc.*, B25, 1–48.

Wilson, E. B. and Worcester, J., 1943. The determination of LD_{50} and its sampling error in bioassay. *Proc. U.S. Natl. Acad. Sci.*, 29, 79–85.

6.11. APPENDIX: THE MAXIMUM-LIKELIHOOD METHOD OF FITTING DOSE–RESPONSE MODELS TO QUANTAL DATA

Assume that the data are of the following form: at each of m dose levels, denoted $d_1, d_2, \ldots, d_m$, we have observed r_i responders out of a total of n_i subjects at risk, $i = 1, \ldots, m$. Assume further that we wish to fit some mathematical model relating dose level to the probability of response which takes the general form,

$$\text{Probability of response at dose } d_i = P(d_i; \theta_1, \ldots, \theta_k) = P_i \qquad (6.33)$$

where $\theta_1, \ldots, \theta_k$ are unknown parameters to be estimated from the observed data. The method of maximum likelihood chooses those values of the θ_i that maximize the likelihood of the observed data.

Assuming that the binomial model holds for each dose level, then the likelihood of observing r_i responders out of n_i subjects at the ith dose level is simply the binomial probability,

$$L(r_i) = \text{Prob}(r_i \text{ out of } n_i) = \binom{n_i}{r_i} P_i^{r_i} (1 - P_i)^{n_i - r_i} \qquad (6.34)$$

where P_i is the unknown response probability at the ith dose. Since the result at any dose level is independent, in the statistical sense, of the results at each of the other doses, the likelihood of the entire set of data can be written as the product of the individual likelihoods,

$$L = L(r_1, \ldots, r_m) = \pi_i L(r_i) = \pi_i \binom{n_i}{r_i} P_i^{r_i} (1 - P_i)^{n_i - r_i} \qquad (6.35)$$

Inserting the dose–response model (equation 6.33) into this likelihood expression gives,

$$L = \pi_i \binom{n_i}{r_i} P(d_i; \theta_1, \ldots, \theta_k)^{r_i} [1 - P(d_i; \theta_1, \ldots, \theta_k)]^{n_i - r_i}$$

The estimates $\hat{\theta}_i$ of the unknown parameters θ_i are obtained by maximizing this expression over all possible values of the θ_i. It is often easier to maximize the logarithm of the likelihood, since maximizing one will maximize the other.

$$l = \log(L) = \Sigma_i \{\log(\tbinom{n_i}{r_i}) + r_i \log[P(d_i; \theta_1, \ldots, \theta_k)]$$
$$+ (n_i - r_i) \log[1 - P(d_i; \theta_1, \ldots, \theta_k)]\} \tag{6.36}$$

One method of obtaining the maximum of a function is to find the values of θ_i such that the first partial derivatives, with respect to the θ_i, are all equal to zero. These partial derivatives are given by

$$\frac{\partial l}{\partial \theta_j} = \Sigma_i \frac{(r_i - n_i P_i)}{P_i Q_i} \frac{\partial P_i}{\partial \theta_j}, j = 1, \ldots, k \tag{6.37}$$

where $P_i = P(d_i; \theta_1, \ldots, \theta_k)$ and $Q_i = 1 - P_i$.

Solving this set of non-linear equations is often computationally difficult and must be accomplished by iteration. The Newton–Raphson iteration technique is one such method that can be used. It can be described as follows:

(1) start with a set of initial estimates of the $\hat{\theta}_j$ denoted by $\hat{\theta}_j^{(0)}$;
(2) for each $\hat{\theta}_j$, find its correction factor denoted by $\Delta_j^{(0)}$;
(3) the new estimates are then given by $\hat{\theta}_j^{(1)} = \hat{\theta}_j^{(0)} + \Delta_j^{(0)}$;
(4) continue this iteration procedure, steps (2) and (3), until the equations (equation 6.37) are all equal or close to zero; in general, the new parameter estimates at the $(i+1)$st stage are obtained from those at the ith stage by the relation, $\hat{\theta}_j^{(i+1)} = \hat{\theta}_j^{(i)} + \Delta_j^{(i)}$.

For the Newton–Raphson iteration method, the correction factors $\Delta_j^{(i)}$ are computed from the second partial derivatives of the log likelihood. These second derivatives are,

$$\frac{\partial^2 l}{\partial \theta_s \partial \theta_t} = \Sigma_i \left[\frac{(r_i - n_i P_i)}{P_i Q_i} \frac{\partial^2 P_i}{\partial \theta_s \partial \theta_t} + \frac{((r_i - n_i P_i) P_i - r_i Q_i)}{P_i^2 Q_i^2} \left(\frac{\partial P_i}{\partial \theta_s}\right) \left(\frac{\partial P_i}{\partial \theta_t}\right) \right]$$
$$s, t = 1, \ldots, k \tag{6.38}$$

Using matrix notation and denoting the vector of first derivatives in equation (6.37) by $M_1^{(i)} = (\partial l/\partial \theta_1, \ldots, \partial l/\partial \theta_k)'$ evaluated at the $\hat{\theta}_j^{(i)}$, $j = 1, \ldots, k$, and denoting the matrix of the second derivatives in equation (6.38) by $M_2^{(i)} = (\partial^2 l/\partial \theta_s \partial \theta_t)$ also evaluated at the $\hat{\theta}_j^{(i)}$, $j = 1, \ldots, k$, then the vector of the ith correction factors, $\Delta^{(i)} = (\Delta_1^{(i)}, \ldots, \Delta_k^{(i)})'$ is obtained from the matrix equation,

$$\Delta^{(i)} = -(M_2^{(i)})^{-1} M_1^{(i)} \tag{6.39}$$

The set of $\hat{\theta}_j$ which solve the equations equation (6.37) are the maximum-likelihood estimates of the parameters. The variances and covariances of these estimates are given by the negative inverse of the matrix of second derivatives

The Statistical Analysis of Dose-Effect Relationships

(equation 6.38) evaluated at the maximum-likelihood estimates $\hat{\theta}_j, j = 1, \ldots, k$. In matrix notation,

$$\Sigma = (\sigma_{st}^2) = -M_2^{-1}, \tag{6.40}$$

where σ_{ss}^2 is the estimated variance of $\hat{\theta}_s$ and σ_{st}^2 is the estimated covariance of $\hat{\theta}_s$ and $\hat{\theta}_t$.

A specific example of this general technique for the log-logistic model follows. The probability of response is given by,

$$P(x_i; \theta_1, \theta_2) = \{1 + \exp[-(\theta_1 + \theta_2 x_i)]\}^{-1},$$

where $x_i = \log(d_i)$. The first and second derivatives of the log likelihood in equations (6.37) and (6.38) are easily obtained by using the relations,

$$\frac{\partial P_i}{\partial \theta_1} = P_i Q_i \text{ and } \frac{\partial P_i}{\partial \theta_2} = P_i Q_i x_i$$

These derivatives turn out to be,

$$\frac{\partial l}{\partial \theta_1} = \Sigma_i (r_i - n_i P_i), \frac{\partial l}{\partial \theta_2} = \Sigma_i (r_i - n_i P_i) x_i$$

$$\frac{\partial^2 l}{\partial \theta_1 \partial \theta_1} = -\Sigma_i n_i P_i Q_i$$

$$\frac{\partial^2 l}{\partial \theta_1 \partial \theta_2} = \frac{\partial^2 l}{\partial \theta_2 \partial \theta_1} = -\Sigma_i n_i x_i P_i Q_i$$

$$\frac{\partial^2 l}{\partial \theta_2 \partial \theta_2} = -\Sigma_i n_i x_i^2 P_i Q_i$$

Table 6.5 Maximum-likelihood Iterative Computation for Rotenone Toxicity Example

Iteration i	Parameter estimates $\hat{\theta}_1^{(i)}$	$\hat{\theta}_2^{(i)}$	First derivatives $\partial l/\partial \theta_1$	$\partial l/\partial \theta_2$	Correction factors $\Delta_1^{(i)}$	$\Delta_2^{(i)}$
0	0	1	−69.474	−88.064	−5.362	2.180
1	−5.362	3.180	13.540	21.742	0.606	−0.152
2	−4.754	3.028	− 0.434	− 0.175	−0.133	0.076
3	−4.887	3.104	− 0.005	0.0002	−0.0021	0.0012
4	−4.889	3.105	0.000	0.000		

The correction factors become

$$\Delta_1 = -[(\Sigma n_i x_i P_i Q_i)(\Sigma(r_i - n_i P_i)x_i) - (\Sigma n_i x_i^2 P_i Q_i)(\Sigma(r_i - n_i P_i))]/D$$

$$\Delta_2 = -[(\Sigma n_i x_i P_i Q_i)(\Sigma(r_i - n_i P_i)) - (\Sigma n_i P_i Q_i)(\Sigma(r_i - n_i P_i)x_i)]/D$$

where $D = (\Sigma n_i P_i Q_i)(\Sigma n_i x_i^2 P_i Q_i) - (\Sigma n_i x_i P_i Q_i)^2$.

The P_i and Q_i are calculated using the current values of $\hat{\theta}_1$ and $\hat{\theta}_2$. Convergence of this iterative procedure depends upon the initial estimates $\hat{\theta}_1^{(0)}$ and $\hat{\theta}_2^{(0)}$, but is normally quite rapid. Table 6.5 shows the computational steps to obtain maximum-likelihood estimates for the rotenone example in Table 6.3. The initial estimates were $\hat{\theta}_1^{(0)} = 0$ and $\hat{\theta}_2^{(1)} = 1$, far from the final estimates, yet only 4 iterations were required. It should also be noted that the estimates obtained from the previous least squares solution agree closely with these maximum-likelihood estimates. This will generally be true for moderately large sized experiments.

SECTION III
EXPERIMENTAL TOXICOLOGY AND FIELD OBSERVATIONS RELATED TO ECOTOXICOLOGY

CHAPTER 7

General Aspects of Toxicology

A. JERNELÖV, K. BEIJER, AND L. SÖDERLUND

Swedish Water and Air Pollution Research Laboratory,
Box 21060, 10031 Stockholm, Sweden

7.1	INTRODUCTION	151
7.2	THE CELL	152
7.3	ENZYMES	154
7.4	METABOLISM AND REGULATORY PROCESSES	156
	(i) Carbohydrate Metabolism	156
	(ii) Respiration	157
	(iii) Lipid Metabolism	157
	(iv) Microsomal Enzyme Systems	158
	(v) Biosynthesis	158
	(vi) Regulatory Processes and Growth	159
7.5	MUTATIONS	160
7.6	TUMOUR INDUCTION	161
7.7	REPRODUCTION AND TERATOGENESIS	163
7.8	IMMUNE RESPONSE	164
7.9	INTERACTIONS OF COMPOUNDS	164
7.10	DISCUSSION	166
7.11	REFERENCES	166

7.1. INTRODUCTION

As a result of new insights into biochemistry, cellular biology, cytogenetics, and pharmacology, increased attention has been paid to the ability of toxic agents, individually or collectively, to induce responses and pathological changes at the cellular level.

A large number of *in vitro* studies on the interaction of chemical agents with cellular organelles, DNA and RNA molecules and individual enzymes or enzyme systems etc., have been undertaken. However, due to the complexity of the living organism, results obtained in such studies do not necessarily explain effects observed in the organism which may be the result of the interference with several biochemical mechanisms. Thus it is important to establish the qualitative and quantitative relationships between effects demonstrated in an *in vitro* experiment and the effect in the organism, which they are supposed to explain. A thorough knowledge of the basic mechanisms by which toxic agents exert their injurious effects is fundamental for an understanding of biological responses of flora and fauna.

7.2. THE CELL

Toxic agents may disrupt the integrity of cellular structures thus impairing vital processes. Inside the cell they may interfere in some way with normal metabolic processes necessary to sustain life.

In order to enter the organism and reach some target organ, any material must penetrate one or more living membranes. The membranes form barriers with specific permeability properties which together with the mediated transport systems control the rate and extent of transfer of substances into and out of a cell and cell organelles. Toxic agents may in different ways modify the permeability of a membrane. A stabilization will occur when the pores are compressed by the penetration of a chemical agent with a decrease in membrane permeability as a result. Some insecticides affect the cell membrane in this way. When a chemical gives rise to an increased permeability a labilization is said to occur. DDT is believed to have this effect. The mediated transfer systems may be disturbed by chemicals interfering with carriers and ATP production. Thiol groups are often involved in the binding of substrates to specific carriers and the transport mechanism may be impaired by toxic agents such as for instance cadmium.

The outer membrane of mitochondria is freely permeable to most low molecular weight solutes including many toxic agents which may interfere with different processes within the organelle, whereas the inner membrane, where the enzymes of electron transport and energy conversion are located, is relatively impermeable and allows only certain substances to pass via a transport system. The permeability can, however, be altered by toxic agents, for example by carbon tetrachloride and some insecticides. Some chemicals produce a swelling of the mitochondria with a resultant unfolding of the membranes and a decrease in ATP production. Cadmium has been found to produce such a swelling in corn, probably through binding to sulphydryl groups on the membrane (Miller *et al.*, 1973).

The lysosomal enzymes responsible for cell autolysis would cause a destruction of most cellular components if they were not separated from the rest of the cell by a membrane. Thus increased permeability or a rupture of the membrane by a toxic agent will injure or even kill the host cell. A well-known example of such an effect is what is called silicosis in humans. The digestive enzymes if released can attack the nucleic acids in the nucleus and cause chromosome damage. An increase of membrane stability so that the release of enzymes is retarded will impede the replacement of 'old cells'. Toxic agents may change the properties of the lysosomal membranes so as to make it less suitable for fusion with other membranes, and thus seriously affect cellular nutrition.

The maintenance of normal cell volume and pressure depends on the sodium—potassium pump. If the pumping is inhibited, sodium will enter the cell along the concentration gradient and water will follow along the osmotic gradient. This will cause a swelling of the cell. The cation pump in cell membranes is a likely point of attack by chemicals in their first interaction with the cell. The pumping may also be inhibited by chemicals interfering with the production of ATP.

Mercury is one toxic agent which may affect living membranes and their functions in several ways. For example, mercuric ions have been shown to cause a decrease in the electrical potential across the membrane of new cells with a resulting loss of cellular potassium and a significant change in nerve conduction.

Bacteria grown in the presence of mercuric chloride have been found to exhibit extensive morphological abnormalities. Significant effects are the numerous structural irregularities associated with cell wall and cytoplasmic membrane synthesis and function (Vaituzis et al., 1975).

Methylmercury is soluble in phospholipids in the central nervous system which leads to membrane lysis specific for those membranes that contain plasmalogens as a major constituent of the phospholipid backbone in membrane structure. *In vitro* experiments have shown that methylmercury can react both catalytically and directly with a group of phospholipids that are important in membrane structure for cells of the central nervous system. It has thus been suggested that this reaction could have a significance in the neurotoxicity of methylmercury (Wood and Seagall, 1974).

The nervous system in particular serves as a target for many toxic agents. The nerve impulses produced by the movement of sodium and potassium ions across the cell membrane may of course be disturbed by chemical agents interfering with the ion pump.

Chemicals may also inhibit or facilitate processes at the synapse involving the transmitter substance (usually acetylcholine, adrenaline, or noradrenaline) and cause disturbances of the synaptic transmission by interfering with the synthesis or storage of the transmitter substance, its release from the synaptic buttons by nerve impulses, its permeability-changing action on the membrane of the next cell or its removal and destruction.

One example of the last process is the action of organophosphorous compounds (used as insecticides and also for instance for fireproofing of textiles) and carbamates which inhibit acetylcholinesterase, the enzyme responsible for the destruction of acetylcholine. Inhibition of the enzyme permits an abnormal accumulation of acetylcholine. This will cause excessive activity of the parasympathetic system, give central nervous system effects and over-reactivity of the voluntary muscles.

Lead may, besides affecting the activity of Na^+-K^+-ATPase, cause demyelination of the axon. Acrylamide is also believed to have this effect. Toxic levels of triethyltin may cause enlarging of the oligodendrocytes, which develop large fluid-filled vacuoles. Extracellular fluid will also accumulate between the myelin rings with a resultant splitting of the sheaths from the axons. Triethyltin is thought to exert its effect by inhibiting the ATPase normally present in astroglia and axonal tubes (Toback, 1965). Hexachlorophene is believed to have similar effects.

7.3. ENZYMES

Enzymes are of supreme importance in biology. They make up the largest and most highly specialized class of proteins. Life depends on a complex network of chemical reactions brought about by specific enzymes. The enzymes are the primary instruments for the expression of genetic action since they catalyse the thousands of chemical reactions that collectively constitute the intermediary metabolism of cells. Clearly any modification of an enzyme pattern may have far-reaching consequences for the living organism.

The consequences of enzyme inhibition for the metabolism will depend on the effects of the inhibition on the enzyme itself. The formation of enzymes is under different kinds of control. Repression and induction of enzyme synthesis by metabolites or substrate is one kind of control. Thus since inhibition of the enzyme will lead to a build-up of substrate and changed levels of metabolites, enzyme formation will be increased and the effects of the inhibition on the overall metabolic function will not be severe. The rate of enzyme formation may also be regulated by the actual level of enzyme present. If the enzyme–inhibitor complex or the inhibitor itself is mistaken for intact enzyme, no more enzyme will be synthesized and the impact of this suppression may be great.

Inhibition of enzymes in the usual sense, by a chemical agent, may be reversible or irreversible, competitive or non-competitive. A reversible inhibition is characterized by an equilibrium between enzyme and inhibitor and will give a definite degree of inhibition depending on inhibitor concentration. Reversible inhibitions are numerous, e.g. the action of cyanide on cytochrome oxidase and of malonate on succinate dehydrogenase.

An irreversible inhibition is characterized by a progressive increase with time, ultimately reaching complete inhibition even with very low concentrations of inhibitor, provided that the inhibitor is in excess of the amount of enzyme present. Examples of irreversible inhibition are the action of cyanide on xanthine oxidase and of the 'nerve gases' on cholinesterases.

Inhibition may involve a direct combination of the inhibitor with the enzyme, but inhibition will also be produced by substances that combine with the substrate, a coenzyme or metal activator.

There are some different mechanisms through which thiol groups of enzymes can be inhibited. Probably the most important is the formation of mercaptides. Heavy metals such as mercury, lead, cadmium, gold, and zinc are examples of metals that will attack thiol groups. Organic compounds of mercury and arsenic also form mercaptides as do trivalent arsenicals. Thus for example, inhibition of succinoxidase by lead has been demonstrated in dogs. The pyruvate oxidase system is inhibited by the action of arsenicals on the dithiol lipoate cofactor. Low concentrations of lead in blood cause a reduction of δ-aminolaevulinic acid dehydrogenase activity in erythrocytes (Prerovska and Teisinger, 1970).

Thiol groups are relatively easily oxidized by a wide variety of oxidizing agents.

For instance, selenium ions oxidize thiol groups rather than complex them and the same can be assumed for cupric ions:

$$4RS^- + 2Cu^{2+} \to 2RSCu + R-SS-R$$

Another possibility is the alkylation of the thiol group by halogenated carbon compounds like ethyl iodoacetate and chloroacetophenone. Certain ions are absolutely necessary for the activity of some enzymes while others are highly toxic to nearly all enzymes (e.g. Ag^+, Hg^{2+} and Pb^{2+}). Some ions are poisons for some enzymes and activators for others, and some may even inhibit an enzyme at one concentration and activate it at another. Particularly with cations, the enzyme is inactive by itself and the requirement for a cation is usually fairly specific; in some cases only one particular cation is effective, in other cases two or three different cations can act.

For instance riboflavinkinase in animal tissues is activated by Mg^{2+}, Co^{2+}, or Mn^{2+}, and inhibited by Ca^{2+}. In plants it is activated by Mg^{2+}, Zn^{2+}, or Mn^{2+}, and inhibited by Hg^{2+}, Fe^{2+}, or Cu^{2+}. Glycerol dehydrogenase in bacteria is activated by NH_4^+, K^+, or Rb^+ and inhibited by Zn^+.

There are several examples of displacement of essential metals by non-essential ones, either in the same or in contiguous periodic groups. The result may produce a conditioned deficiency as in chlorosis in plants which is caused by iron deficiency or by the displacement of iron by other metals, e.g. nickel or copper, in chlorophyll with an impairment of photosynthesis as a result.

This displacement is dependent on the strength of the individual ligand complexes, generally following the extended Irving–Williams series

$$Ca^{2+} < Mg^{2+} < Fe^{2+} < Co^{2+} \ll Zn^{2+} \ll Ni^{2+} < Cu^{2+}$$

(which is independent of ligand type), and also on the concentrations of ligands and metal ions present in the system. It is also dependent on the relative strength of the complexes involved. The effect of this may be illustrated by the following example: According to the Irving–Williams series, cupric ions should replace ferrous ions in ligand complexes. However, this will not happen when the ferrous ion is complexed by an aromatic nitrogen-type ligand, where the ferrous ion is abnormally stabilized, resulting in a very small difference in stability constants for these types of complexes with cupric and ferrous ions, compared to those of almost any other complex possible in the system (Williams, 1953).

Zinc occurring in carbonic anhydrase will be replaced by administered cupric ions, with the consequent loss of activity for the enzyme (Williams, 1953). Cadmium inhibits peptidase and enhances esterase activity by displacing zinc from its nitrogen–sulphur ligands in carboxypeptidases (Perry *et al.*, 1955). Large doses (0.25% in diet) of pure zinc salts retarded the growth of rats and caused hypochromic anaemia, probably resulting from displacement of copper and iron (Underwood, 1962). Nickel has the ability to displace beryllium from alkaline phosphatase and reactivate the enzyme (Schroeder *et al.*, 1962). The feeding of tungstate ion in large quantities has caused displacement of molybdate in the body of experimental animals (De Renzo, 1962).

Metal ions in the active site of enzymes can be inhibited by substances forming stable complexes with the metal. Cyanide, hydrogen sulphide, azide, and carbon monoxide are examples of such compounds. Hydrogen sulphide and cyanide inhibit many of the enzymes which contain iron or copper as an essential part of their catalytic activity. They are, for instance, powerful inhibitors of the respiration of many tissues through interaction with the cytochrome oxidase system. Carbonic anhydrases both in blood and plants are inhibited by cyanide, azide, and hydrogen sulphide.

Cyanide may inhibit enzymes by several different mechanisms, such as combining with essential metals in the enzyme or removing a metal ion thus leaving the enzyme as an inactive complex, combining with carbonyl groups in the enzyme, a cofactor, a prosthetic group or the substrate, or by acting as a reducing agent breaking disulphide bonds. Carbon monoxide will only inhibit enzymes containing iron or copper and particularly those that react directly with oxygen, combining with the reduced form of the enzymes. Cyanide combines most stably with the oxidized forms.

In the formation of avian eggshell, carbonic anhydrase is believed to be necessary to supply the carbonate ions required for calcium carbonate deposition. It has been shown that the DDT metabolite DDE and also PCB may inhibit the formation of carbonic anhydrase and other transport enzymes in the shell gland and may thus be contributing to thinness.

7.4. METABOLISM AND REGULATORY PROCESSES

Metabolism, the sum of all enzymatic reactions occurring in the cell, can be disturbed by a great number of toxic agents, in many different ways. Chemicals may interfere with the enzymes essential for various processes. Hormones and other regulatory systems may be affected, resulting in an uncontrolled metabolic rate. Toxic agents may also cause genetic mutations which can result in a failure to synthesize an enzyme in its active form. Such defects, if they are not lethal, may result in an accumulation or excretion of the normal substrate of the defective enzyme.

(i) Carbohydrate Metabolism

One of the major biochemical lesions arising from the action of toxic compounds is the disturbance of carbohydrate metabolism. Since carbohydrate catabolism provides energy for other metabolic processes, deviations from the normal metabolic pattern might be expected to result in impairment of respiratory chain reactions. The kinds of deviation are numerous as are the possible mechanisms through which they are mediated. Only a few examples are given here.

In glycolysis the phosphorylation of glucose to yield glucose-6-phosphate is catalysed by two types of enzymes, hexokinase and glucokinase. Hexokinase may be inhibited by certain sulphydryl reagents, especially arsenicals. Hexokinase has also been shown to be completely inhibited by copper in sheep (Agar and Smith, 1973). The conversion of 3-phosphoglycerate to 2-phosphoglycerate is catalysed by the enzyme phosphoglyceromutase. This enzyme may be inhibited by mercury ions in heart muscle (Diederich et al., 1970). The enzyme enolase catalyses the reaction in which 2-phosphoglycerate is transferred to phosphoenolpyruvate. Enolase may be inhibited by inorganic fluoride which forms a ternary complex with the essential activators Mg^{2+} and inorganic phosphate. This complex competes with Mg^{2+} for the active site (Smith, 1970). In the last step of glycolysis, pyruvate is reduced to lactate. This reaction is catalysed by lactate dehydrogenase, and this enzyme is inhibited by oxalate, a structural analogue of pyruvate.

Also, the tricarboxylic acid cycle may be disturbed in several ways. For example, the pyruvate dehydrogenase complex is characteristically inhibited by trivalent arsenicals or by arsenite, which can react with both thiol groups to yield an inactive cyclic derivative. A metabolite, namely fluorocitrate, of fluoroacetate, used as a rodenticide and also synthesized by certain plants growing on fluorine-rich soils, inhibits the metabolism of citric acid by aconitase. Thus the citric acid cycle is blocked and citrate is accumulated in the tissues.

(ii) Respiration

Biological oxidation proceeds through a sequence of intermediary carriers transferring electrons from substrates to molecular oxygen. The respiratory chain is of vital significance for supplying energy to living cells. The electron transport along this chain may be inhibited at specific sites by different toxic agents. For instance, cyanide, azide, and CO may inhibit the electron transport between cytochrome a_3 and O_2. Antimycin A and 2-alkyl-4-hydroxyquinoline-N-oxide may inhibit the electron transport between cytochrome b and cytochrome c_1. Rotenone and barbiturates may inhibit the electron transport between flavoprotein and coenzyme Q. Oxidative phosphorylation, through which energy released from the exergonic catabolic process is conserved for driving a multitude of endergonic biological processes, is tightly coupled to the respiratory chain. There are several toxic agents that do not inhibit respiration directly but prevent the associated phosphorylations. These agents are called uncoupling agents. Substituted phenols such as 2,4-dinitrophenol and pentachlorophenol belong to this group, as do triethyltin and triethyllead.

(iii) Lipid Metabolism

Disturbances of lipid metabolism resulting in impaired liver function are caused by many toxic agents, e.g. lindane, cobalt salts, and selenium. One example of this is that chronic exposure to both DDT and dieldrin has been shown to increase significantly lipogenesis in fish, and their effects are additive.

(iv) Microsomal Enzyme Systems

The multienzyme system located in hepatic microsomal fractions is responsible for the metabolism of, e.g. steroid hormones, fatty acids, alkylpurines, thyroxine, and bioactive amines, and also for the biotransformation of foreign compounds. This system is subject to modification by a vast number of chemical agents as well as by various physiological conditions such as hormonal derangement. The alteration in microsomal metabolic function is of two types – stimulation or inhibition.

The pattern of microsomal enzyme depression varies with the inhibitory agent and may be mediated through a variety of mechanisms. Indirect mechanisms like generalized hepatotoxicity with consequent damage to endoplasmic membrane structures may also be involved in chemically induced inhibition of microsomal metabolism.

Among the chemicals of environmental importance capable of impairing hepatic microsomal biotransformations are organophosphate insecticides, carbon tetrachloride, carbon disulphide, and carbon monoxide.

Microsome enzyme induction by organochlorines and by polyaromatic hydrocarbons is exhibited by a variety of species including mammals, fish, and birds. Also toxic elements, such as Ni, Cr, Br, Cd, Pb, in trace amounts affect the liver microsomal enzyme system.

For example, chlordane, DDT, dieldrin, and PCB's have been noted to bring about an accelerated hydroxylation of androgens, estrogens, and glucocorticoids in several animal species thus causing a more rapid deactivation of these steroid hormones.

(v) Biosynthesis

Protein synthesis can be disturbed on many levels by a variety of mechanisms, either by affecting the nucleic acid metabolism or structure, or in the protein-forming system itself. Toxic agents acting directly on ribosomes, RNA, enzymes or coenzymes may also have a drastic influence on protein synthesis.

For example, all aminoacyl-tRNA synthetases have thiol groups in their active sites and are therefore, of course, especially sensitive to sulphur-attacking toxicants.

It is thought that carbon tetrachloride exerts its toxic effect on protein synthesis on single-unit ribosomes (Farber *et al.*, 1971). The inhibition of amino acid incorporation in microsomes by ethionine is thought to be connected with a deficiency of available ATP in the cell. Trapping of cellular adenine and a diminution in the rate of ATP synthesis is brought about by the replacement of methionine by ethionine with the formation of *s*-adenosylethionine (Farber *et al.*, 1971).

All biosynthetic processes can of course be disturbed in one way or another. The inhibition of carbon dioxide fixation — the synthesis of organic compounds in photosynthetic plants — by zinc and cadmium, which has been demonstrated for spinach, is an example from the plant world (Hampp et al., 1976).

(vi) Regulatory Processes and Growth

In general, the rate of catabolism is controlled by the second-to-second need for energy. The rate of biosynthesis of cell components is also adjusted to immediate needs.

The regulatory system may be disturbed by toxic agents in many ways. The structure or activity of regulatory enzymes may be altered and the synthesis, storage, release, or sequestration of hormones may be affected. For example, the release of peptide hormones such as insulin is strongly dependent on calcium ions. Magnesium ions can block the secretion of these systems.

Disturbance of the regulatory system may have serious effects on the organism. The rate of different metabolic processes may become uncontrolled. Such effects may be lethal for the organism. For example, when the herbicide 2,4-D penetrates the leaves of a plant, the result is a violent and uncontrolled growing to death. In this example growth is affected. All chemicals interfering with metabolic pathways may affect the growth.

The thyroid has many regulatory functions in all vertebrates. Its main function is control of metabolic rate. It also plays a major role in the control of growth rate and is essential for reproduction. It is believed that faulty thyroid activity and changes in vitamin A storage may be the cause of sublethal effects by organochlorine pesticides and PCB, which have structural similarities to thyroid hormones. This subject has been reviewed recently by Jefferies (1975). For example, in one experiment with pigeons an increase in metabolic rate was shown to occur with the administration of low doses of DDT and at higher doses a decrease was observed.

The pesticide dieldrin has been shown to decrease the growth of fawns that had parents whose diet incorporated dieldrin. Dieldrin was passed across the placenta to the fawns, and also was secreted with the mother's milk.

D-Threose-2,4-diphosphate has been found to inhibit the activity of growth hormone and thereby the growth of certain microorganisms.

Also the growth of aquatic invertebrates has been found to be impeded by exposure to several pesticides.

Especially the impact of atmospheric pollutants such as ozone, sulphur dioxide, and nitrogen oxides adversely affects plant growth (Mudd and Kozlowski, 1975). For instance, using duckweed, carnations, corn, petunias, marigolds, chrysanthemums, and turf grasses, a reduction in growth rate, stem elongation, leaf area, general plant size, top and root weight, fruit and seed set, and floral productivity was shown, when the plants were grown in air with ozone levels ranging from 0.1 to 10 ppm (Feder, 1970).

7.5. MUTATIONS

Mutations occur spontaneously in nature in all living organisms, but they are also artificially induced. Ionizing radiation induces mutations and chromosome aberrations in all kinds of cells. Chemicals, however, may act as powerful mutagens in one kind of organism or type of cell and remain without effect in others.

Two main categories of molecular mechanisms lead to point mutations during DNA replication: base substitution and intercalation. Base substitutions may be brought about by alkylating agents, for instance diethyl sulphate and mustard gas (β, β'-dichlorodiethyl sulphide) alkylate guanine, and will result in the incorporation of an incorrect amino acid into the protein coded for, or possibly the synthesis of the protein may be stopped at a premature stage. This may be of little consequence if the correct amino acid(s) is unimportant for the activity of some enzyme, but on the other hand, it may cause severe damage to the enzyme activity if this is not the case. Alkylating agents such as mustard gas with two or more functional groups produce extensive crosslinking of guanine moieties in the opposite strands of the DNA molecule, so that they cannot separate for replication.

The intercalation of a chemical substance between base pairs results in a distortion of the sugar-phosphate backbone of the DNA molecule producing deletion or insertion of nucleotide bases. This will have a drastic effect on the activity of the gene in that the grouping together of the bases three by three will be shifted up or down along the DNA chain. The amino acid sequence of a protein synthesized will thus, from the point of mutation on, be completely changed, and this will lead to a completely non-functional protein unless the mutation affects only the very end of the protein.

A large number of chemicals cause breaks in the DNA chain and the chromosomes. When a chromosome is broken, the ends are usually fused together through special enzymes and the break will pass unnoticed. It may happen, though, that the chromosome is not repaired due to some fault in the enzyme system, leading to structural changes. Such changes may be due to translocations or inversion of fragments or the duplication of fragments. Dicentric bridges may be formed between different chromosomes in close proximity. Many chromosome aberrations will lead to cell death, or grave abnormalities may arise in the offspring if reproductive cells are thus affected.

The distribution of chromosomes at cell division may be disturbed, leading to different numbers of chromosomes in the daughter cells. Some metal compounds have been shown to have this effect, and organic mercury compounds are particularly powerful agents effective at very low concentrations.

Most chromosome mutations in reproductive cells will cause abnormalities or lethality in the offspring. The effects of mutations in somatic cells and disturbances of mitosis are not very well known or easily predictable, however. In the organism a faulty cell is most likely to be selected against, and replaced by a healthy one. Some data, however, indicate a connection between genetic defects in somatic cells and

General Aspects of Toxicology 161

the formation of tumours. Thus if a chemical is shown to cause mutations there is reason to suspect that it may cause cancer as well.

Many biocides have been tested for their mutagenicity and some have been shown to be mutagenic. For instance the insecticidal organophosphate ethyl esters can act as alkylating agents and are powerful mutagens. The fungicidal action of benzimidazole carbamic acid methyl ester (BCM) is thought to be due to some interference with spindle formation.

There are both 'natural' and artificial food additives that have been shown to be mutagenic. Aphlatoxin has been demonstrated to be a mutagen and it is also suspected of causing liver tumours. Cycasin is, after being metabolized in the intestine, a potent methylating agent.

Examples of artificial food additives with mutagenic and carcinogenic action are cyclamates, nitrite, and nitrofurazone.

In vitro experiments with DNA nucleosides and nucleotides have shown that Cu^{2+} binds to the $N_{(1)}$ position of cytidine and to $N_{(7)}$ in adenosine and thymine. Copper also binds phosphate, cleaving the phosphodiester bonds in polynucleotides (Eichhorn *et al.*, 1966). The interaction of silver ion with guanosine has been investigated and it is found to form a bond at the $N_{(7)}$ position (Tu and Reinosa, 1966). The manganous ion seems to have a marked preference for association with the guanine ring (Anderson *et al.*, 1971); it also interacts with the phosphate groups. Such ligand complexes of metal ions with nucleotide bases will interfere with hydrogen bonding between bases in the DNA molecule.

In experiments with human diploid cells, namely leukocytes and fibroblasts, chromosome alterations were not observed with salts of cadmium, cobalt, nickel, iron, selenium, vanadium, and mercury. Chromosome breaks were shown to be induced by arsenic and tellurium salts, however. The chromatid breaks were frequent and often located close to the end of a chromosome chain, and cells with several damaged chromosomes were common.

Arsenic is known to block sulphydryl groups and it has been suggested that arsenic may inhibit DNA repair enzymes. The uptake of phosphate into DNA is reduced in the presence of arsenate ion which is thought to be incorporated into DNA in lieu of phosphate thereby forming weak bonds between the DNA strands.

Surveys on the mutagenic effects of environmental contaminants may be found in reviews and monographs (e.g. Fishbein *et al.*, 1970; Voegel, 1970; Hollaender, 1971).

7.6. TUMOUR INDUCTION

Cancer can be induced by chemical carcinogens, oncogenic virus, or ionizing radiation. Chemicals are the most likely major cause of human cancer and environmental pollutants are thought to be responsible for the increasing tumour frequency observed in fish from contaminated waters.

It seems that carcinogenic agents are generally chemically reactive electrophilic compounds, such as the alkylating agents. These electrophilic reagents interact with a variety of informational macromolecules, for instance the proteins and nucleic acids of the cell. The reactive sites on DNA and RNA molecules have been shown to be the nucleotide bases and in proteins they seem to be tryptophan, tyrosine, methionine, and histidine.

It has been shown that even though procarcinogens are generally not mutagenic, their electrophilic metabolites are almost invariably potent mutagens. Alkylating agents are both highly mutagenic and potent carcinogens. Several agents — initiators, promoters, co-carcinogens etc., will often interact in the production of cancer.

There are non-carcinogenic chemicals that augment the effect of a primary carcinogen. This promotion involves the growth and development of dormant or latent tumour cells, resulting from the interaction of the primary carcinogen and specific receptors in susceptible cells — the initiation step. The promotion could be described as any situation yielding an increased growth rate of dormant cells, that is, hastening the mitotic rate.

The work that led to the important distinction between tumour initiation and tumour promotion described the increase in tumour yield and shortened latent period caused by, for example, croton oil applied subsequently to a dose of a carcinogenic polycyclic aromatic hydrocarbon (Sall et al., 1940; Berenblum, 1941).

Sulphur, sulphur dioxide, and some other compounds; dodecane, aldehydes, and phenols all stimulate the effects of polycyclic aromatic hydrocarbons and other carcinogens.

There are two types of chemical carcinogens. The first consists of those that are direct acting and do not require metabolic activation. Metabolic conversion of such compounds leads to detoxification and loss of effectiveness. These compounds usually act at the point of application. Examples are some active halogen derivatives (e.g. methyl iodide), strained lactones, unsaturated larger ring lactones, epoxides, imines, alkyl sulphate esters (e.g. dimethyl sulphate), and nitrogen mustard gas derivatives.

Most of the chemical carcinogens in the environment however, belong to the second type, which require some sort of activation — they must first be converted either directly or by chemical activation to an electrophilic compound with suitable properties of stability and reactivity. Many of these activation reactions consist of biochemical oxidation or hydroxylation. This type of chemical carcinogen can be divided into different groups, such as polycyclic aromatic hydrocarbons, aromatic amines and azo compounds, N-nitrosamines and amides, and metal compounds etc. They all give rise to tumours under certain conditions and at specific sites. The carcinogenic action of polycyclic aromatic hydrocarbons seems to be related to their binding to DNA and it has been postulated that an epoxide in the K-region formed through oxidation by microsomal enzymes may be the metabolically activated form responsible for the carcinogenic action (Boyland, 1950; Sims and Grover, 1974).

To induce cancer, aromatic amines need first to undergo N-hydroxylation. They are then changed into the highly reactive N-sulphate which reacts with proteins and guanine in nucleic acids. (Miller and Miller, 1974a). The chemical interaction of azo dyes and aromatic amines with tissue constituents has been extensively investigated (Miller and Miller, 1974b). For instance the carcinogenic nitrosamines are transformed either enzymatically or non-enzymatically into agents that donate methyl and ethyl radicals to RNA and DNA (Swann and Magee, 1968). Biochemical activation of nitrosamines appears to take place in all species so far studied. Many potent carcinogens are found among the organic nitroso compounds (Magee and Barnes, 1967). Nitrosamines can be formed *in vivo* in the gastrointestinal tract by reaction of nitrite ion and secondary or tertiary amines and amides, causing tumours in other parts of the organism (Sander and Bürkle, 1969). It has been shown that nitrosamines may pass the placental barrier to the foetus and give rise to the development of tumours after birth (Ivankovic and Preussmann, 1970).

Aphlatoxin, a mycotoxin from *Aspergillus flavus*, and cycasin, a plant toxin from cycad tree fern (methylazo methyl-β-D-glucoside) are examples of carcinogenic substances which occur naturally (Miller, 1974).

Metals and their compounds induce cancer after a long latent period during which the metals probably initiate irreversible cellular changes necessary for the subsequent development of tumours. It seems that they give rise to cancer at the point of application in most instances. Inhalation of arsenic compounds can result in tumour formation. Inhalation of nickel carbonyl has been shown to give rise to nasal sinus- and lung cancer. Chromium salts have been shown to be the cause of lung cancer in workers employed in the manufacture of dichromate from the ore.

7.7. REPRODUCTION AND TERATOGENESIS

Foreign compounds may interfere with the organs of reproduction so that the formation of their product is inhibited. Thus in mammals, chemicals may act to prevent implantation of the fertilized egg, causing its expulsion from the uterus. Or they may interfere with the normal development of the placenta, for instance by alteration of the enzymatic functions, or interfere with the transfer of nutrients. Several hormones are involved in ovulation and spermatogenesis and there are many direct and indirect effects of toxic agents that may disturb the hormonal balance. Compounds of very diverse structure may affect the reproductive system which is very sensitive to toxic agents.

Metals may cause disturbances of the reproductive functions. Cadmium, for example, can cause testicular damage in several species of animals, probably due to interference with the seminiferous epithelium and Leydig cells resulting in testicular degeneration.

Certain insecticides, notably the organochlorine compounds, have been shown to affect the male and female reproductive systems. Several workers have shown the connection between pesticide residues and viability of fish eggs. The organochlorines are lipid soluble and may thus be accumulated in fish eggs and lead to the death of fry as the yolk-sac is absorbed at a critical stage of growth.

Pesticides may influence fish reproduction by inducing abortion, thus DDT, DDE, methoxychlor, aldrin, dieldrin, endrin, toxaphene, heptachlor, and lindane have all been shown to cause some degree of abortion in mosquito fish. The prostatic uptake of testosterone is decreased in male mice by continuous administration of dieldrin.

DDT has structural similarities with the synthetic oestrogen diethyl stilboestrol and oestrogenic activity has been shown to be affected by DDT in rats.

For many species of birds the organochlorines cause severe damage to the reproductive system with a decrease in egg production, thinning of eggshells, and reduced hatchability and fledgling success. In a second generation of pheasants whose parents received 4–6 mg of dieldrin per week for thirteen weeks, fertility and hatchability of eggs were found to be significantly lowered even though they only received the toxin through the egg.

The offspring may be born malformed mentally or physically as a result of chemical action on the reproductive system of either parent (a mutation) or on cell differentiation in the conceptus (a teratogenic change) or by toxic actions of chemicals on the developing organs of the foetus which can result in growth retardation or in degenerative toxic effects similar to those seen in the postnatal animal.

7.8. IMMUNE RESPONSE

The most powerful allergens are chromium, beryllium, nickel, and cobalt but many other industrial and environmental pollutants are also strongly allergenic. Some examples to illustrate other types of immune response will be given.

It is thought that bone marrow damage may be due to an allergic reaction involving antibodies to the precursor cells and a sensitizing chemical. This is suspected in some cases involving chloramphenicol.

Peripheral destruction of red cells may involve an allergic mechanism (auto-immune haemolytic anaemia) via sensitization by a chemical such as acetanilide.

Several environmental contaminants, e.g. lead, cadmium, mercury, DDT, and PCB, have been found to be synergistic to infectious agents through an immuno-suppressive action. Thus chronic exposure to lead has been shown to cause a significant decrease in antibody synthesis, particularly IgG globulin (Koller and Kovacic, 1974).

7.9. INTERACTIONS OF COMPOUNDS

The presence of large numbers of toxicants in the environment makes it necessary to consider interactions between such substances. Thus the concurrent presence of two or more foreign compounds in the organism often yields a toxic response which deviates from a simple additive one.

The interacting effect is said to be additive or substitutive, when the combined effects of two or more toxicants equal that expected when considering the toxicants individually. However, the additive effect does not imply a strict summation of the toxicity of the two substances. If the effect is greater than additive, it is said to be synergistic and means that one toxicant will potentiate the effect of the other. If the effect is less than additive, it is said to be antagonistic (see also Chapter 9).

As the toxicity of a compound depends largely on the efficiency of its biotransformation any interference with metabolism is likely to influence its toxic potential. Altered sensitivity of receptor sites, hormonal intervention, and competition for binding sites of biological receptors are other mechanisms which may contribute to potentiating or antagonistic effects.

A direct chemical interaction of exogenous compounds, which often results in a decreased response, is still another mechanism. One example of this is the precipitation of calcium fluoride which is not readily resorbed, upon the ingestion of sodium fluoride together with food rich in calcium. Often toxic agents concentrate at sites other than the site of toxic action. They may be stored in plasma proteins, in proteins in liver and kidney, in fat and in bone. Lead, for example, is stored in bone but exerts its toxic action in soft tissues. Displacement is another type of interaction. One chemical agent may displace another of similar structure from its binding site and thereby alter its toxicity. The binding capacity of the receptor may be influenced producing the same result. Thus lead stored in bone may be mobilized with a concomitant rise in the serum level of lead. Suppression or induction of microsomal enzyme systems can explain many of the metabolic interactions arising from the combined presence of different chemicals. The consequences of a suppression are usually a potentiation of the toxicity of foreign compounds. Formulations for this topic are to be found in Chapter 9.

Simultaneous intake of certain organophosphorus esters has been observed to result in potentiation of toxicity. Phosphothioates such as malathion are rendered much more toxic in combination with e.g. EPN (*O*-ethyl-*O-p*-nitrophenyl phenyl phosphoethionate). EPN competes with malathion for its hydrolysing enzyme thus impeding hydrolysis and resulting in a prolonged retention of malathion within the organism.

The increased toxicity of insecticides achieved by the addition of so called pesticidal synergists results from the inhibitory action of the synergists upon the microsomal mixed-function oxidase system. They have been developed to synergize insecticidal toxicity of pyrethrins but they have been shown to potentiate the toxicity of a number of other insecticides such as carbamates and DDT, and to be active in mammals as well as insects. Piperonyl butoxide, the most commonly used synergist also causes the formation of toxic metabolites of freons to occur. In insecticide spray bottles, 'Freons' occur together with the pyrethrins and the synergist.

7.10. DISCUSSION

Based on a knowledge of the biochemistry and physiology of the organisms and on previous experience of disturbances caused by toxic substances, various bioassay tests have been and can be worked out.

Obviously the total number of tests that would have to be performed in order to evaluate possible disturbances to any part of the metabolic system is enormous.

Sometimes arguments of resemblance and analogies can give some guidance as to what to test.

As an attempt to reduce the number of tests and yet to cover a variety of biochemical and physiological processes, bioassays are sometimes carried out to check net effects on functions composed of a large number of processes. Such functions may be reproduction or growth rate.

A further argument for such an approach is that even with the knowledge of how a substance will affect each process by itself, it may not be possible to predict the total effect on the function.

Regulating systems or feed-back or feed-forward type may have a major influence on the overall result.

It should also be kept in mind that the response of an individual organism provides only one part of the information required to predict effects on an ecological or population level.

Knowledge about the reaction of one species to a given chemical does not tell whether that species will increase or decrease in an exposed ecosystem (except when the result is death or total sterility).

The reaction of other species — competing ones, predators on prey — must also be known as the interactions between organisms often is a major regulating factor.

Obviously there is also a need for knowledge of the nature of these interactions.

Thus in addition to bioassays based on single biochemical or physiological reactions and those based on complex functions at the individual level, there is a need for testing procedures at higher levels of organization such as test ecosystems and computer-based mathematical models.

7.11. REFERENCES

Agar, N. S. and Smith, J. F., 1973. Effect of copper on red cell glutathione and enzyme levels in high and low glutathione sheep (37095). *Proc. Exp. Biol. Med.*, **142**, 502–05.

Anderson, J. A., Kuntz, G. P. P., Evans, H. H., and Swift, T. J., 1971. Preferential interaction of manganous ions with the guanine moiety in nucleosides, dinucleoside monophosphates and deoxyribonucleic acid. *Biochemistry*, **10**, 4368–74.

Berenblum, I., 1941. Co-carcinogenetic action of croton resin. *Cancer Res.*, **1**, 44–8.

Boyland, E. 1950. The biological significance of the metabolism of polycyclic compounds. *Biochem. Soc. Symp.*, **5**, 40. Cited in *Chemical Carcinogenesis*

(Eds. P. O. P. Ts'o and J. A. DiPaolo), Part A, Marcel Dekker, New York, 1974, p. 157.
De Renzo, E. C., 1962. Molybdenum. In *Mineral Metabolism: An Advanced Treatise* (Eds. C. L. Comar and F. Bronner), Vol. 2, Pt. B. Academic Press, New York, pp. 483–98.
Diederich, D. A., Khan, A., Santos, I., and Grisolia, S., 1970. The effects of mercury and other reagents on phosphoglycerate mutase-2,3-diphosphoglycerate phosphatase from kidney, muscle and other tissues. *Biochim. Biophys. Acta*, **212** 441–9.
Eichhorn, G. L., Clark, P., and Becker, E. D., 1966. Interactions of metal ions with polynucleotides and related compounds. VII. The binding of copper (II) to nucleotides, nucleosides and deoxyribonucleic acids. *Biochemistry*, **5**, 245–53.
Farber, E., Liang, H., and Shinozuka, H., 1971. Dissociation of effects on protein synthesis and ribosomes from membrane changes induced by carbon tetrachloride. *Amer. J. Pathol.*, **64**, 601–22.
Feder, W. A., 1970. *Chronic Effects of Low Levels of Air Pollutants upon Floricultural and Vegetable Plants in the North-East, Ann. Rept. Publ. Health Serv. Contr.*, No. PH22-68-39.
Fishbein, L., Flann, W. G., and Falk, H. L., 1970. *Chemical Mutagens: Environmental Effects on Biological Systems*, Academic Press, New York.
Hampp, R., Beulich, K., and Ziegler, H., 1976. Effects of zinc and cadmium on photosynthetic CO_2-fixation and Hill activity of isolated spinach chloroplasts. *Z. Pflanzenphysiol.*, **77**, 336–44.
Hollaender, A. (Ed.), 1971. *Chemical Mutagens: Principles and Methods for Their Detection*, Vols. 1 and 2, Plenum Press, New York.
Ivankovic, S. and Preussmann, R., 1970. Transplacental generation of malignant tumours after oral administration of ethylurea and nitrite to rats (in German). *Naturwissenschaften*, **57**, 460–1.
Jefferies, D. J., 1975. The role of the thyroid in the production of sublethal effects by organochlorine insecticides and polychlorinated biphenyls. In *Organochlorine Insecticides: Persistent Organic Pollutants* (Ed. F. Moriarty), Academic Press, London, pp. 131–230.
Koller, L. D. and Kovacic, S., 1974. Decreased antibody formation in mice exposed to lead. *Nature*, **250**, 148–50.
Magee, P. N. and Barnes, J. M., 1967. Carcinogenic nitroso compounds. *Adv. Cancer Res.*, **10**, 163–246.
Miller, E. C. and Miller, J. A., 1974a. Biochemical mechanisms of chemical carcinogenesis. In *The Molecular Biology of Cancer* (Ed. H. Busch), Academic Press, New York, pp. 377–402.
Miller, E. C. and Miller, J. A., 1974b. The presence and significance of bound amino azo dyes in the livers of rats fed p-dimethylaminoazo benzene. *Cancer Res.*, **7**, 468–80.
Miller, J. A., 1974. Naturally occurring substances that can induce tumors. In *Toxicants Occurring Naturally in Foods*, Chmn. F. M. Strong. Natl. Acad. Sci., Washington, D.C., pp. 508–49.
Miller, R. J., Bittell, J. E., and Koeppe, D. E., 1973. The effect of cadmium on electron and energy transfer reactions in corn mitochondria. *Physiol. Plant*, **28**, 166–71.
Mudd, J. B. and Kozlowski, T. T. (Eds.), 1975. *Responses of Plants to Air Pollution*, Academic Press, New York.
Perry, H. M., Teitlebaum, S., and Schwartz, P. L., 1955. Effects of antihypertensive

agents on amino acid decarboxylation and amine oxidation. *Fed. Proc.*, **14**, 113–4.
Prerovska, I. and Teisinger, J., 1970. Excretion of lead and its biological activity several years after termination of exposure. *Brit. J. Ind. Med.*, **27**, 352–5.
Sall, R. D., Shear, M. J., Leiter, J., and Perrault, A., 1940. Studies in carcinogenesis. XII. Effect of the basic fraction of creosote oil on the production of tumours in mice by chemical carcinogens. *J. Natl. Cancer Inst.*, **1**, 45–55.
Sander, J. and Bürkle, G., 1969. Induction of malignant tumours in rats by simultaneous feeding of nitrite and secondary amines (in German). *Z. Krebsforsch.*, **73**, 54–66.
Schroeder, H. A., Balassa, J. J., and Tipton, I. H., 1962. Abnormal trace metals in man: nickel. *J. Chronic Dis.*, **15**, 51–65.
Sims, P. and Grover, P. L., 1974. Formation and reactions of epoxy derivatives of aromatic polycyclic hydrocarbons. In *Chemical Carcinogenesis* (Eds. P. O. P. Ts'o and J. A. DiPaolo), Part A, Marcel Dekker, New York, pp. 237–47.
Smith, F. A. (Ed.), 1970. *Pharmacology of Fluorides*, Springer-Verlag, Berlin, pp. 48 and 98.
Swann, P. F. and Magee, P. N., 1968. Nitrosamine-induced carcinogenesis: the alkylation of nucleic acids of the rat by *n*-methyl-*n*-nitrosourea, dimethylnitrosamine, dimethylsulfate and methyl methane sulfonate. *Biochem. J.*, **110**, 39–47.
Toback, R. M., 1965. The relationship between adenosine triphosphatase activity and triethyltin toxicity in the production of cerebral edema of the rat. *Amer. J. Pathol.*, **46**, 245–62.
Tu, A. T. and Reinosa, J. A., 1966. The interaction of silver ion with guanosine, guanosine monophosphate and related compounds. Determination of possible sites of complexing. *Biochemistry*, **5**, 3375–83.
Underwood, E. J., 1962. *Trace Elements in Human and Animal Nutrition*, 2nd edn., Academic Press, New York.
Vaituzis, Z., Nelson, J. D., Wan, L. W., and Colwell, R. R., 1975. Effects of mercuric chloride on growth and morphology of selected strains of mercury-resistant bacteria. *Appl. Microbiol.*, **29**, 275–86.
Voegel, F. (Ed.), 1970. *Chemical Mutagenesis in Mammals and Man*, Springer-Verlag, Berlin.
Williams, R. J. P., 1953. Metal ions in biological systems. *Biol. Rev., Cambridge Phil. Soc.*, **28**, 381–412.
Wood, J. M. and Seagall, H. J., 1974. Reaction of methyl mercury with plasmalogens suggests a mechanism for neuro-toxicity of metalalkyls. *Nature*, **248**, 456–8.

CHAPTER 8

*Terrestrial Animals**

F. MORIARTY

Monks Wood Experimental Station, Institute of Terrestrial Ecology, Huntingdon, U.K.

8.1 INTRODUCTION	169
8.2 AMOUNTS OF POLLUTANTS WITHIN ORGANISMS	169
(i) Absorption	175
8.3 TOXICITY	176
8.4 SUBLETHAL EFFECTS	179
8.5 EFFECTS ON POPULATIONS	180
(i) Pesticides and the peregrine falcon (*Falco peregrinus*)	181
8.6 CONCLUSIONS	184
8.7 REFERENCES	184

8.1. INTRODUCTION

Pollutants matter because of their biological effects, and the type and magnitude of these effects depend on the degree of exposure that individual organisms receive. Terrestrial animals will be used to illustrate four related themes: the intake, distribution, and loss of pollutants by organisms; the causes of death; the effects of sublethal exposures; and the effects on populations. Most of the examples come from insects, birds, and mammals, and refer to the effects of insecticides, polychlorinated biphenyls (PCBs) and metals, but the ideas that these examples illustrate should apply to a much wider range of habitats, organisms, and pollutants. Some aspects of pollutants in terrestrial ecosystems, especially the effects of organochlorines on avian reproduction, have recently been reviewed by Stickel (1975).

8.2. AMOUNTS OF POLLUTANTS WITHIN ORGANISMS

There is an enormous amount of data published on results from chemical analyses of plants and animals, but much is of doubtful scientific value. Although the details of chemical analysis for pollutants can be complicated and difficult, it is too easy to collect many biological specimens and have them analysed for a range of pollutants. The precise questions to be answered need to be clearly formulated at the beginning (Holden, 1975).

*© F. Moriarty, 1978

Some pollutants, such as ozone, are so transitory that it is not possible to detect them in animal tissues — one can only detect the effects. However, for pollutants that are more persistent it is useful to know the amounts that are present within animals: it gives some measure of past exposure, and it can indicate the likelihood of biological effects.

The compartmental model (Atkins, 1969), developed extensively in physiology and pharmacology, appears to be the most useful approach developed so far for the quantitative study of pollutants within organisms (Moriarty, 1975a). A compartment is defined as a quantity of pollutant that has uniform kinetics of transformation and transport, and whose kinetics are different from those of all other compartments. A whole animal is envisaged as consisting entirely of compartments, which are linked in a mamillary system: the peripheral compartments are all linked to the central compartment, but not with each other (Figure 8.1).

The central compartment, compartment 1, always represents the blood, haemolymph, extracellular fluid or transport system. This is the only compartment that receives pollutant from, or returns it to, the exterior. This makes reasonable physiological sense. A fish, for example, is likely to take up (indicated by R) most of any pollutant into its bloodstream either through its gills or from food in its gut. Excretion, via the kidney (indicated by k_{01}), occurs from the blood. Compartment 2 might represent the liver, the principal site of metabolism. Other tissues of the body may be represented by additional compartments.

The model assumes that the rate at which pollutant leaves any compartment is

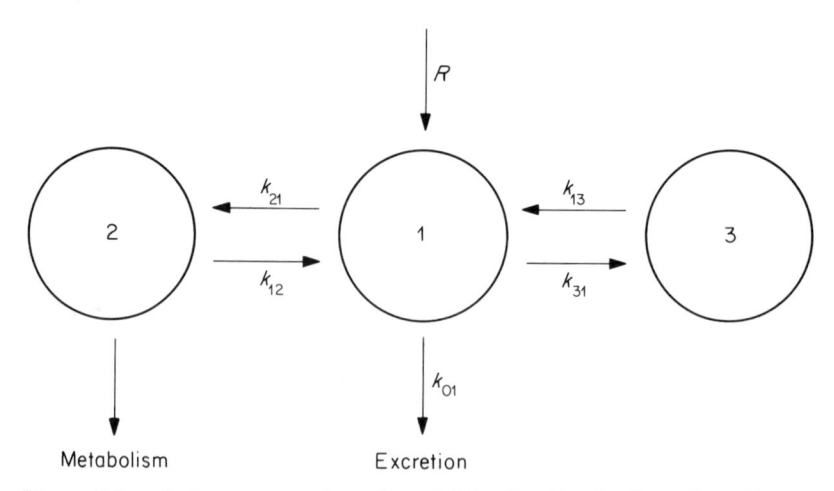

Figure 8.1 A three-compartment model for the distribution of a pollutant within a vertebrate. Pollutant is absorbed into the blood (compartment 1) at a rate R, most metabolism occurs in the liver (compartment 2), and rates of transfer between compartments are indicated by the rate constants (k)

directly proportional to the amount of pollutant in that compartment. The rate constants (fraction of pollutant transferred per unit of time) are indicated by k, with suitable subscripts.

In theory, different cell constituents should probably be represented by distinct compartments. In most cases two or three compartments have sufficed to describe the distribution of pollutants within entire organisms. There are at least two reasons for this:

(1) Analytical results usually refer to, at least, whole tissues, so that smaller compartments become meaningless.
(2) Even if there is reason to suppose that there are many compartments and there are analytical data with which to test this supposition, in practice the residual variation in even very good data makes it difficult to estimate parameters on retention or excretion for a model with more than two or three compartments.

In spite of these severe practical limitations, the compartmental model has proved a very useful way of describing experimental results. Equations derived from these models consist of a series of exponential terms. Figure 8.2 shows as an example the loss of dieldrin from blood (after cessation of uptake), which has been described by a two-term exponential equation. The two exponential constants (0.535 and 0.0529) are derived, theoretically, from the three rate constants for the two compartments. In practice of course they are estimated directly by fitting an equation to the data points with a least squares procedure. In practical terms these data show that, for the first 71 days after exposure, the dieldrin in rats' blood could be considered in two parts. That part that is lost more rapidly has a half-life of 1.3 days ($\log_e 2/0.535$) while the part lost more slowly has a half-life of 13.1 days. The exponential coefficients indicate that, when exposure stopped, the rapid component was almost twice as large as the slow component (542 : 298, or about 9 : 5). The higher this ratio, the more rapidly most of the HEOD disappears. It should be noted that, unless the simplest one-compartment model is used, it is meaningless to talk of a pollutant's half-life in organisms — the proportion lost from the organism in unit time is not constant.

Similar equations can be derived for the amounts of pollutant found within a compartment during chronic exposure (Figure 8.3). This example shows clearly that the equation for a one-compartment model gives a poor fit to the data for amounts of dieldrin in the blood, whereas a two-compartment model gives a good fit. The compartment model assumes that the rate constants for transfer of pollutant between compartments have the same values during and after exposure. There is at present little experimental evidence either way on this point.

This model implies that, if exposure continues for long enough, then a steady state will be reached, in which the amounts of pollutant within each compartment

172   *Principles of Ecotoxicology*

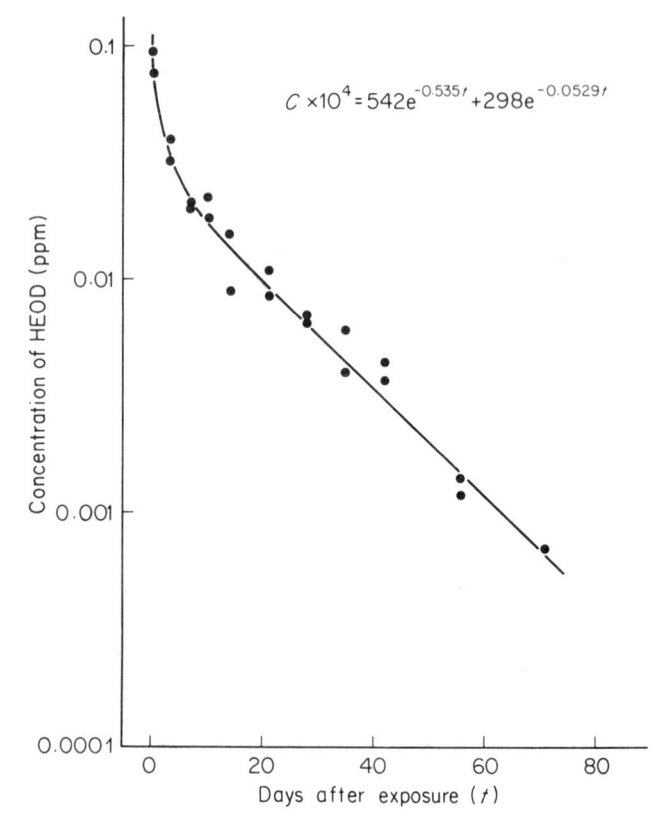

$$C \times 10^4 = 542e^{-0.535t} + 298e^{-0.0529t}$$

Figure 8.2   Decrease in the concentration (C) of dieldrin in rats' blood during the first 71 days after exposure. Data fitted to an equation with two exponential terms (data and equation from Robinson *et al.*, 1969). (Reproduced by permission of Academic Press, London, from Moriarty, 1975a)

stay constant. At this stage the rate of intake equals the rate of loss. There is considerable evidence to support this. In theory, the amount of pollutant in a compartment, once the steady state has been reached, should be proportional to the degree of exposure and inversely proportional to the rate of loss from the compartment. This does seem to be true as a first approximation (see Figure 8.4). This is potentially of great value for any monitoring system, but several points should be noted:

(1) In some of the few experiments where exposure has continued for long enough, the steady state, apparently reached, does not continue and the amounts in the compartment have increased again (see Figure 8.5). This

$$C = 689 - 234e^{-0.400t} - 470e^{-0.00774t}$$

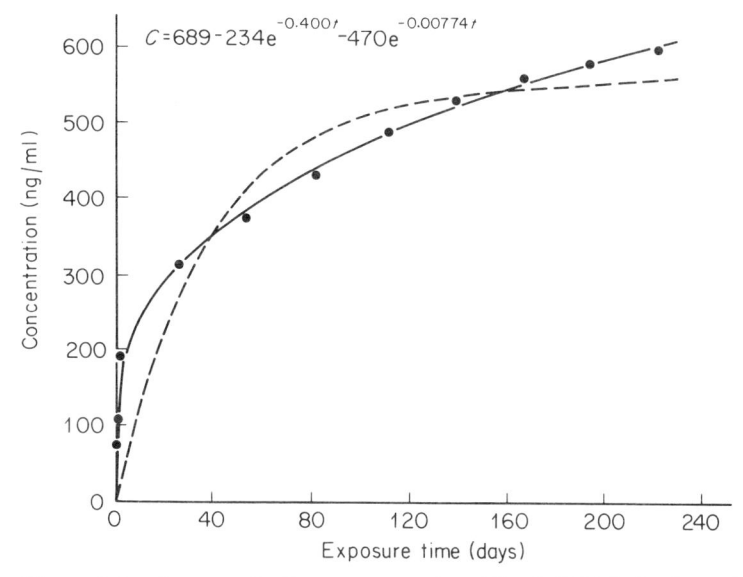

Figure 8.3 Increase in the concentration (C) of dieldrin in sheeps' blood while ingesting 2 mg dieldrin/kg body weight/day. ———, line derived from equation with two exponential terms; ———, line derived from equation with one exponential term (data from Davison, 1970). (Reproduced by permission of Academic Press, London, from Moriarty, 1975a)

may result from some change in metabolism possibly due to the presence of the dieldrin or to some assumption in the model which is false, but at present we have no information on possible explanations.

(2) Specimens from the field will not have experienced a constant or standard physical environment, nor, in many instances, will exposure have been constant. Many physiological events, such as breeding activity (Anderson and Hickey, 1976) and hibernation (Jefferies, 1972), quite apart from catastrophes and inclement conditions (Clark, 1975), may well affect pollutant levels.

The crux of the compartmental model is the use of rate constants to quantify the movement of pollutants into, within, and out of an organism. Future work will doubtless modify this model, but it seems most unlikely that this central feature can be discarded.

Another discussion of retention equations and the changes in body and organ content with time is to be found in Chapter 5.

The application of this model to calculation of the amounts of a persistent pollutant in each level of a food chain, provides an important advance over earlier

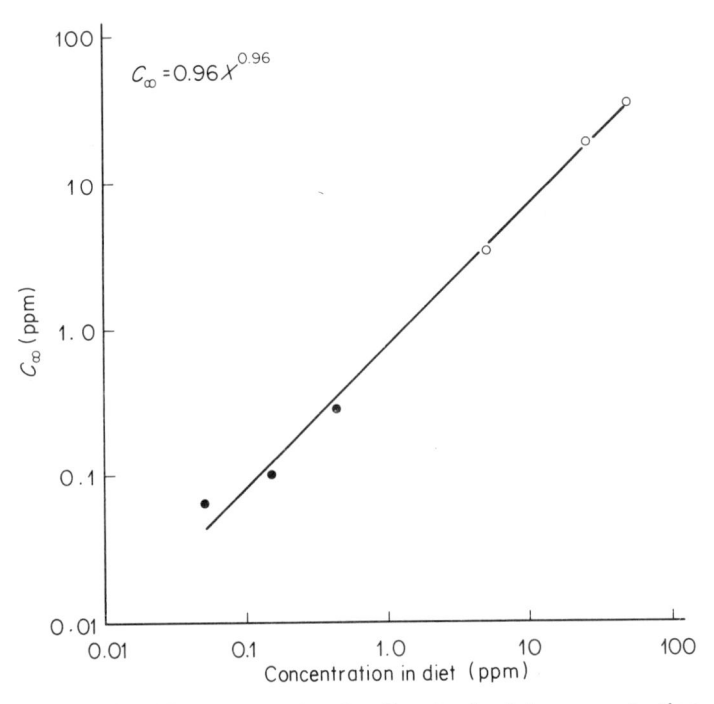

Figure 8.4 Linear regression for the steady-state concentration (C_∞) of p,p'-DDT in eggs of white leghorn hens on the concentration of p,p'-DDT in the diet (X). Both scales are logarithmic. ●, values calculated from data of Cummings et al. (1966); ○, values calculated from data of Cecil et al. (1972). (Reproduced by permission of Academic Press, London, from Moriarty, 1975a)

The equation shown on the graph is $C_\infty = 0.96 X^{0.96}$.

concepts. One of these, that persistent pesticides accumulate and concentrate along food chains, is too simple and does not take account of the physiological processes operating at each trophic level. The compartmental model approach is clearly to be preferred (Moriarty a, in press) over that of Hamelink et al. (1971) and Neely et al. (1974) who have proposed that for pesticides in aquatic habitats, partition coefficients will indicate the amounts to be found in animals. The uptake and retention of molecules depends not only on passive diffusion but also on processes of active transport, metabolism, and excretion, which vary with different species (Walker, 1975).

The great potential value of the compartmental model is that, by estimating rate constants or parameters derived from them, it should enable us to measure the differences between species in the way they take in, accumulate, distribute, and get rid of pollutants. It should also help us assess the more intractable problems of interactions between pollutants and the effects of other variables. There are reviews

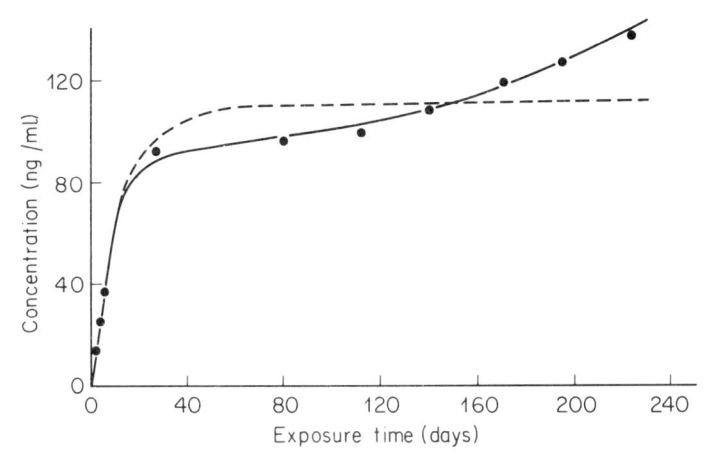

Figure 8.5 Changes in the concentration of dieldrin in sheeps' blood while ingesting 0.5 mg dieldrin/kg body weight/day, ———, line derived from equation with two exponential terms; ————, line derived from equation with one exponential term (data from Davison, 1970). (Reproduced by permission of Academic Press, London, from Moriarty, 1975a)

of the available data for metals (Task Group, 1973) and for organochlorine insecticides (Moriarty, 1975a).

(i) Absorption

Intake from the physical environment deserves special mention. Intake, be it through the epidermis, respiratory organs, or gut, depends very much on the precise form of the pollutant — both its chemical 'species' and the degree and type of physical aggregation. Much of our information relates to man.

Aerial pollutants can occur either as gases or associated with particulate matter. The ease with which gases reach the lungs' alveoli depends greatly on their water solubility: those that are highly water-soluble will dissolve readily in the mucous membranes before the alveoli are reached. For those gases which do reach the alveoli, lipid solubility probably aids penetration through the membrane into the lung tissue and thence into the blood. For example, mercury vapour is non-polar, is not dissolved in the nasopharyngeal or tracheobronchial tracts of man, but about 80% of that inhaled is absorbed through the alveoli (Kudsk, 1965).

Deposition and retention of aerial pollutants from particulate matter depend greatly on the size of the particles. Particles may settle onto respiratory surfaces by any of three distinct processes:

(1) Impaction, the most important process for particles larger than a few microns, occurs where the airstream changes course abruptly.

(2) Settlement by gravitational forces, which of course depends on particle mass and shape. For particles of about unit density, this process is important for particles whose diameter is within the range of about 0.5–5.0 microns.

(3) Diffusion, which is important for the smaller particles.

The lung model for Reference Man described in Chapter 5 shows that inhaled particles may be deposited in the nasopharyngeal, the tracheobronchial, or the alveolar regions of the lung. In each of these regions the pollutant may be either absorbed into the extracellular fluids or transported up to the pharynx in mucus propelled by ciliary action. After reaching the pharynx the material is swallowed into the digestive tract. Material taken up from the alveoli (the most important site of absorption) may go to the bloodstream or be retained in the broncho-pulmonary lymph nodes.

Ingestion of pollutants can obviously occur with food and drink, but inhaled particulate matter may also be transferred to the gut in mucus. In qualitative terms, several pathways exist. Some may pass straight through the gut, unaltered and unabsorbed, some may be metabolized by the gut microflora, while some may enter the gut wall, by endocytosis or in a chemical form suitable for penetration of the gut wall. Absorption through the gut wall does not necessarily indicate transfer to the whole organism. Dichlorobiphenyl for example may pass to the liver, and then be excreted in the bile back into the gut (Iatrapoulos *et al.*, 1975).

8.3. TOXICITY

Pollutants matter because of their many possible effects on individual organisms (see Chapter 7). The most dramatic effect is of course death. Death is the end result of many disrupted bodily functions, but these are usually secondary effects from a single initial biochemical change. This 'site of action' may be dispersed throughout the body, and it can be important, if one wishes to relate amount of pollutant to degree of effect, to have precise knowledge of the site of action. There are two reasons for this.

Firstly, unless the critical organ is known it may prove difficult, or impossible, to relate degree and type of exposure, and amount of pollutant within the body, to the likelihood of adverse biological effects. For example, different compartments can approach their steady-state concentrations at very different speeds, so that the amounts of pollutant in one tissue or organ may give very little indication of the amount in another (Table 8.1).

Secondly, if an enzyme, or group of enzymes, that is widespread throughout the body is inhibited by a pollutant, results may appear contradictory unless the critical organ is known. The organophosphorus insecticides provide a good example. They have been known for a long time to inhibit esterases, and acute toxicity is due to acetylcholinesterase inhibition. It proved difficult to correlate degree of acetyl-

Table 8.1 Amounts of Dieldrin in the Blood, Liver, and Fat of Rats Fed on a Diet containing 50 ppm Dieldrin. Results for liver and fat also expressed as a ratio of that found in the blood (data from Deichmann et al., 1968)

Days fed	Concentration of dieldrin (ppm)			Ratio of concentration to that in blood	
	blood	liver	fat	liver	fat
0	0.001	0.00	0.00		
1	0.047	1.01	6.42	25	160
2	0.080	1.97	9.74	25	122
4	0.121	2.63	25.00	22	208
9	0.243	6.22	98.95	26	412
16	0.261	9.56	174.19	36	669
31	0.237	12.70	280.25	53	1182
45	0.255	8.55	196.10	33	754
60	0.262	7.44	148.22	29	570
95	0.193	4.77	130.77	25	688
183	0.268	8.15	184.75	30	684

cholinesterase (AChE) inhibition with the onset of death (e.g. Burt et al., 1966), but the discrepancies appear to have been resolved now that it has been discovered that there are several isozymes of AChE (Tripathi and O'Brien, 1973). The housefly contains four isozymes of AChE in the head, and another three different isozymes in the thorax. Groups of houseflies were given the LD_{50} dose of four insecticides, and the minimal activity estimated for each of the seven isozymes (Table 8.2). It can be seen that isozyme 7 always loses virtually all of its activity, even in the survivors, so it cannot be the relevant site of action. Most of the other isozymes are inhibited to different degrees by the different insecticides, but isozyme 5, in the thorax, has a constant degree of inhibition, of just over 80%. These data suggest that this isozyme is the critical site of action.

Care is needed when applying the compartmental model to analytical results from an animal that has received enough pollutant to seriously impair its normal functions. At the present stage of the model's development it is usually assumed that the animal is in a state of equilibrium. However, an animal near to a lethal exposure will often depart from such a state. Thus it has been found many times that animals killed by insecticides have reduced fat reserves. The highest concentrations of organochlorine insecticides occur in fatty tissues, so that if these fat reserves start to be mobilized, the concentrations of insecticide in all tissues are likely to increase suddenly. This will include the critical organ, the brain. It follows that if we know that 10 ppm of an insecticide in the brain indicates that the animal is likely to, or has, died from that insecticide, and a wild population has say, a steady-state concentration of 2 ppm of insecticide in the brain, we cannot deduce

Table 8.2 The Minimal Percentage Activity of Seven Acetylcholinesterase Isozymes in the Housefly after Exposure to the LD$_{50}$ Dose of Four Insecticides. (Reproduced by permission of Academic Press, New York from Tripathi and O'Brien, 1973)

Body region	Isozyme	Activities following treatment with				Range of minimal activities
		malaoxon	paraoxon	diazinon	dichlorvos	
Head	1	18	45	54	39	36
Head	2	42	53	28	72	44
Head	3	51	78	32	57	46
Head	4	28	67	67	70	42
Thorax	5	15	18	22	21	7
Thorax	6	5	38	1	16	37
Thorax	7	1	1	1	1	0.0

that the population's exposure is one-fifth of that needed to kill it. It may in fact be very near the critical level at which fat will be mobilized, brain levels rise, and the animals die (Jefferies and Davis, 1968).

It is important to distinguish between pollutants whose toxicity is intrinsic to one of the elements contained within the molecule, and those whose toxicity resides in the precise molecular configuration. Mercury, for example, has many inorganic and organic chemical 'species', some of which are much more toxic than others, but the toxicity depends in all instances on the presence of the mercury atom, which is, for practical purposes, indestructible. p,p'-DDT is also regarded as a persistent pollutant, but its toxicity resides in the whole molecule, which can in time break down into non-toxic components. The critical organ can change according to chemical 'species' for the elemental pollutants. With mercury, methylmercury affects the central nervous system, several other forms of mercury affect the kidney, while the vapour of elemental mercury affects both these and the peripheral nervous system.

The classical measure of a pollutant's toxicity is the LD_{50}, but this is of limited value. Strictly, this is the result of a bioassay, and although the results are reproducible, they have limited relevance to the problems of pollution. Other criteria, such as changes of behaviour or enzyme activity can be, and often are, used but these may more conveniently be considered as sublethal effects.

8.4. SUBLETHAL EFFECTS

Numerous types of sublethal effects have been described, and those for pesticides have been reviewed several times. Moriarty (1969) has reviewed them for insects, while Jefferies (1975) has made the most recent review for vertebrates. A few general points merit attention here.

Death is obvious. Sublethal effects may be very subtle and not at all obvious. So we are often faced with the question: have we missed observing a subtle sublethal effect? There is probably no complete answer to this question, because it is usually difficult to prove a negative. The usual, empirical, approach is to look for effects on functions, such as growth and reproduction. The alternative is to look for effects on specific systems, such as enzyme activity. This is most useful when some clues exist about the pollutant's mode of action. Then tests are sometimes done *in vitro*, rather than *in vivo*, but extrapolation of results to the whole organism may be difficult. For example, inhibition of carbonic anhydrase has been suggested as the cause of thin eggshells in birds exposed to DDT. *In vitro* tests have demonstrated inhibition of carbonic anhydrase activity by DDT, but this appears to be an artefact, caused by precipitation of DDT occluding the enzyme from solution (Pocker *et al.*, 1971).

With both approaches there is a need to distinguish between a deleterious effect and an adaptive response. Animals do respond to their environment — they would not survive if they didn't, and it can be difficult to decide when an effect has become deleterious. Enzyme induction for instance is an adaptive response, and occurs in response to a wide range of liposoluble foreign compounds, including organochlorine insecticides. It can, however, have unfortunate consequences. Wergedal and Harper (1964) fed one group of rats on a high protein diet, and another group on a low protein diet. These diets were given for from five to twenty days, after which the rats were starved for twelve hours. Each rat was then injected with glycine. All the rats on the high protein diet died from ammonia poisoning. Their diet had induced the formation of enzymes that enabled them to metabolize amino acids more rapidly. Ammonia is one of these metabolites, which is normally metabolized to urea before excretion. The high protein diet meant that the injected glycine was metabolized more rapidly, and the twelve hours starvation meant that there was a shortage of the intermediate compounds needed to convert ammonia to the less toxic urea. By contrast, half the rats fed a low protein diet survived the injection of glycine.

This was a rather unnatural situation, but does illustrate an important point. An adaptive response does not necessarily increase the individual's chances of survival, it just increases the probability of survival in normal environmental conditions. That is an ecological question that can be difficult to answer.

One of the greatest deficiencies in current toxicity testing is the paucity of information on the effects of very long low-level exposures. Long-term experiments are of course routine in testing for carcinogenic and mutagenic action on mammals, but not for other sublethal effects. To put it in its simplest form, very rarely do we try to determine how far below a lethal exposure will still produce deleterious effects.

8.5. EFFECTS ON POPULATIONS

Apart from our own species and some domesticated species, we are not usually concerned with the effects of pollutants on individual organisms. Rather we are concerned with the effects on populations of individuals. This introduces an extra order of complexity: we cannot automatically assume that because, in the field, a pollutant kills some individuals from a population, therefore the population will become smaller. Most of our knowledge on this topic relates to the effects of insecticides on insects, where resurgence is well known (Dempster, 1975). Other possible results include increases in the populations of other species, replacement of the affected species by another, and resistance.

Sublethal effects also can affect population size, by their effects on either survival ability or reproductive ability. No field studies have been made so far in which the separate quantitative consequences of the individual lethal and sublethal effects of a pollutant have been estimated. It is therefore difficult to estimate the significance of sublethal effects alone, distinct from lethal effects, for population dynamics.

There have been many studies of the ecological effects of pollutants, especially the unexpected consequences of pesticide usage, on wildlife (three examples taken more or less at random: Grolleau and Giban, 1966; Borg et al., 1970; Craig and Rudd, 1974). However, study of effects on populations is difficult. The communities within which populations exist are usually too complex to permit sufficient resources for a strictly experimental approach, with control and experimental areas and adequate replication — the work on *Tribolium* species is a notable, but special, exception (Park, 1962). Because of the difficulties in understanding these complex situations, conclusions have usually to rely heavily on experience and judgment, and the most effective way to illustrate the difficulties of interpretation is to consider a specific example in some detail.

(i) Pesticides and the Peregrine Falcon (*Falco peregrinus*)

The peregrine falcon, widespread throughout Eurasia and North America, feeds almost entirely on live birds, caught in flight. It is the largest native falcon in Great Britain, where it is one of the few wildlife species for which extensive and acceptable data on past population size exist. The total number of breeding pairs in Great Britain appears to have been remarkably stable over the last four centuries. This species is at the end of a food chain, and the number of breeding pairs appears to be limited in part at least by the occurrence of suitable breeding sites — the nest or eyrie is usually on a steep rock face, often in the same spot year after year. Ratcliffe (1970) discovered that during the years 1947 and 1948 an index of eggshell thickness (eggshell weight/egg length x egg breadth) decreased significantly by, on average, 19.1% (Figure 8.6). Before 1947 there were no geographical variations in this index, but after that date eggshells from the eastern and central Scottish Highlands decreased in thickness by only 4.4%. There has been much controversy about the causes and significance of this phenomenon (Moriarty, 1975b). The fundamental problem is that the field data are essentially of correlations between different measurements, and to deduce cause and effect from correlations is always difficult. However, it does seem highly probable that DDT, or its metabolite DDE, or both, initiated the thinning of eggshells. More recently, other organochlorine insecticides, and other pollutants too, may have been partly responsible. Whether that diagnosis be right or wrong, a marked and widespread change occurred among the individuals in a field population, for which a reasoned argument can be advanced to suggest pollution as the cause, but it has been impossible to prove beyond all doubt that insecticides were the cause.

One important piece of evidence would be an experiment in which, when a pollutant is administered:

(1) amounts in the eggs are comparable to those found in field specimens;

(2) eggshells are comparably thinner.

The results of many experiments have been published, for a range of pollutants and species, but Lincer (1975) has been the first to satisfy these two criteria

182   *Principles of Ecotoxicology*

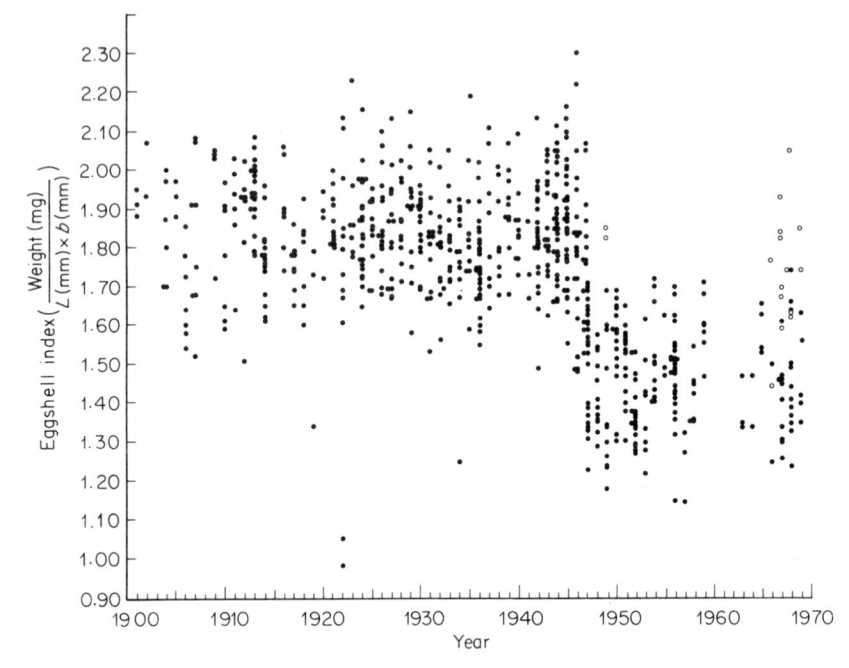

Figure 8.6 Changes from 1901–1969 of the eggshell index for the peregrine falcon in Great Britain. ○, eggshells from the central and east Scottish Highlands; ●, eggshells from other districts. (Reproduced by permission of Blackwell Scientific Publications Ltd., from Ratcliffe, 1970)

(Figure 8.7). He worked on a related species, *Falco sparverius*, for which, in New York state, DDE was the most abundant insecticide found in eggs, although it must be noted that the available data suggest that PCBs were present in similar concentrations.

Ratcliffe discovered this phenomenon of thin eggshells as an indirect result of complaints by owners of racing pigeons that the peregrine falcon was exacting a heavy toll of their birds. In the event Ratcliffe found the complaint was ill founded; in fact the peregrine falcon had all but disappeared as a breeding species from southern England. Nobody had been aware of this.

Several theories have been advanced to explain how eggshells are thinned (Cooke, 1973). The most favoured theory is perhaps that an enzyme in the shell gland is inhibited: either carbonic anhydrase (Bitman *et al.*, 1970; Peakall, 1970), or Ca-ATPase (Miller *et al.*, 1976).

The effect of thin eggshells on the population size is again not simple to understand. Eggshells of the peregrine falcon in Great Britain have been thinner from 1947/8 onwards. Correlated with this, and presumably as a result, the proportion of eggs broken in the eyrie rose from 4 to 39%. It was another 8–10

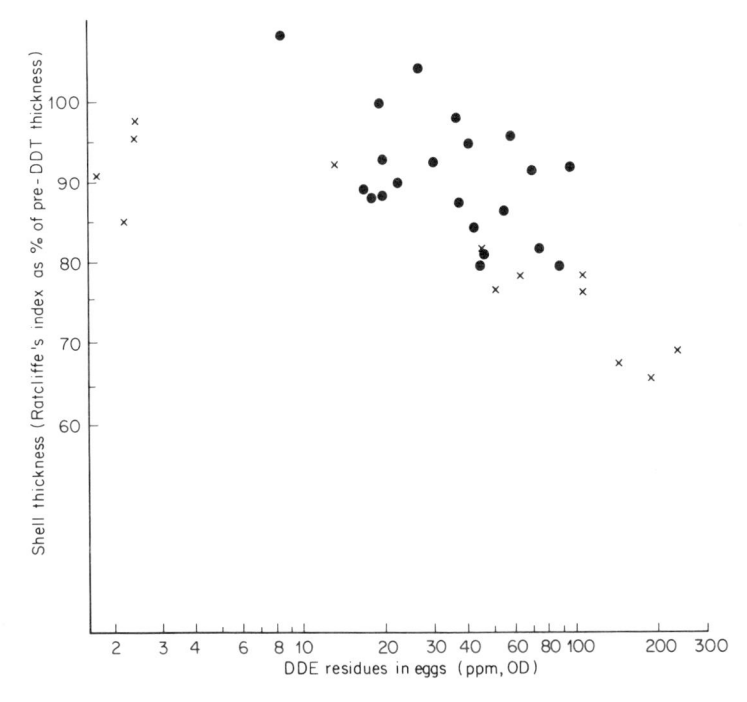

Figure 8.7 Relationship between mean clutch shell thickness and DDE residue of kestrel eggs collected in Ithaca, New York, during 1970 (●) and same relationship experimentally induced with dietary DDE (X). (Reproduced by permission of Blackwell Scientific Publications Ltd., from Lincer, 1975)

years however, before the number of peregrine falcons declined. This drop in the size of the breeding population was associated with the use of dieldrin as a seed dressing. Dieldrin too may cause thin eggshells, but it also killed many pigeons in the spring, when they ate the newly sown dressed cereal seed. It is estimated that a peregrine falcon could acquire a lethal dose from eating two or three heavily contaminated pigeons.

To summarize, this incident illustrates:

(1) Even an obvious complex of effects, such as thin eggshells, reduced breeding success and reduced population, in a very 'popular' species, can escape notice.

(2) It is difficult, particularly after the event, to decide on causes and effects.

(3) The detailed mechanism of the effect on the individual is almost impossible to predict, and difficult to discover, when the site of action is unknown;

organochlorine insecticides are usually considered to interfere with the transmission of nerve impulses (Narahashi, 1971), whereas this presumably is not the primary lesion that causes thin shells.

(4) The ultimate effects on the population, and even more on the community, are again not simple. The reduced breeding success of the peregrine falcon had no apparent effect on population size, although it has been suggested for the golden eagle (*Aquila chrysaëtos*) that a similar reduction in breeding success would result initially in an ageing breeding population. As this is a long-lived species the impact of reduced breeding success would be slow to appear in the population (Dempster, 1975).

8.6. CONCLUSIONS

We have become aware only during the last two decades of the need to consider the possible ecological effects of pollutants. It should be clear that we lack both toxicological and ecological understanding. Current practice, exemplified in particular by regulatory procedures for pesticides, consists of a modicum of toxicology, field trials and safety factors (Moriarty, 1977). Present testing methods are relevant and important, but if we are to increase our predictive abilities we must also, as has been emphasized by many advisory committees, undertake some long-term research projects. Important areas for research include:

(1) Development of the compartmental model as a means of understanding the factors that determine the amount and distribution of pollutants within animals.

(2) Understanding of pollutant modes of action, and the consequent sublethal effects on the whole animal's functions.

(3) Studies of relatively simple ecosystems, with simulations of significant aspects in laboratory experiments, in an attempt to improve our ability to understand and predict ecological effects.

8.7. REFERENCES

Anderson, D. W. and Hickey, J. J., 1976. Dynamics of storage of organochlorine pollutants in herring gulls. *Environ. Pollut.*, **10**, 183–200.

Atkins, G. L., 1969. *Multicompartment Models for Biological Systems*, Methuen, London, xiii, 153 pp.

Bitman, J., Cecil, H. G., and Fries, G. F., 1970. DDT-induced inhibition of avian shell gland carbonic anhydrase: a mechanism for thin eggshells. *Science*, **168**, 594–6.

Borg, K., Erne, K., Hanko, E., and Wanntorp, H., 1970. Experimental secondary methylmercury poisoning in the goshawk (*Accipiter g. gentilis*). *Environ. Pollut.*, **1**, 91–104.

Burt, P. E., Gregory, G. E., and Molloy, F. M., 1966. A histochemical and electrophysiological study of the action of diazoxon on cholinesterase activity and nerve conduction in ganglia of the cockroach *Periplaneta americana* L. *Ann. Appl. Biol.*, **58**, 341–54.

Cecil, H. C., Fries. G. F., Bitman, J., Harris, S. J., Lillie, R. J., and Denton, C. A., 1972. Dietary p, p'-*DDT*, o, p'-DDT or p, p'-DDE and changes in eggshell characteristics and pesticide accumulation in egg contents and body fat of caged white leghorns. *Poult. Sci.*, **51**, 130–8.

Clark, D. R., 1975. Effect of stress on dieldrin toxicity to male redwinged blackbirds (*Agelaius phoeniceus*). *Bull. Environ. Contam. Toxicol.*, **14**, 250–66.

Cooke, A. S., 1973. Shell thinning in avian eggs by environmental pollutants. *Environ. Pollut.*, **4**, 85–152.

Craig, R. B. and Rudd, R. L., 1974. The ecosystem approach to toxic chemicals in the biosphere. In *Survival in Toxic Environments* (Eds. M. A. Q. Khan and J. P. Bederka), Academic Press, New York. pp. 1–24.

Cummings, J. G., Zee, K. T., Turner, V., Quinn, F., and Cook, R. E., 1966. Residues in eggs from low level feeding of five chlorinated hydrocarbon insecticides to hens. *J. Assoc. Off. Anal. Chem.*, **49**, 354–64.

Davison, K. L., 1970. Dieldrin accumulation in tissues of the sheep. *J. Agric. Food Chem.*, **18**, 1156–60.

Deichmann, W. B., Dressler, I., Keplinger, M., and MacDonald, W. E., 1968. Retention of dieldrin in blood, liver, and fat of rats fed dieldrin for six months. *Ind. Med. Surg.*, **37**, 837–9.

Dempster, J. P., 1975. Effects of organochlorine insecticides on animal populations. In *Organochlorine Insecticides: Persistent Organic Pollutants* (Ed. F. Moriarty), Academic Press, London, pp. 231–48.

Grolleau, G. and Giban, J., 1966. Toxicity of seed dressings to game birds and theoretical risks of poisoning. *J. Appl. Ecol.*, **3**, (suppl.), 199–212.

Hamelink, J. L., Waybrant, R. C., and Ball, R. C., 1971. A proposal: Exchange equilibria control the degree chlorinated hydrocarbons are biologically magnified in lentic environments. *Trans. Am. Fish. Soc.*, **100**, 207–14.

Holden, A. V., 1975. Monitoring persistent organic pollutants. In *Organochlorine Insecticides: Persistent Organic Pollutants* (Ed. F. Moriarty), Academic Press, London, pp. 1–27.

Iatropoulos, M. J., Milling, A., Müller, W. F., Nohynek, G., Rozman, K., Coulston, F., and Korte, F., 1975. Absorption, transport and organotropism of dichlorobiphenyl (DCB), dieldrin, and hexachlorobenzene (HCB) in rats. *Environ. Res.*, **10**, 384–9.

Jefferies, D. J., 1972. Organochlorine insecticide residues in British bats and their significance. *J. Zool.*, **166**, 245–63.

Jefferies, D. J., 1975. The role of the thyroid in the production of sublethal effects by organochlorine insecticides and polychlorinated biphenyls. In *Organochlorine Insecticides: Persistent Organic Pollutants* (Ed. F. Moriarty), Academic Press, London, pp. 131–230.

Jefferies, D. J. and Davis, B. N. K., 1968. Dynamics of dieldrin in soil, earthworms, and song thrushes. *J. Wildl. Manage.*, **32**, 441–56.

Kudsk, F. N., 1965. Absorption of mercury vapour from the respiratory tract in man. *Acta Pharmacol. Toxicol.*, **23**, 250–62.

Lincer, J. L., 1975. DDE-induced eggshell-thinning in the American kestrel: a comparison of the field situation and laboratory results. *J. Appl. Ecol.*, **12**, 781–93.

Miller, D. S., Kinter, W. B., and Peakall, D. B., 1976. Enzymatic basis for DDE-induced eggshell thinning in a sensitive bird. *Nature (London)*, **259**, 122–4.

Moriarty, F., 1969. The sublethal effects of synthetic insecticides on insects. *Biol. Rev.*, **44**, 321–57.

Moriarty, F., 1975a. Exposure and residues. In *Organochlorine Insecticides: Persistent Organic Pollutants* (Ed. F. Moriarty), Academic Press, London, pp. 29–72.

Moriarty, F., 1975b. *Pollutants and Animals. A Factual Perspective*, Allen and Unwin, London, 140 pp.

Moriarty, F., (a) In press. Prediction of ecological effects by pesticides. In *Ecological Effects of Pesticides* (Eds. K. Mellanby and F. H. Perring), Academic Press, London.

Moriarty, F., 1977. Ecotoxicology: some complexities of effects on ecosystems. In *The Evaluation of Toxicological Data for the Protection of Public Health* (Eds. W. J. Hunter and J. G. P. M. Smeets), Pergamon Press, Oxford. pp. 281–7.

Narahashi, T., 1971. Effects of insecticides on excitable tissues. *Adv. Insect Physiol.*, **8**, 1–93.

Neely, W. B., Branson, D. R., and Blau, G. E., 1974. Partition coefficient to measure bioconcentration potential of organic chemicals in fish. *Environ. Sci. Technol.*, **8**, 1113–5.

Park, T., 1962. Beetles, competition and populations. *Science, N.Y.*, **138**, 1369–75.

Peakall, D. B., 1970. p,p'-DDT: effect on calcium metabolism and concentration of estradiol in the blood. *Science, N.Y.*, **168**, 592–4.

Pocker, Y., Beug, W. M., and Ainardi, V. R., 1971. Carbonic anhydrase interaction with DDT, DDE, and dieldrin. *Science, N.Y.*, **174**, 1336–9.

Ratcliffe, D. A., 1970. Changes attributable to pesticides in egg breakage frequency and eggshell-thickness in some British birds. *J. Appl. Ecol.*, **7**, 67–115.

Robinson, J., Roberts, M., Baldwin, M., and Walker, A. I. T., 1969. The pharmacokinetics of HEOD (dieldrin) in the rat. *Food Cosmet. Toxicol.*, **7**, 317–32.

Stickel, W. H., 1975. Some effects of pollutants in terrestrial ecosystems. In *Ecological Toxicology Research. Effects of Heavy Metal and Organohalogen Compounds* (Eds. A. D. MacIntyre and C. F. Mills), Plenum Press, New York and London, pp. 25–74.

Task Group on Metal Accumulation, 1973. Accumulation of toxic metals with special reference to their absorption, excretion and biological half-times. *Environ. Physiol. Biochem.*, **3**, 65–107.

Tripathi, R. K. and O'Brien, R. D., 1973. Effect of organophosphates *in vivo* upon acetylcholinesterase isozymes from housefly head and thorax. *Pestic. Biochem. Physiol.*, **2**, 418–24.

Walker, C. H., 1975. Variations in the intake and elimination of pollutants. In *Organochlorine Insecticides: Persistent Organic Pollutants* (Ed. F. Moriarty), Academic Press, London, pp. 73–130.

Wergedal, J. E. and Harper, A. E., 1964. Metabolic adaptations in higher animals. IX. Effect of high protein intake on amino nitrogen catabolism *in vivo*. *J. Biol. Chem.*, **239**, 1156–63.

CHAPTER 9

Aquatic Animals

P. D. ANDERSON AND S. D'APOLLONIA

Department of Biological Sciences, Concordia University, 1455 de Maisonneuve Blvd. W. Montreal, P.Q. H3G 1M3

9.1 INTRODUCTION	187
9.2 SUBLETHAL TOXICITY	188
9.3 PHYSIOLOGICAL SYSTEMS FOR SUBLETHAL TESTS OF POLLUTANTS	191
(i) Osmoregulation	191
(ii) Respiration	192
(iii) Blood	194
(iv) Locomotion	195
(v) Growth	198
(vi) Reproduction	198
9.4 TERATOGENESIS	199
9.5 MUTAGENESIS	201
9.6 CARCINOGENESIS	202
9.7 MULTIPLE TOXICITY	204
(i) Presumptive Mechanisms of Interaction	206
(ii) Multiple Toxicity Models – based on Interactions at the Dynamic Level	207
(iii) Multiple Toxicity Models – based on Interactions at the Kinetic Level	212
(iv) Application of Multiple Toxicity Models	213
(v) Cancer	213
(vi) Water Quality Management	213
9.8 REFERENCES	214

9.1. INTRODUCTION

Natural waters are the ultimate recipients of much of the chemical wastes generated by man's industrial, agricultural, and domestic activities and released into the geosphere. Wastes enter the aquatic environment directly, by point source dumping, or indirectly, by rain, snow, or fall-out and by surface water run-off and ground-water leachings. Aquatic ecosystems, to varying degrees, are adaptable; they have a variety of physical, chemical, and biological mechanisms by which wastes may be assimilated without serious implications to endemic biota. When chemical contaminants reach levels in excess of the assimilative capacity of receiving waters they affect the survival, reproduction, growth, and movement of organisms. As a result the distribution and abundance of populations change with a possible alteration in the energy budget of the ecosystem. The end effect is observed as perturbations in community structure and successional patterns. Man's economic, recreational, and agricultural interests in aquatic ecosystems may be jeopardized by pollution. Furthermore, man's health may be harmed by toxic contaminants upon their uptake from drinking water and from the consumed flesh of aquatic organisms.

Chemical wastes can impair the natural state of an ecosystem through either nutrient enrichment or toxic processes. Enrichment by man-mobilized organics and minerals may promote excessive growth rates in autotrophs whose abundance then alters environmental modifying factors such as oxygen concentration and temperature. The resulting conditions may be unfavourable to other organisms which either migrate or perish often to be replaced by less desirable species. A change in the nutritional balance of an ecosystem can increase the susceptibility of certain organisms to the toxic action of chemical wastes. In these instances, nutrient enrichment may be regarded as an accessory factor in toxicity (Fry, 1947). Moreover, the incomplete catabolism of excess nutrients by decomposers can, in itself, be toxicogenic. The technology is, however, available to regulate effectively the releases of biostimulatory wastes at their major sources, e.g. in municipal and industrial effluents. This technology coupled with recent methods of monitoring eutrophication (Vollenweider and Dillon, 1974) offers a promising solution to the problems created in aquatic ecosystems by biostimulatory wastes.

The management of toxic wastes entering natural waters would appear more difficult if only because poisonous contaminants often arise from multiple or diffuse sources. The seriousness of this form of pollution is compounded by the facts that toxicants may be highly potent and selective as lethal and sublethal agents, resistant to denaturation by environmental and biological agents, altered by environmental and biological agents to more toxic forms, transported widely by environmental and biological vectors and accumulated by organisms whereby the levels of a toxicant undergo magnification within the food chain.

The following account provides approaches deemed to be useful in assessing sublethal toxicity with particular reference to teratogenesis, mutagenesis, and carcinogenesis. Furthermore, a rationale is presented for the investigation of the potentially harmful additive and synergistic interactions of pollutant mixtures to aquatic fauna.

It will be noticed that the tests of sublethal effects described in the sequel apply to individual organisms or to small groups studied in the laboratory. At present these are the principal means of assessing the effects of environmental pollutants in spite of the fact that the methodology for applying laboratory results to field situations is in serious need of development. Some discussion of this is to be found in Section IV.

9.2. SUBLETHAL TOXICITY

The deleterious effects of chemical pollutants on aquatic ecosystems may result not only from toxicity that directly causes the death of an organism, but also from a variety of sublethal effects. Sublethal impairment of an animal's development or its capacity to perform and adapt can reduce the chances for survival and the potential for growth and reproduction. These constraints, though possibly slight for individual members of a population, may be highly detrimental to that population's long-term ecological success as measured by its abundance and distribution. The consequences may be as severe as extinction. Ultimately, such changes in populations lead to perturbations of community structural and successional patterns.

This subtle yet insidious pattern of sublethal toxicity makes the assessment of levels below which the respective chemical contaminants do not pose a toxic hazard to an ecosystem, an immensely complex problem. Adequate monitoring of natural ecosystems to measure ecologically significant levels (i.e. levels that are detrimental under field conditions) usually results in severe disruption of the community under study. The only practical recourse in water quality management is to devise laboratory screening tests from which are extrapolated pollutant levels that are deemed 'safe' for aquatic ecosystems. The task in this latter approach is to decide what level of impairment observed under laboratory conditions would constitute a detrimental effect under field conditions.

The popular use of 'application factors' (5–10% or the 48-hour LC_{50} for non-persistent pollutants and 1–10% of the 48-hour LC_{50} for persistent pollutants, Sprague, 1971) as a means of predicting 'safe' levels derived from measurements of lethality has been criticized by many pollution biologists. They argue that the assumptions in the application factor approach are not supported empirically and advocate as an alternative the use of chronic, sublethal tests. However, as of 1972 fewer than 40% of the water quality criteria for toxicants set by the Environmental Protection Agency of the United States were based on sublethal tests (Sprague, 1976).

Sublethal studies conducted in the laboratory are amenable to predicting 'safe' or ecologically insignificant levels of toxicants if responses are quantified and dose–response relationships are established. By converting these numerical data to a quantal (all or none) form, the effective concentration (EC) for a given sublethal response can be computed for various percentages of the test organisms (Figure 9.1). Sprague (1971) has proposed that the EC's be used to set water quality criteria for safeguarding aquatic organisms. Also recommended as a measure for predicting 'safe' levels is the incipient threshold concentration computed by plotting median effective concentrations (EC_{50}) against time (Figure 9.2).

With a few notable exceptions, such analyses of sublethal toxicity have not been applied. This scarcity of quantitation and the lack of predictive models that can be used to correlate sublethal impairment and population hazard are perhaps the major deficiencies in the field of sublethal toxicity and its application to environmental assessment.

Another important problem in the design of sublethal bioassays is the choice of test organisms. Test species, representative of the ecosystem to which the eventual

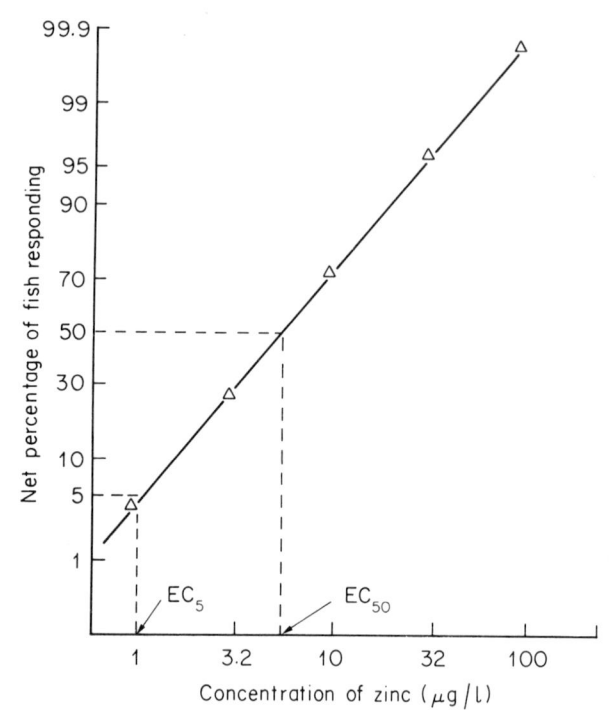

Figure 9.1 Example of probit analysis applied to sublethal studies of avoidance behaviour in trout. The technique allows the computation of the median effective concentration (EC_{50}) and the effective concentration for 5 per cent of the test population (EC_5). (Reprinted with permission from *Water Res.*, 2, 367. Sprague, J. B. Avoidance reactions of rainbow trout to zinc sulphate solutions. © 1968, Pergamon Press, Ltd.)

criteria are to be applied, should be selected for their sensitivity, availability, and position in the food chain. For marine environments, Eisler (1972) listed teleosts, crustaceans, and molluscs in order of increasing tolerance to lethal levels of several organochlorine and organophosphate insecticides. It cannot be assumed that this order of sensitivity would be exemplified at sublethal levels; pollutants may have modes of toxicity that are highly selective and therefore particularly potent to only certain organisms. In consideration of the extreme variety in organisms and toxicants in aquatic ecosystems, neither one screening test nor one test species should be relied upon for the establishment of criteria to protect against sublethal impairment. It is suggested that knowledge of modes of action of toxicants may prove useful in predicting sensitive organisms in an aquatic ecosystem. This proposal is based on the assumption that cells having similar functions and similar metabolic pathways in various organisms are similarly affected by a given chemical entity (Loomis, 1974).

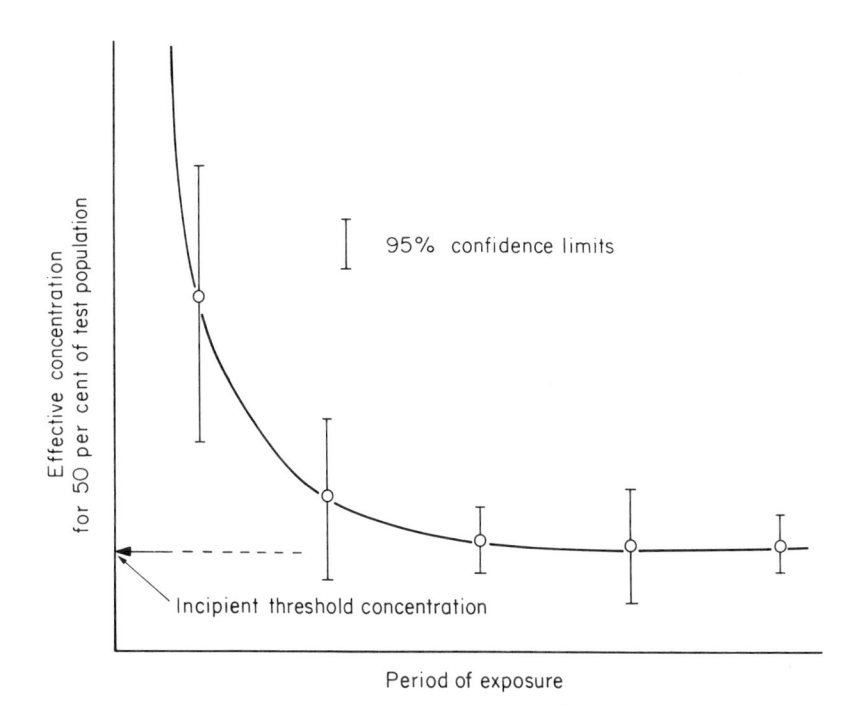

Figure 9.2 Illustration of a sublethal toxicity curve derived by plotting EC_{50} values against the period of exposure. The incipient threshold concentration represents the point at which the curve becomes asymptotic to the time scale. (Reproduced by permission of Ann Arbor Science Publishers, Inc., from Brown, 1973)

Organisms have evolved intricate regulatory mechanisms vital to survival, growth, and reproduction within a hostile environment. Toxic damage to these physiological systems, although not necessarily lethal under laboratory conditions, often produces impairment under natural conditions. The following section offers a brief survey of physiological systems deemed to have ecological significance for aquatic organisms and some functional laboratory tests used to evaluate the sublethal toxicity of discrete chemical toxicants as they apply to fish. The principles may be applied to other organisms. It is not a complete review of the topic; further information may be found in recent reviews (Sprague, 1971; Mitrovic, 1972; Rosenthal and Alderdice, 1976).

9.3. PHYSIOLOGICAL SYSTEMS FOR SUBLETHAL TESTS OF POLLUTANTS

(i) Osmoregulation

Aquatic organisms often expend considerable energy regulating the concentrations of osmotically active constituents in body fluids. For example, freshwater fish are hyper-osmoregulators and must counteract the tendency to lose ions and gain water. Thus, ions are selectively taken up at the gill epithelium and retained by the urinary bladder epithelium. Ionic regulation across a concentration gradient imposes the largest energy expenditure on standard metabolism in freshwater fish (Renfro et al., 1973). Gills, a primary site of osmoregulation, as well as respiration, may be highly vulnerable to lesions because they are in immediate contact with aquatic toxicants. These considerations suggest that inquiries into the sublethal effects of pollutants should include tests for osmoregulatory damage.

There is considerable evidence that certain heavy metals (e.g. mercury, zinc, and copper) cause structural gill damage (Skidmore, 1970; Mitrovic, 1972) and interfere with ion transport across gill epithelium (Lewis and Lewis, 1971; Renfro et al., 1973). Toxicant-induced disruptions in osmoregulation are reflected in alterations in the ion content of blood (Lewis and Lewis, 1971; O'Conner and Fromm, 1975) or in changes in the rate of urine flow (Lloyd and Orr, 1969; Swift and Lloyd, 1974).

Renfro et al. (1973) have shown that both mercuric chloride and methylmercury bind to gill tissue where they inhibit sodium uptake in kilifish. The rates of mercury clearance from gill tissue correlate with the disappearance of this sodium uptake inhibition. Furthermore, there was a dose-related decrease in sodium–potassium-activated ATPase activity in bladder homogenates isolated from fish exposed to varying concentrations of methylmercury.

This enzyme is found in high concentrations in all epithelial membranes involved in active ion exchange. In anadromous fish adapting to sea water, the activity of sodium–potassium-activated ATPase increases in proportion to sodium excretion (Giles and Vanstone, 1976). Consequently, this period of adjustment may be particularly sensitive to mercurial contaminants and other osmoregulatory antagonists which inhibit sodium–potassium-activated ATPase (Davis and Wedemeyer, 1971; Janicki and Kinter, 1971).

Osmoregulatory variables, such as plasma ion concentrations, rates of urine flow, and activities of ion transport enzymes, would appear to offer sensitive systems for the quantitative determination of effective sublethal concentrations of many toxicants.

(ii) Respiration

Fish have evolved ventilatory, circulatory, and haematological systems for the uptake and transport of oxygen to cellular oxidation sites. These systems are integrated to deliver an oxygen supply to tissues sufficient for oxidative phosphorylation and other energy requirements of the cells. Homeostatic mechanisms are present to regulate oxygen tension at cellular respiratory sites independent (within set limits) of varying respiratory stress. These mechanisms may be either short-term responses (e.g. increased heart rate, increased blood pH) or long-term adaptations (e.g. increased haematocrit). All these require an additional expenditure of energy which, under certain conditions (e.g. food shortage, low ambient oxygen concentration) or during times of increased energy demand, the organism cannot supply. Although respiratory distress imposed by a pollutant may have a detrimental effect on fish there are very few studies in which a sublethal effect has been proved to cause population perturbations. The effects of pollutants on ventilation and gas transport have been investigated.

(a) Ventilation

The muscular expansion and contraction of the buccal and opercular cavities maintains a flow of water over gill surfaces in teleost fishes. These respiratory movements are under physiological control of the respiratory centre and are modified by both internal (oxygen concentration, carbon dioxide concentration, pH) and external (temperature, oxygen depletion) factors (Shelton, 1971). Under normal circumstances the ventilatory pump is coupled to the cardiac pump, so that there is an adequate residence time of blood in the gills for oxygen exchange. Generally, an increase in ventilation acts to increase the oxygen supply at the gills at a low additional energy cost (Cech and Wohlschlag, 1973). Large increases in respiratory movements however may result in a lower oxygen-carrying capacity of the blood with resulting tissue anoxia (Cech et al., 1976).

A number of techniques have been used to monitor respiratory movements (Heath, 1972). Implanted cannulae, appropriately positioned pressure tranducers, and the physical separation of inspired and expired water allow ventilatory volume, ventilatory frequency, and opercular and buccal pressures to be monitored simultaneously under conditions of least stress (Davis and Randall, 1973). In this way alterations in respiratory movements have been recorded in response to sublethal concentrations of a variety of toxicants (Cairns and Sparks, 1971; Davis, 1973; Lunn et al., 1976). Although alterations in respiratory movements are relatively easy to monitor, modifying conditions are difficult to control. Activity, time of day, internal and external oxygen and carbon dioxide concentration, and temperature modify ventilation rate. Furthermore the required anaesthetic and surgery are stressors that are manifested by alterations in ventilatory responses (Houston et al., 1971). Perhaps for these reasons quantitative studies of the effect of toxicants on respiratory activity and the possible detrimental effect of this sublethal response on fish, have not been made.

A more promising respiratory activity for sublethal studies is 'coughing' (the reversal of water flow over gills, presumably to clear them). Schaumberg et al. (1967) have suggested that the cough reflex be used as an indicator of sublethal toxicity. An increased cough reflex has been correlated with low levels of pesticides (Lunn et al., 1976), copper (Sellers et al., 1975), and kraft mill effluent (Davis, 1973). An elevated cough frequency interferes with oxygen uptake at the gills and reduces the efficiency of oxygen uptake (Davis, 1973) and if continued might cause tissue anoxia and respiratory collapse. Although alterations in cough reflex are relatively easy to monitor using either implanted cannulae and pressure transducers, or surface electrodes, fish quickly accommodate to the toxicant and the cough rate returns to normal. Thus the relevance of this sublethal response to populations' impairment can be questioned.

(b) Gas Transport

The amount of oxygen reaching tissue sites is a function of not only the oxygen-carrying capacity of the blood but also the blood flow. The rate of blood flow is modulated by peripheral vascular resistance to flow (measured by blood pressure) and cardiac output (heart rate x stroke volume). Although both arterial and venous blood pressures are easily monitored the effects of a toxicant on blood pressure, to our knowledge, have not been measured. Heart rate and amplitude of contraction are under nervous control and can be measured by either electrocardiography or by the methods used to monitor respiratory movements and oxygen uptake. A number of cardiovascular functions can be measured directly (e.g. blood pressure, heart rate, oxygen consumption); cardiac output and therefore stroke volume can also be determined from the Fick relationship (Cech and Wohlschlag, 1973; Davis and Randall, 1973). Although studies have shown that heart rate is altered by a number of toxicants (Pfuderer and Francis, 1973; Lunn et al., 1976), the response is transitory and thus can only be used for short-term experiments; furthermore, the possible long-term detrimental effects of a decreased heart rate have not been established.

(iii) Blood

Blood forms a unique compartment between the external and internal tissue compartments in animals: all substances that enter or leave tissues, be they nutrients, toxicants, excretory products, or gases, pass through blood. Most toxic substances alter blood composition either directly (e.g. erythrocyte destruction) or indirectly by altering osmotic and ion regulation, gas uptake, bilirubin formation, and catabolism. Because of the crucial role of blood in oxygen uptake many pollutants cause anaemia and tissue hypoxia. These sublethal effects can not be considered a successful acclimation to an environmental change because if they persist they result in reduction in the survival potential of the fish (Sprague, 1971).

Haematological analysis is an established procedure for the diagnosis of disease and poisonings in medicine. Blood is relatively easy to obtain, store, analyse, and quantitate. Analyses for haematocrit, haemoglobin content, iron content, ion composition, or serum protein analysis offer quick screening methods for assessing the health of fish (Wedemeyer and Chatterton, 1970; Blaxhall and Daisley, 1973).

Haematological analysis can also be used to determine the incipient lethal concentration of a toxicant (McLeay, 1973). For example, Buckley et al (1976) assayed haemoglobin content and the percentage of mature erythrocytes in coho salmon exposed to residual chlorine for twelve weeks. A linear dose—response relation was established between both these variables and total residual chlorine. McKim *et al.* (1970) demonstrated statistically significant alterations in a number of blood variables in brook trout to short-term (6-day) copper exposure. The copper concentration, however, resulted in the death of 40% of the fish, scarcely a sublethal effect.

(iv) Locomotion

Most non-sedentary aquatic animals depend on swimming to maintain their position against a current, to obtain food, to escape predators, and to reach spawning beds. Thus a toxicant that interferes with swimming is likely to have far-reaching effects. Swimming represents the integrated result of many physiological processes including behaviour, sensory perception, and muscular activity. These three aspects of locomotion are discussed below.

(a) Behaviour

Fish have a number of behavioural patterns which enable them to select environments favourable to survival and reproduction. These behavioural patterns provide useful measures of sublethal toxicity because they represent the integrated result of many biochemical and physiological processes. Furthermore, even under laboratory conditions, behavioural patterns remain consistent with those in the field and behavioural measurements often require less physical manipulation of the test organism than physiological measurements.

Two approaches used to monitor the effect of a toxicant on behaviour are avoidance studies in a 'free-choice' situation by unconditioned fish and impairment of a conditioned response in trained fish.

In avoidance studies, fish are introduced into an experimental tank in which there is either a continuous gradient from water to toxicant (Ishio, 1965) or a differential partition of a toxicant and water (Sprague, 1968). The distribution of fish in the tank or the time spent in the uncontaminated water are the indices of the avoidance reaction.

Fish have been shown to avoid sublethal concentrations of detergents (Abel, 1974), heavy metals (Sprague, 1968; Ishio, 1965), and insecticides (Hansen *et al.*, 1972) at sublethal concentrations; goldfish avoid fenitrothion at a concentration two orders of magnitude below the incipient lethal level (Scherer, 1975). Westlake and Kleerekoper (1974) have shown that the extent of avoidance and, conversely, attraction to copper is dependent on the slope of the copper ion gradient rather than the absolute copper concentration. Thus a steep gradient may elicit avoidance while a shallow gradient may elicit attraction.

In this latter experiment the bimodal response to the rate of change in the concentration of a toxicant presents a problem in applying the data to quantal analysis, but Sprague (1968) was able to employ quantal analysis in estimating the median effective concentration of zinc that elicits an avoidance reaction in fish

(Figure 9.1). Avoidance by salmon of polluted segments of their migratory routes to spawning areas did have far-reaching effects on their population (Saunders and Sprague, 1967).

A phenomenon related to avoidance reaction is temperature preference. There is evidence that pollutants like DDT alter temperature preference in fish (Anderson, 1971).

The effect of a pollutant on the ability of fish to be conditioned (or learn) can also be used as a measure of sublethal toxicity. Fish can be trained either by reward for performance (e.g. feeding) or by punishment for failure to perform (e.g. electric shock, Warner et al., 1966). This experimental approach has shown that a number of insecticides at sublethal concentrations reduce both learning and retention of a conditioned response in fish (Warner et al., 1966; Hatfield and Johansen, 1972; Davy et al., 1973). It is highly probable that learning impairment has a detrimental effect on such complex behaviour patterns as migration and reproduction. However the ecological significance of alterations in learning in fish has not been studied. Preliminary evidence would suggest that salmon completely recovered their learning ability within 1 week following exposure to 1000 ppm fenitrothion (Hatfield and Johansen, 1972).

(b) Swimming Performance

Brett (1967) and Sprague (1971) have suggested that swimming performance be used as a criterion in determining sublethal impairment. The critical swimming speed (the fastest swimming speed reached) or the maximum sustained swimming speed (the maximum swimming speed maintained for an hour) can be determined by forcing fish to swim against an applied water current in order to maintain a stationary position. The water current is increased at a fixed rate and the swimming speeds (fish lengths/sec) determined from the water current. The design of the swimming chamber is critical as turbulence, drag, and temperature fluctuations must be minimized. The food ration and respiratory ability of the fish also influence swimming performance (Howard, 1975). Despite the difficulties due to fish variability and equipment design, cyanide (Neil, 1957), fenitrothion (Peterson, 1974), hydrogen sulphide (Oseid and Smith, 1972), and bleached kraft pulpmill effluent (Howard, 1975) have been shown to impair swimming performance.

A more promising bioassay for a sublethal effect related to swimming performance is the determination of the critical rate of rotation of a rotating water mass at which fish cannot compensate for torque (Lindahl et al., 1977). Using this technique, Bengtsson (1974) showed that exposure to zinc at levels as low as 0.06 ppm in the water for 190 days significantly reduced the ability of minnows to compensate for torque. This sublethal effect occurs at a concentration of zinc lower than that for any other sublethal impairment reported.

A similar effect of methylmercury has been reported by Lindahl and

Schwanbom (1971a,b). With roach, the 'critical rate of rotation', which measures the ability of the fish to compensate for torque, was inversely proportional to the methylmercury content of the fish muscles.

(c) Perception

Sense perception is thought to play a crucial role in many activities vital to the survival of fish. For example, the homing abilities of migrating fish are mediated through their olfactory, visual, and auditory senses (Hoar, 1975). Impairment of these sensory functions by chemical toxicants would have adverse consequences for fish populations. Furthermore, because many sensory receptors are in immediate contact with the ambient environment they are prime target sites for uptake of toxicant and for subsequent impairment. Considering the importance of olfaction in fish biology, Hara *et al.* (1976) and Bardach *et al.* (1965) studied the effects of a number of pollutants on olfactory response. The alterations in response to a standard stimulus, serine, caused by a pollutant can be detected by recording the electrical responses of the olfactory bulb. After exposure to sublethal concentrations of copper or mercury the electrical activity generated by the olfactory bulbs of anaesthetized immobilized rainbow trout in response to serine was depressed (Figure 9.3). Such depression in olfactory response can be quantified and is dose-dependent.

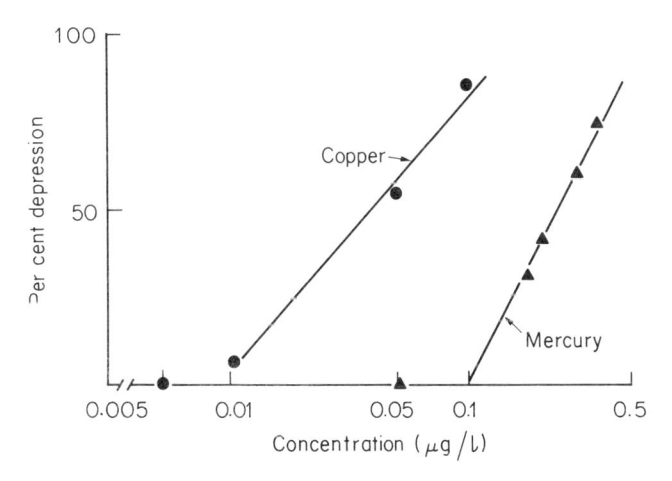

Figure 9.3 Relationship between per cent depression of olfactory response and the ambient concentration of discrete solutions of mercury and copper at 4 hours exposure. (Reproduced by permission of Minister of Supply and Services, Canada, from Hara *et al.*, 1976)

(v) Growth

All organisms must continually repair and replace bodily structures to survive. To increase in size, organisms must obtain energy and materials from the environment in excess of minimum requirements. Growth is dependent on the metabolic state of the organism, thus on its age, on its physiological state, and on environmental controlling and limiting factors (Fry, 1947). Both natural and pollutant stressors may affect food conversion in animals. Pollutants can be an indirect stress on growth by altering the quality and quantity of available food and a direct stress by increasing the metabolic cost of other life functions (Warren, 1971).

Growth rate was initially suggested as a good stress indicator (Sprague, 1971), but proper execution of growth studies, particularly over long periods of time, requires careful control of ration level and locomotory activity. Furthermore, the ration level must be such that the growth rate does not become asymptotic during the experimental period (Warren, 1971). For these reasons, certain results of studies on the influence of toxicants on growth are contradictory (Mount and Stephen, 1969; McKim and Benoit, 1971; Lett et al., 1976). A complicating factor is that many pollutants, especially heavy metals, initially reduce appetite so that less food is eaten (Lett et al., 1976). These latter authors found that after prolonged exposure to toxicants fish exhibit an increased appetite. Consequently, in the long-term studies there is no growth retardation in test fish.

Recently, the effectiveness of growth studies for the establishment of 'safe' levels has been challenged (Sprague, 1976). Nevertheless, Webb and Brett (1973) showed measurements of growth rate and conversion efficiency to be more sensitive indicators of the sublethal effects of sodium pentachlorophenate than swimming performance.

Warren and Davis (1971), through the use of artificial streams, have added a promising new dimension for investigating effects of pollutants on intertrophic relationships that govern growth and production of fish. Such experimental approaches may provide an effective intermediary in extrapolations from laboratory to field conditions.

(vi) Reproduction

The biogenic potential of a population may be adversely affected by a pollutant in a number of ways:

(1) physiological stresses may reduce gametogenesis and therefore egg and sperm production and fertilization success;
(2) behavioural alterations may reduce spawning or parental care of eggs and thus reduce the number of fertilized eggs and their rate of hatching;

(3) egg or sperm viability may be reduced;

(4) teratological effects on the developing embryo may reduce hatching success and larval survival.

Regardless of the mode of action, pollutants affecting reproduction reduce the number of viable offspring.

The reproduction process appears to be highly sensitive to the adverse effects of certain pollutants (Brungs, 1969). Laboratory test systems are now available for the quantitative evaluation of not only survival and growth but also reproduction of fish during chronic exposure to pollutants (Benoit, 1975).

These latter experiments, although informative, are time requiring and expensive. Lesniak (1977) applied light- and electron microscopy as tools for the quantitative measurements of oocyte retardation in rainbow trout, exposed to cyanide during short-term (two weeks) experiments. This approach may offer a satisfactory alternative to the *in vivo* reproductive test.

Two special aspects of reproductive impairment, teratogenesis and mutagenesis, are considered separately in the following two sections.

9.4. TERATOGENESIS

All chemical contaminants are probably toxic to embryos and larvae at some level, but only certain substances impair the ontogenic processes of sequential differentiation, growth and metamorphosis at levels that are normally tolerated by later stages of the life cycle. The adverse effects of such selectively acting compounds – herein designated as embryotoxicants – can range from a simple retardation in developmental rate through a variety of distortions in form (teratogenesis) or function in embryos (Rosenthal and Alderdice, 1976). Whether the action of these toxicants is lethal by contributing directly to the death of an organism during ontogeny or sublethal by reducing indirectly the survival potential of an organism at some subsequent stage in its life, the consequence of embryotoxicity can be a serious reduction in the distribution and abundance of animal populations. It follows that a check list in the toxicity screening of environmental contaminants should include, in conjunction with reproduction surveys, tests identifying and quantifying teratogenic and other forms of selective embryotoxicity.

Whereas certain species appear to serve as sensitive indicators of the potentially adverse effects of pollutants on juvenile and adult organisms inhabiting different trophic levels (Reish, 1972), evidence would suggest that species genotype is also a significant determinant of susceptibility in embryotoxicity (FDA, 1970). This specificity is credited to the tendency of teratogens to act on, through or in conjunction with, unstable genetic loci found only in certain species. As a result there may be limited predictive value in using indicator organisms to estimate the relative hazard of teratogens and other embryotoxicants to a broad spectrum of animal groups. Rosenthal and Alderdice (1976), who tabulated the responses of eggs and larvae of marine fish to a wide selection of stressors including pollutants, suggested that embryotoxicity often involves a limited number of similar physiological and biochemical mechanisms common to many species. By our definition the pollutants that these authors selected to illustrate their rationale were not 'true' teratogens or embryotoxicants because deleterious effects were caused at concentrations equal to or greater than levels affecting adult and juvenile organisms (McKee and Wolf, 1963). The problem of whether embryotoxicants can be characterized by studies involving limited kinds of test organisms would appear unresolved.

Egg capsules, jelly-like coatings and extraembryonic membranes are effective screening devices which often impede the access of pollutants to the developing organism. The high resistance of eggs and embryos to the toxic effects of many pollutants may be explained in part by the protection provided by these coatings. Developing organisms may however be more vulnerable to contaminants absorbed from yolk deposits which occurred during oogenesis; this possibility should be considered in the design of experiments in teratology (Davis, 1972). Studies of teratogenesis should also include complete life-cycle examinations because deleterious effects may be initiated well in advance of physiological or structural manifestations of toxicity (FDA, 1970).

Certain events in the time course of development are particularly vulnerable to the action of teratogens. These 'sensitive' periods include gastrulation, early organogenesis, hatching, and metamorphosis (Rosenthal and Alderdice, 1976). Sufficient periods of toxicant exposure prior to or during these sensitive periods may be necessary for effects to be manifested (e.g. Anderson and Battle, 1967). The cell proliferation stage, i.e. cleavage, that immediately follows fertilization in the indeterminate type of vertebrate development is known to be relatively insensitive to toxic insult. Embryonic development of crustaceans and molluscs exhibits determinant cleavage where, unlike the former omnipotent pattern, each daughter cell or blastomere is the sole primordium of major organ systems in the adult organism (Balinsky, 1975). The implications of this pronounced difference in the pattern of ontogeny between major groups of aquatic organisms in reference to their respective susceptibility to toxicants require investigation.

Amongst the various agents reported by FDA (1970) as possible teratogens are classes of chemicals known to pollute aquatic ecosystems. They include alkylating agents, azo dyes, salicylates, and analogues and antimetabolites of nucleic acid metabolism. Rosenthal and Alderdice (1976) described the embryotoxicity of certain heavy metals (particularly cadmium), DDT, dinitrophenol, cyanide, oil dispersants and extracts, benzene and vinyl chlorides, amongst others, but failed to demonstrate, as suggested previously, that embryonic development was more sensitive to these latter toxicants than physiological, structural, or behavioural properties of adult or juvenile organisms.

As observed by Davis (1972), few studies have explored the teratogenic effects of interacting pollutants in aquatic organisms even though the importance of such investigations has been enunciated in mammalian research (Wilson, 1964).

9.5. MUTAGENESIS

The list of aquatic pollutants suspected of being mutagens has increased dramatically in the last three decades. Amongst the more prominent contaminants attributed with mutagenic capabilities are nitrites, certain pesticides, alkylating agents, commercial solvents (Sanders, 1969). Somatic mutations may contribute to premature senescence and may initiate neoplastic growths within an organism's life span. Gametic mutations may be inherited and thereby cause at some point during subsequent generations either death or a reduced potential for survival (Durham and Williams, 1972). The implication that mutagens are adversely affecting the health of contemporary and future generations of aquatic organisms can not be dismissed. Mutagenicity testing should become an integral part of screening programmes for the assessment of the toxic hazard that chemical contaminants represent to aquatic biota.

Definitive mutagenic tests such as specific locus, backcross, and genetic loading studies are tedious, time consuming, and expensive (FDA, 1971). The only reasonable alternative in accomplishing the task of surveying potential mutagens is to employ rapid 'indicator' tests, the significance of which is then extrapolated to field conditions.

The kinds and numbers of indicator tests required for deducing mutagenic activity are still in a stage of research and development. Nevertheless most investigators advocate a multi-component or tier approach which screens for evidence of DNA damage, chromosomal aberrations, metabolite mutagenicity, and mutation transmissibility by gametes to subsequent generations (Flamm, 1974).

For a study of DNA damage, bacterial cultures are often employed on the assumption that genetic material is similar in all organisms and that a chemical which is mutagenic in one species is likely to be mutagenic in others (Ames et al., 1973). Either cell cultures or *in vivo* cytogenetic examination appear effective for chromosomal aberration studies. These investigations are complicated by the fact that chromosomal breakage and repair occurs spontaneously in organisms, the rate of which can be altered by hormones and background radiations (Chu, 1971; Schmid, 1973). Host-mediated assays, in which an organism is administered both a candidate mutagen and then, by another route, indicator bacteria, are carried out to detect the mutagenic action of metabolites while at the same time providing the advantages of a bacterial assay system (Legator and Malling, 1971). Heritability of mutations caused by pollutants can be investigated through dominant lethal tests with the recommendation that studies span several generations (Epstein, 1973). Extrapolations from these tests on microorganisms to whole animals must be made with caution because false negatives and false positives have occurred in all these tests.

The use of bacteria in tests for environmental mutagens and in screening of environmental carcinogens is mentioned in Chapter 13.

A different approach has been developed by Osterman-Golkar (1975) who treated *E. coli* with several alkylating agents and found a good correlation between the degree of alkylation and mutation rate. The amount of *in vivo* alkylation of haemoglobin in mice was used as a method for monitoring the dose of alkylating agent received by the animals (Osterman-Golkar *et al.*, 1976).

9.6. CARCINOGENESIS

The epidemiological studies presently available are not sufficient to allow an assessment of the hazard that chemical pollutants as oncogenic agents represent to the welfare of aquatic organisms. However, given that carcinogens initiate lesions by reacting with DNA (Kotin, 1976) and considering the similarity of genetic material in all organisms, some impression of the impact of carcinogens on aquatic populations may be gained from the extensive data on man. The World Health Organization (WHO, 1964) has estimated that up to 85% of all human tumours are caused by factors related to man's environment. Ninety per cent of these oncogenic factors are considered to be chemicals (Boyland, 1969). The National Institute for Occupational Safety and Health in the U.S.A., as of 1974, implicated more than 1,300 chemical pollutants as candidate carcinogens for man. Their threat to the health of man is immense in view of the fact that twenty five per cent of all deaths in the United States are accredited to cancer (Lassiter, 1976). Because many of the above candidate carcinogens are common contaminants of natural waters, one can only infer that aquatic biota are being jeopardized by pollutants causing cancer (Stich *et al.*, 1975).

There is some evidence to suggest that tumours occur with a higher incidence in aquatic organisms inhabiting polluted, rather than pristine, waters. Brown *et al.* (1973) after an extensive five-year survey found a 4.4% incidence of neoplasms in fish taken from a polluted watershed. This frequency of tissue lesions was significantly different from the 1.03% occurrence of tumours identified in similar fish species captured from uncontaminated waters. In a polluted habitat, catfish, *Ictalurus nebulousus*, which are bottom dwelling and therefore likely to contact high concentrations of chemical contaminants, had a high frequency of hepatomas (12.2%). Harshbarger (1974) also observed the highest incidence of neoplasms (epidermal papillomas and carcinomas) in benthic fish, e.g. catfish, from polluted rivers. Although neoplasms are extremely rare in aquatic invertebrates, particularly molluscs (Wolfe, 1974), sarcomas, with an average incidence of 12%, were detected in oysters, *Ostrea lurida* and mussels, *Mytilus edulis* residing in an estuary receiving pulpmill effluent (Farley, 1974). Carcinomas were fatal to 80% of the afflicted detritus-feeding clams, *Macoma balthica*, found in a watershed receiving agricultural wastes (e.g. pesticides, herbicides, and ammonia).

In all the epidemiological studies listed above, the epizootic occurrence of tumours varied seasonally and the possibility that viruses were acting as oncogenic agents can not be ruled out. Viruses have been associated with the occurrence of fish tumours for years (Nigrelli, 1952; Mawdesley-Thomas, 1971). Mulcahy (1974) suggested a viral cause in malignancies (12.5%) afflicting certain genetic stocks of northern pike, *Esox lucius*, caught in natural waters of Ireland. Nevertheless, chemical carcinogenesis has been demonstrated empirically in fish. For example, salmonoids, more than any other group of animals, are susceptible to the hepatoma-inducing action of aflatoxins (Wales and Sinnhuber, 1972). Khudoley (1972) produced liver tumours in guppies, *Poecilia reticulata*, through exposure to aminoazobenzene derivatives.

One reasonable explanation for these epidemiological results is that the higher tumour incidences were promoted by the combined action of pollutant chemicals and viruses (see section on multiple toxicity). Kirschbaum *et al.* (1940) demonstrated enhanced incidences and growth rates of tumours caused by viruses in the presence of chemicals. Furthermore, Kotin and Wisely (1963) suggested that interactions between viruses that are not considered oncogenic and chemical carcinogens may be important in the occurrence of neoplasms. In any case, future definitive tests for carcinogens must consider the viral background and the natural ecological settings of aquatic organisms.

Ames *et al.* (1973) have devised a rapid, sensitive, and inexpensive test for screening candidate carcinogens on the assumption that most chemical carcinogens

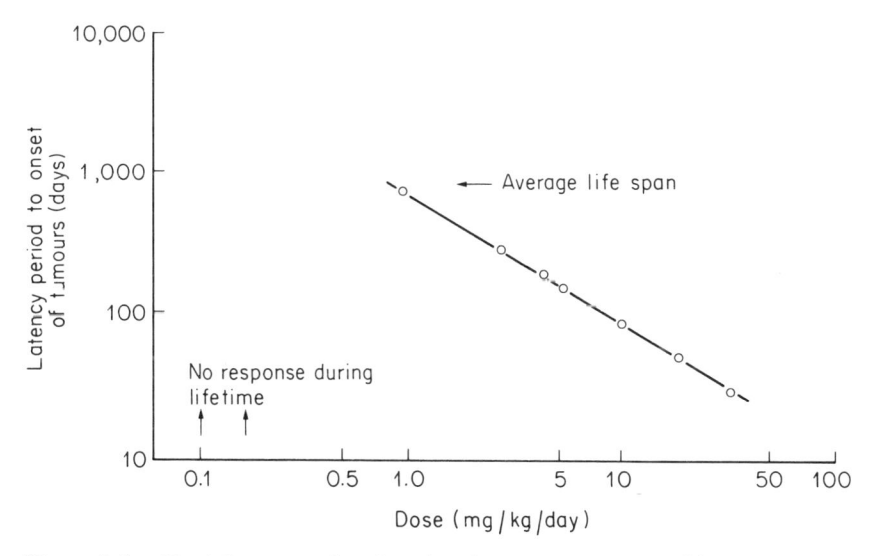

Figure 9.4 Toxicity curve showing the time to occurrence of hepatomas as a function of dose, in rats administered *p*-dimethylaminoazobenzene. (Reproduced with permission from National Academy of Sciences/National Research Council, 1959)

cause tumours by somatic mutation. The test uses a special set of bacterial strains combined with liver homogenates for carcinogen activation. Whether adaptations of this test system would be effective for identifying oncogenic agents in aquatic organisms — given that many tumours may be the consequence of interactions between viruses and chemicals — is a question requiring investigation.

With a knowledge of the dose—response relation for a carcinogen, definitive tests should be designed to employ high dosages with at least one likely to yield a maximum incidence of tumours thereby reducing latency periods to within manageable time periods (Durham and Williams, 1972). Other considerations in the design of definitive tests are the choice of short-lived species, the inclusion of both sexes, the route of administration, the presence of viruses, bacteria and parasites, and the possibilities of cocarcinogens (FDA, 1971).

In accordance with the relation between dose and response depicted in Figure 9.4, Evans (1966) showed that the time of appearance of radium-produced bone cancers in humans was inversely related to dose. When the dose of radium was small enough the latent period was so long that it exceeded the life expectancy. This resulted in an apparent zero incidence of bone cancer; a phenomenon that he called 'practical threshold'.

9.7. MULTIPLE TOXICITY

In receiving waters multicontaminant pollution by the biota appears to be the rule rather than the exception. *In situ* chemical monitoring of aquatic ecosystems has confirmed the virtual ubiquity of pollutant mixtures in both the ambient environment and tissues of organisms (FAO, 1972; Kerr and Vass, 1973). The facts gathered to date show that unique forms of toxicity can be ascribed to the concurrent and sequential exposure of organisms to two or more pollutants (Sprague, 1970).

Types of multiple toxicity particularly hazardous to aquatic fauna are characterized by:

(1) an effect greater than that predicted on the basis of the potency of each component of a mixture;

(2) an effect different from and therefore more toxic or insidious than that predicted by a knowledge of the toxicity of individual constituents of a mixture.

Organisms are not safeguarded against either of these effects by water quality standards which set permissible levels based on single toxicants. Because the movement of pollutants from various diffuse or point sources gives rise to 'coincidental' mixtures in the ecosystem, multiple toxicity risks can not be completely avoided by the application of water quality standards based on bioassay criteria for complex effluents monitored 'at the pipe'. There is a need to understand the mechanisms of multiple toxicity and to derive approaches which quantify the effects. Such knowledge would provide the authorities, who are charged with the responsibility of determining valid water quality criteria and standards, with a rationale that adequately estimates the adverse effects of toxicant mixtures.

Multiple toxicity is the consequence of interaction between constituents of chemical mixtures. This interaction can be either chemical or physiological and can occur at one or more of three phases of pollutant movement and activity in the ecosystem as illustrated in Figure 9.5.

Chemical interaction involves the mutual influence between pollutants that results in, for example, new compounds, complexes, chelates, or modifications in valency. One would expect this form of interaction to occur primarily in the environmental phase, although incidences of chemical combinations are known to occur within the organism (Gaddum, 1957). Physiological interactions can occur in the dynamic phase by altering the sequence of events commencing with, and following from, the binding of a toxicant to the target tissue. Within this sequence,

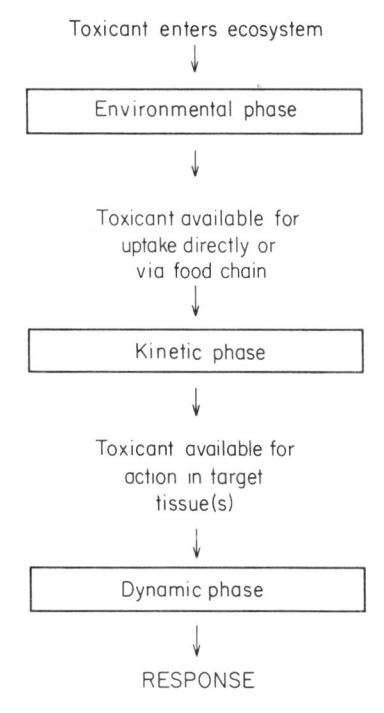

Toxicant enters ecosystem
↓
Environmental phase
↓
Toxicant available for uptake directly or via food chain
↓
Kinetic phase
↓
Toxicant available for action in target tissue(s)
↓
Dynamic phase
↓
RESPONSE

Figure 9.5 Phases of interactions between chemical constituents of pollutant mixtures (modified from Ariens, 1972)

it is useful to distinguish between the processes of affinity (i.e. events in binding with tissue receptors) and intrinsic activity or efficacy (i.e. events initiated by binding). Physiological interactions also occur in the kinetic phase by altering mechanisms of toxicant uptake, distribution, deposition, degradation, and excretion. Kinetic processes determine the concentration of a contaminant or its metabolite(s) available in body compartments, e.g. blood, tissue and excreta, as a function of time and dosage.

The following discussion is limited to multiple toxicity caused by physiological interactions*. A rationale is presented in an attempt to stimulate and guide further research in the field of multiple toxicity.

(i) Presumptive Mechanisms of Interaction

Possible modes of physiological interaction which create hazardous forms of multiple toxicity are proposed to occur as follows:

In the dynamic phase between,

(a) pollutant constituents which act at the same site(s) in target tissue(s);

(b) pollutant constituents which act at different sites possibly in different target tissues but which contribute to a common adverse response;

(c) pollutant constituents where one is normally inactive as a toxic agent but in combination changes the response of an organism to one or more of the other toxic constituents;

(d) pollutant constituents that mutually produce a toxic response different from the response induced by each toxicant alone.

In the kinetic phase between,

(e) pollutant constituents which alter toxicant availability to the target tissue;

(f) pollutant constituents which enhance or induce (e.g. by the mixed oxidase system in liver) the production of metabolites more toxic than the original pollutants.

---

*An understanding of the problems of chemical interaction may be achieved through precise analytical knowledge of pollutant kinds, forms, and quantities that result from reciprocal actions between constituents, either in receiving waters or in target organisms. It is obviously essential to have this information before proceeding with a quantitative and qualitative assessment of physiological interactions.

(ii) Multiple Toxicity Models — based on Interactions at the Dynamic Level

(a) Strict Addition

The simplest form of pollutant interaction occurs when constituents of a pollutant mixture have qualitatively similar toxic effects and these effects combine additively, although the concentration or amount of one or another required to produce a given effect may be quite different. Thus if C_{s1}, C_{s2}, etc., are concentrations of pollutants 1, 2, etc., which produce identical effects, i.e. if

$$E(C_{s1}) = E(C_{s2})$$

where $E(C_{s1})$ = the biological effect of pollutant P_1 at concentraction C_{s1} then it is also true that

$$E(\tfrac{1}{2}C_{s1} + \tfrac{1}{2}C_{s2}) = E(C_{s1}) = E(C_{s2}) \tag{9.1}$$

and more generally

$$E\left(\frac{1}{n}\sum_{j=1}^{n} C_{sj}\right) = E(C_{s1}) = \ldots = E(C_{sn}) \tag{9.2}$$

The concept of 'Toxic Units' (Sprague, 1970) concerns the calculation of a quantity

$$q = \sum_{j=1}^{n} \frac{C_j}{C_{sj}} \tag{9.3}$$

where C_j is the actual concentration of pollutant j, and the comparison of this quantity to unity. If $q = 1$, we would predict that such a combination of toxic substances would lead to the specified effect, even though the concentration of any one may be below toxic levels. This concept has been used to compare multiple-pollutant effects in a variety of systems (Anderson and Weber, 1975). Simply stated, equations (9.1) to (9.3) mean that a similarly acting constituent contributes to its mixtures in proportion to its relative potency. Expressed in another way, the contribution of any constituent to the toxic effect of a mixture may be calculated by multiplying its concentration by its relative potency. Empirically the respective contributions in dosage of each toxicant are added to achieve a quantitative measure of the potency of the mixture. This criterion of dose summation inherent in equations (9.1) to (9.3) has come to be known as strict addition.

Bliss (1939) discussed a way of examining strict additivity in the case where one of the pollutants (in fact, by assumption, all of them) gives a linear graph when probit response was plotted against the logarithm of the dose. In that case one pollutant would have a dose—response curve of the form,

$$Y = a + b \log X_1 \tag{9.4}$$

where

Y = probit of response
X_1 = concentration of pollutant no. 1

If additivity is expected, then a second pollutant could be combined with the first in a proportion π, so that

$$Y = a + b \log[\pi X_1 + (1 - \pi)\rho_{21} X_1] \tag{9.5}$$

that is, an amount π of a solution of pollutant X_1 and an amount $(1 - \pi)$ of a solution of pollutant no. 2 of concentration $X_2 = \rho_{21} X_1$, would have the same effect. If now $\pi = 0$, the equation becomes

$$Y = a + b \log \rho_{21} + b \log X_1 \tag{9.6}$$

i.e. an equation which has the same slope but is laterally shifted. Figure 9.6 illustrates this methodology applied to the multiple toxicity of nickel and copper.

Strict addition may not occur between toxicants which have a similar action. If the response is larger than that expected for the algebraic summation of dosages (i.e. $q > 1$ in equation 9.3), then the effect is termed supra-additive — a form of synergism; if the response is less than predicted for additivity (i.e. $q < 1$ in equation 9.3), then the effect is termed infra-additive — a form of antagonism*. The degree of this displacement from strict addition is often illustrated as isobols, curves for equal biological response, as seen in Figure 9.7. A relative potency factor (q observed/q expected) has been suggested as a measure of the supra- and infra-additive effects (Anderson and Weber, 1975).

(b) Response Addition

Several authors (Bliss, 1939; Plackett and Hewlett, 1952; Finney, 1971) have advanced a quantitative approach for the assessment of such an interaction between pollutant constituents which act at different sites but contribute to a common response. In accordance with their model, toxic constituents, although acting on different systems, may contribute to a common response only if their respective concentrations are equal to or greater than the threshold computed from quantal response curves for single toxicants. This mechanism of physiological interaction is called independent action (Bliss, 1939).

---

*Synergism and antagonism were terms originally synonymous with supra- and infra-additive effects as defined above (Gaddum, 1953). However, through general usage the definitions of these terms have been broadened to mean respectively *any* effects of a mixture greater and lesser than the toxicity or potency estimated on the basis of the effects of its constituents taken individually. To avoid confusion, this chapter will follow the suggestion of Ariens and Simonis (1964) and qualify the type of synergistic and antagonistic effects, e.g. supra-additive synergism.

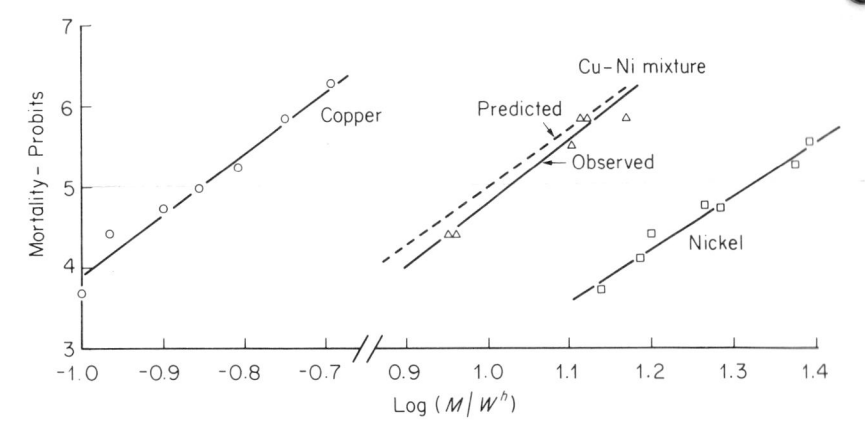

Figure 9.6 Linear regressions for discrete solutions and for mixtures of copper and nickel. The observed toxicity curve for the mixtures is not significantly different from that predicted in accordance with the Bliss (1939) model of similar action. (Reproduced by permission of International Heavy Metals Conference Committee, from Anderson and Weber, 1975)

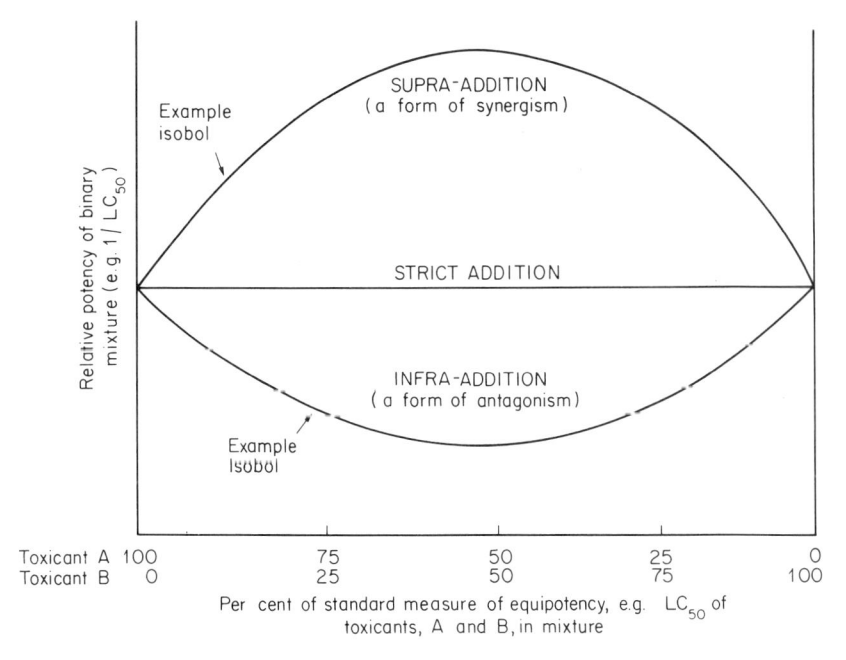

Figure 9.7 Possible types of responses which can occur between two hypothetical toxicants, A and B, which have similar actions. (Reproduced by permission of W. B. Saunders Co., Philadelphia, Pa., from Warren, 1971)

At the outset, it is generally not known whether tolerances (LD_{50}s) to different toxicants are correlated. Therefore, it is not meaningful to attach great significance to parameters, such as regression coefficients, that arise from the study of dose—response curves for one pollutant at a time. One can, however, compare the response to a combination of doses with what would be expected if they acted independently. For a quantal response to two toxicants, this is calculated as follows (P_r = probability):

$$P(\text{response}) = 1 - P_0 \text{ (no response)}$$
$$= 1 - (1 - P_1)(1 - P_2)$$
$$= P_1 + P_2 - P_1 P_2$$

where P_1 and P_2 are the probabilities of exhibiting a response to pollutant 1 or 2 at concentration X_1 or X_2 respectively.

For several toxicants, the expanded form is,

$$P_m = 1 - (1 - P_1)(1 - P_2) \ldots (1 - P_n) \tag{9.7}$$

where

P_m = proportion of individuals responding to a mixture
$P_1, P_2 \ldots P_n$ = proportion of individuals responding to pure solutions of each constituent at concentrations $X_1, X_2 \ldots X_n$ respectively
(Finney, 1971)

To find an example of the application of this method to aquatic organisms, refer to Anderson and Weber (1975). These authors have suggested that the criterion of independent action be called response addition*.

This model makes possible the prediction of effects of mixtures that, by our original definition, are not hazardous. The common effect in each is never greater than that predicted on the basis of the potency of each component of a mixture. In this case water quality standards which establish safeguards against individual, independently acting toxicants also protect organisms against their combined effect in mixtures. Nevertheless equation (9.7) provides a useful quantitative approach for screening combinations of toxicants for those constituents which would not create, through interaction, hazardous forms of multiple toxicity. Further research may identify patterns of independent action by which a mixture's effect is greater than that predicted by equation (9.7).

Another formulation for calculating the total effect of several pollutants in several effluents has been given by Esvelt et al. (1973).

---

*This item is introduced to maintain coherence of thought whereby strict and response addition represent the empirical aspects of the mechanisms of similar and independent action respectively. This terminology avoids inference to a knowledge of mode of action which may be unknown or unattainable.

(c) Sensitization and Potentiation

Theoretically, synergistic interaction can occur as a consequence of a non-toxic pollutant promoting either the binding or the toxic action of another toxic pollutant. Antagonistic interactions occur when either of these is inhibited.

It may be possible to separate, empirically, affinity from intrinsic interactions on the assumption that in the former the non-toxic component acts prior to or concurrently with the toxic agent whereas in the latter the non-toxic agent may act after the binding of the toxicant to its site of action. This temporal distinction is deemed useful for the development of quantitative methods for the assessment of this type of interaction. Ariens (1972) identified synergism due to the former affinity-related interactions as sensitization, and those due to the latter type of interaction as potentiation.

A relative measure of the joint potency of constituents interacting in this manner can be obtained through the use of isobols as demonstrated in Figure 9.8 (Hewlett, 1969).

Over a certain range an increase in the dose of B, the potentiating or sensitizing agent, would result in a corresponding decrease in the dose (M) of toxicant A required to induce a particular standard of effect, e.g. LC_{50}. Eventually, a level of B is reached beyond which no further synergism occurs. As seen in Figure 9.8, this

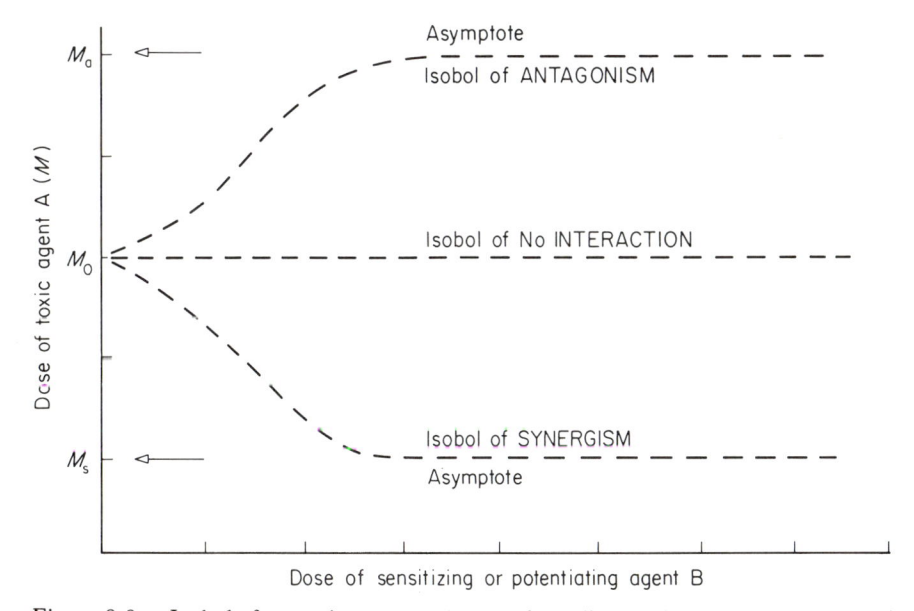

Figure 9.8 Isobols for a mixture consisting of a pollutant A, an active toxicant if applied singly, and a pollutant B, a non-toxicant but which antagonizes or synergizes the response to pollutant A. (Reproduced by permission of The Biometric Society, from Hewlett, 1969)

asymptote allows the computation of a minimum effective dose M_s of toxicant A. The ratio M_0/M_s is a relative measure of the maximum potency of B as a synergistic agent in a mixture containing A. The ratio M_0/M_a measures agent B's potency as an antagonist.

(d) Permissive Synergism

There is recent evidence that pollutants can interact to produce an effect different from those of the individual toxicants. For example, Crocker *et al.* (1974) suggested that Reye's syndrome was an expression of ammonia toxicity generated by the simultaneous or sequential interaction of general toxic agents. The same symptoms of toxicity were not evident in studies of the poisonous constituents taken individually. The proposal was that the ammonia problem arose in the presence of the mixture due to one poison's atttack on the liver which was then unable to detoxify at an adequate rate the increased levels of ammonia generated as a by-product of another toxic agent's action elsewhere in the body. We were unable to find evidence of aquatic studies to represent this model.

(iii) Multiple Toxicity Models – based on Interactions at the Kinetic Level

(e) Uptake Interactions

These interactions can create enhanced, and consequently hazardous, effects through an increased availability of one or more toxicants at their respective target sites. For example, Bingham and Falk (1969) showed a 1,000-fold increase in the potency of the carcinogen, benzo(a)pyrene, when administered in solution in *n*-dodecane rather than in toluene. Similarly, oil solvents enhanced the toxicity of DDT, BHC, toxaphene, and chlordane to aquatic organisms (Cope, 1971). In both instances, the accelerated rate of uptake of the toxicant promoted by the second pollutant augmented the magnitude of response beyond that predicted.

(f) Interactions at Sites of Detoxification

Evidence exists that synergism can occur due to interactions between toxicants at sites of detoxification. Piperonyl butoxide suppresses allethrin denaturation by inhibiting mixed oxidase systems in liver, thus enhancing the toxicity of allethrin (Menzie, 1972). On the other hand, certain combinations of non-carcinogenic and carcinogenic hydrocarbons resulted in a reduced incidence of tumours (Falk *et al.*, 1968). A similar mechanism of antagonism was shown to occur between two carcinogens (Gelboin, 1967). A measure of protection against the lethal toxicity of chlordane and parathion was gained by organisms exposed to mixtures including chlorinated hydrocarbons (Triolo and Coon, 1966). Thus both synergistic and antagonistic interactions can occur at the kinetic level.

(iv) Application of Multiple Toxicity Models

Often, the suggestion that a pollution-related disease is linked to the action of more than one chemical contaminant comes only from extensive epidemiological studies (Koeman and van Genderen, 1972; Brown et al., 1973). This limitation may explain the dearth of knowledge and quantitative methodology relative to most of the models proposed. However this paucity of information should not negate the probable occurrence and hazard of these interactions in the aquatic environment. Below are two examples of the applicability of this rationale to cancer and to water quality management.

(v) Cancer

Kotin (1976) proposed that most, if not all, forms of chemical carcinogenesis are potentially triggered by a contaminant which binds covalently to the genetic material (DNA) of the cell. Subsequent development of neoplastic tissue, however, depends upon a number of intrinsic factors, such as DNA repair mechanisms and abnormal protein production and activity in the cell, as well as extrinsic factors, such as the competence of the immunological response and the nutritional and hormonal state of the organism. Kotin (1976) termed the events involving carcinogen–DNA binding, the initiation phase, and those which followed, the promotion phase. Since the classical study of Berenblum and Shubik (1947), it has been shown that the promotion phase of certain forms of cancer is enhanced by a synergizing agent either administered concurrently with or subsequently to the carcinogen (Weisburger et al., 1965; Bingham et al., 1976). These synergizing agents are collectively called cocarcinogens and are believed not to be carcinogenic in themselves. It appears that certain carcinogen–cocarcinogen relations fit the model of sensitization and potentiation.

If the assumption is correct, then standard measures of the relative potency of cocarcinogens as synergizing agents could be obtained through the use of Hewlett's isobol method. It may also be useful to explore the nature of cocarcinogens as either potentiating or sensitizing agents in accordance with the model. Of interest in the pursuit of this distinction in cocarcinogenic action is knowledge of the degree of specificity which each form displays in this collaboration with carcinogens. The threat of cocarcinogens to the health of the ecosystem is deemed to increase with decreasing specificity of their synergistic action in carcinogenesis.

(vi) Water Quality Management

Of the multiple toxicity models proposed, strict addition and various synergistic forms of interaction meet the original criteria of what constitutes a hazard in the aquatic environment.

The standard based on the toxic unit principle is calculated as follows, (Esvelt et al., 1973)

$$T_{cr} = \frac{T_{c1}Q_1 + T_{c2}Q_2 + \ldots T_{cn}Q_n}{Q_t} = 0.05 \text{TU} \qquad (9.8)$$

where

T_{cr} = total allowable relative concentration in receiving waters

$T_{c1} - T_{cn} = \dfrac{C_{actual}}{LC_{50}}$ for each contaminant respectively

0.05 = application factor

Q_t = total effluent flow = sum of $Q_1, Q_2 \ldots Q_n$

TU = toxic unit = 1 (Seba, 1975)

and may prove to be satisfactory as a measure of protection for the health of aquatic biota in receiving waters of strictly additive toxicants. However, the eventual applicability of this standard requires knowledge supporting its inference that toxic constituents which obey the strictly additive criterion at the lethal level interact by the same mechanism at the sublethal level.

Some authors (Sprague, 1970; Seba, 1975) have proposed that strict addition methodology adequately quantifies the multiple toxicity of most pollutant mixtures. However, even a cursory survey of toxicology literature reveals a diversity of effects — and presumably modes of action — between poisonous pollutants studied singly. One might reasonably expect that this complexity would be compounded for the toxicity of mixtures and that the adverse effects of combinations of toxicants could not be simply characterized by the mechanism of similar action (i.e. strict addition). In fact, recent investigations of the effects of pollutant mixtures on aquatic biota report a variety of lethal and sublethal response patterns indicating not only strict addition but also response addition, synergisms, and antagonisms, (Sprague, 1970; LaRoche et al., 1973; Halter and Johnson, 1974; Roales and Perlmutter, 1974; Anderson and Weber, 1975, 1976; Hutchinson and Czyrska, 1975; Macek, 1975; Sellers et al., 1975; Reinbold and Metcalf, 1976; Statham and Lech, 1976; Herbes and Beauchamp, 1977). These latter reports do not support reliance on a single water quality standard for the safeguarding of organisms against toxic mixtures. Additional research is required, particularly in the area of synergistic interactions.

9.8. REFERENCES

Abel, P. D., 1974. Toxicity of synthetic detergents to fish and aquatic invertebrates. *J. Fish Biol.*, **6**, 279–98.

Ames, B. N., Lee, F. D., and Durston, W. E., 1973. An improved bacterial test system for the detection and classification of mutagens and carcinogens. *Proc. U.S. Natl. Acad. Sci.*, **70**, 782–6.

Anderson, J. M., 1971. Assessment of the effects of pollutants on physiology and behaviour. *Proc. Roy. Soc. London*, B**177**, 307–20.

Anderson, P. D. and Battle, H. I., 1967. Effects of chloramphenicol on the development of the zebrafish, *Brachydanio rerio*. *Can. J. Zool.*, **45**, 191–204.

Anderson, P. D. and Weber, L. J., 1975. The toxicity to aquatic populations of mixtures containing certain heavy metals. *Proc. International Conference on Heavy Metals in the Environment*, **2**, 933–53. Institute for Environ. Studies, Univ. of Toronto, Toronto.

Anderson, P. D. and Weber, L. J., 1976. The multiple toxicity of certain heavy metals: additive actions and interactions. In *Proc. of Workshop on Toxicity to Biota of Metal Forms in Natural Water* (Eds. R. W. Andrew, P. V. Hodson, and D. E. Konasewich), Publ. Int. Joint. Comm. Res. Ad. Bd. Windsor, Canada, pp. 263–82.

Ariens, E. J., 1972. Adverse drug interactions. *Proc. Eur. Soc. Study Drug Tox.*, **13**, 137–63.

Ariens, E. J. and Simonis, A. M., 1964. A molecular basis for drug action. *J. Pharm. Pharmacol.*, **16**, 137–57.

Balinsky, B. I., 1975. *An Introduction to Embryology*, 4th ed., W. B. Saunders Co., Philadelphia, U.S.A., 648 pp.

Bardach, J. E., Fujiya, M., and Hall, A., 1965. Detergents: effects on the chemical senses of the fish *Ictalurus malates* (le Sueur). *Science, N.Y.*, **148**, 1605–7.

Bengtsson, B. E., 1974. The effects of zinc on the ability of the minnow, *Phoxinus phoxinus* L., to compensate for torque in a rotating water-current. *Bull. Environ. Contam. Toxicol.*, **12**, 654–8.

Benoit, D. A., 1975. Chronic effects of copper on survival growth and reproduction of the bluegill (*Lepomis macrochirus*). *Trans. Am. Fish Soc.*, **104(2)**, 353–8.

Berenblum, I. and Shubik, P., 1947. A new quantitative approach to the study of the stages of chemical carcinogenesis on the mouse skin. *Brit. J. Cancer*, **1**, 383–91.

Bingham, E. and Falk, H., 1969. Environmental carcinogenesis. Threshold concentrations of carcinogens and cocarcinogens. *Arch. Environ. Health*, **19**, 779–83.

Bingham, E., Niemeier, R. W., and Reid, J. B., 1976. Multiple factors in carcinogenesis. *Ann. N.Y. Acad. Sci.*, **271**, 14–21.

Blaxhall, P. C. and Daisley, K. W., 1973. Routine haematological methods for use with fish blood. *J. Fish. Biol.*, **5**, 771–81.

Bliss, C. I., 1939. The toxicity of poisons applied jointly. *Ann. Appl. Biol.*, **36**, 385–615.

Boyland, E., 1969. The correlation of experimental carcinogens and cancer in man. *Prog. Exp. Tumour Res.*, **11**, 222–34.

Brett, J. R., 1967. Swimming performance of sockeye salmon (*Oncorhynchus nerka*) in relation to fatigue time and temperature. *J. Fish. Res. Bd. Canada*, **24**, 1731–41.

Brown, E. R., Hazdra, J. S., Keith, L., Greenspan, I., Kwapinski, J. B. G., and Beamer, P., 1973. Frequency of fish tumours found in a polluted watershed as compared to nonpolluted Canadian waters. *Cancer Res.*, **33**, 189–98.

Brown, U. N., 1973. Concepts and outlook in testing the toxicity of substances to fish. In *Bioassay Techniques and Environmental Chemistry* (Ed. E. Glass) Ann Arbor Science Publishers Inc., Ann Arbor, Mich., pp. 73–95.

Brungs, W. A., 1969. Chronic toxicity of the fathead minnow, *Pimephalus promelas* (Rafinesque). *Trans. Am. Fish Soc.*, **98**, 272–9.

Buckley, J. A., Whitmore, C. M., and Matsuda, R. I., 1976. Changes in blood chemistry and blood cell morphology in Coho salmon (*Oncorhynchus kisutch*) following exposure to sublethal levels of total residual chlorine in municipal wastewater. *J. Fish. Res. Bd. Canada*, 33, 776–82.

Cairns, J. Jr. and Sparks, R. E., 1971. The use of bluegill breathing to detect zinc. Environment Protection Agency, *Water Pollution Control Research Series 18050 EDQ*, 45 pp.

Cech, J. J. Jr., Bridges, D. W., Rowel, D. M., and Balzer, P. J., 1976. Cardiovascular responses of winter flounder, *Pseudopleuronectes americanus* (Walbaum), to acute temperature increase. *Can. J. Zool.*, 54, 1383–8.

Cech, J. J. Jr. and Wohlschlag, D. E., 1973. Respiratory responses of the striped mullet, *Mugil cephalus* (L.) to hypoxic conditions. *J. Fish. Biol.*, 5, 421–8.

Chu, E. H. Y., 1971. Induction and analysis of gene mutations in mammalian cells in culture. In *Chemical Mutagens, Principles and Methods for their Detection* (Ed. A. Hollaender), Vol. 2, Plenum Press, New York, pp. 411–44.

Cope, O. B., 1971. Interactions between pesticides and wildlife. *Ann. Rev. Ent.*, 16, 325–63.

Crocker, J. F. S., Ozere, R. L., Rozee, K. R., Digout, S. C., and Hutzinger, O., 1974. Insecticide and viral interactions as a cause of fatty visceral changes and encephalopathy in the mouse. *Lancet*, 2, 22–4.

Davis, C. C., 1972. The effects of pollutants on the reproduction of marine organisms. In *Marine Pollution and Sea Life* (Ed. M. Ruivo), FAO, Fishing News (Books) Ltd., London, England, pp. 305–11.

Davis, J. C., 1973. Sublethal effects of bleached kraft pulp mill effluent on respiration and circulation in Sockeye salmon (*Oncorhynchus nerko*). *J. Fish. Res. Bd. Canada*, 30, 369–77.

Davis, J. C. and Randall, D. J., 1973. Gill irrigation and pressure relationships in rainbow trout *(Salmo gairdneri)*. *J. Fish. Res. Bd. Canada*, 30, 99–104.

Davis, P. W. and Wedemeyer, G. A., 1971. NaK-activated ATPase inhibition in rainbow trout: a site for organochlorine pesticide toxicity? *Comp. Biochem. Physiol.*, 40B, 823–7.

Davy, F. B., Kleerekoper, H., and Matis, J. H., 1973. Effects of exposure to sublethal DDT on the exploratory behaviour of goldfish *(Carassius auratus)*. *Water Resources Res.*, 9, 900–5.

Durham, W. F. and Williams, C. H., 1972. Mutagenic, teratogenic and carcinogenic properties of pesticides. *Am. Rev. Ent.*, 17, 123–48.

Eisler, R., 1972. Pesticide-induced stress profiles. In *Marine Pollution and Sea Life* (Ed. M. Ruivo) FAO, Fishing News (Books) Ltd., London, England, pp. 229–33.

Epstein, S. S., 1973. The use of dominant-lethal test to detect genetic activity of environmental chemicals. *Environ. Hlth. Perspectives*, 6, 23–7.

Esvelt, L. A., Kaufman, W. J., and Selleck, R. E., 1973. Toxicity assessment of treated municipal wastewaters. *J. Water Poll. Control Fed.*, 45, 1558–72.

Evans, R. D., 1966. The effect of skeletally deposited α-ray emitters in man. *Br. J. Radiol.*, 39, 881–95.

Falk, H. L., Kotin, P., and Thompson, S., 1968. Inhibition of carcinogenesis: the effect of polycyclic hydrocarbons and related compounds. *Arch. Environ. Health*, 9, 169–79.

FAO., 1972. Food and Agriculture Organization of the United Nations. *Marine Pollution and Sea Life* (Ed. M. Ruivo), Fishing News (Books) Ltd., London, England, 624 pp.

Farley, C. A., 1974. Epizootic neoplasia in bivalve mollusks. *Abs. Symp. Conf. XIth Int. Cancer Congress, Florence*, pp. 219–20.

FDA, 1970. Food and Drug Administration Advisory Committee on Protocols for Safety Evaluation: panel on reproduction report on reproduction. Studies in the safety evaluation of food additives and pesticide residues. *Toxicol. Appl. Pharmacol.*, 16, 264–96.

FDA, 1971. Food and Drug Administration Advisory Committee on Protocols for Safety Evaluation: panel on carcinogenesis report on cancer testing in the safety evaluation of food additives and pesticides. *Toxicol. Appl. Pharmacol.*, 20, 419–38.

Finney, D. E., 1971. *Probit Analysis*, 3rd edn., Cambridge Univ. Press, London, England, 333 pp.

Flamm, W. G., 1974. A tier system approach to mutagen testing. *Mutat. Res.*, 26, 329–33.

Fry, F. E. J., 1947. Effects of the environment on animal activity. Univ. Toronto Studies Biological Series 55, *Ontario Fish Res. Laboratory Publ.*, 68, 62 pp.

Gaddum, J. H., 1953. *Pharmacology*, 4th edn., Oxford Univ. Press, London, 562 pp.

Gaddum, J. H., 1957. Drug antagonism. *Pharmacol. Rev.*, 9, 211–8.

Gelboin, H. V., 1967. Carcinogen, enzyme induction and gene action. *Adv. Cancer Res.*, 10, 1–81.

Giles, M. and Vanstone, W. E., 1976. Changes in oubain-sensitive adenosine triphosphatase activity in gills of Coho salmon (*Oncorhynchus kisutch*) during parr-smolt transformation. *J. Fish. Res. Bd. Canada*, 33, 54–62.

Halter, M. T. and Johnson, H. E., 1974. Acute toxicity of a polychlorinated biphenyl and DDT alone and in combination to early life stages of Coho salmon, *Oncorhynchus kisutch*. *J. Fish. Res. Bd. Canada*, 31(9), 1543–7.

Hansen, D. J., Matthews, E., Nall, S. L., and Dumas, D. P., 1972. Avoidance of pesticides by untrained mosquitofish, *Gambusia affinis*. *Bull. Environ. Contam. Toxicol.*, 8, 46–51.

Hara, T. J., Law, Y. M. C., and MacDonald, S., 1976. Effects of mercury and copper on the olfactory response in rainbow trout *Salmo gairdneri*. *J. Fish. Res. Bd. Canada*, 33, 1568–73.

Harshbarger, J. C., 1974. Integumentary papillomas and carcinomas in fish. *Abs. Symp. Conf. XIth Int. Cancer Congress, Florence*, pp. 218–9.

Hatfield, C. T. and Johansen, P. H., 1972. Effects of four insecticides on the ability of Atlantic salmon parr (*Salmo salar*) to learn and retain a simple conditioned response. *J. Fish. Res. Bd. Canada*, 29(3), 315–21.

Heath, A. G., 1972. A critical comparison of methods for measuring fish respiratory movements. *Water Res.*, 6, 1–7.

Herbes, J. F. and Beauchamp, J. J., 1977. Toxic interaction of mixtures of 2 coal conversion effluent components (resorcinol and 6-methyl quinoline) to *Daphnia magna*. *Bull. Environ. Contam. Toxicol.*, 17(1), 25–32.

Hewlett, P. S., 1969. Measurement of the potencies of drug mixtures. *Biometrics*, 25, 477–87.

Hoar, W. S., 1975. *General and Comparative Physiology*, 2nd edn., Prentice-Hall Inc., New Jersey, 848 pp.

Houston, A., Madden, J. A., Woods, R. S., and Miles, H. M., 1971. Some physiological effects of handling and TMS anesthetization upon brooktrout *Salvelinis fontinalis*. *J. Gen. Physiol.*, 56, 342–58.

Howard, T. E., 1975. Swimming performance of juvenile Coho salmon (*Oncorhynchus kisutch*) exposed to bleached kraft pulpmill effluent. *J. Fish. Res. Bd. Canada*, 32, 789–93.

Hutchinson, T. C. and Czyrska, H., 1975. Heavy metal toxicity and synergism to floating aquatic weeds. *Verh. Internat. Verein. Limnol.*, 19, 2102–11.

Ishio, S., 1965. Behaviour of fish exposed to toxic substances. In *Advances in Water Pollution Research* (Ed. O. Jaag), Vol. 1, Pergamon Press, London, pp. 19–33.

Janicki, R. H. and Kinter, W. B., 1971. DDT inhibits NaK Mg-ATPase in the intestinal mucosae and gills of marine teleosts. *Nature New Biol.*, 233, 148–9.

Kerr, S. R. and Vass, W. P., 1973. Pesticide residues in aquatic invertebrates. In *Environmental Pollution by Pesticides* (Ed. C. A. Edwards), Plenum Press, London, pp. 134–80.

Khudoley, V. V., 1972. Induction of liver tumours in aquarium guppies with some nitrogen compounds. *J. Ichthyol.*, 12, 319–24.

Kirschbaum, A., Strong, L. C., and Gardner, W. U., 1940. Influence of methyl cholanthrene on age incidence of leukemia in several strains of mice. *Proc. Soc. Exptl. Biol. Med.*, 45, 287–9.

Koeman, J. H. and van Genderen, H., 1972. Tissue levels in animals and effects caused by chlorinated hydrocarbon insecticides, chlorinated biphenyls and mercury in the marine environment along the Netherlands coast. In *Marine Pollution and Sea Life* (Ed. M. Ruivo), FAO, Fishing News (Books) Ltd., London, England.

Kotin, P., 1976. Dose response relationship and threshold concepts. *Ann. N.Y. Acad. Sci*, 271, 22–8.

Kotin, P. and Wisely, A. V., 1963. Production of lung cancer in mice by inhalation exposure to influenza virus and aerosols of hydrocarbons. *Prog. Exp. Tumour Res.*, 3, 186–215.

LaRoche, G., Gardner, G. R., Eisler, R., Jackim, E. H., Yevish, P. P., and Zaroogian, G. E., 1973. Analysis of toxic response in marine poikilotherms. In *Bioassay Techniques and Environmental Chemistry* (Ed. G. E. Glass), Ann Arbor Science Publishers Inc., Ann Arbor, Mich., pp. 199–216.

Lassiter, D. V., 1976. Prevention of occupational cancer – toward an integrated program of governmental action. *Ann. N.Y. Acad. Sci.*, 271, 214–5.

Legator, M. S. and Malling, H. V., 1971. The host-mediated assay, a practical procedure for evaluating potential mutagenic agents in mammals. In *Chemical Mutagens, Principles and Methods for their Detection* (Ed. A. Hollaender), Vol. 2, Plenum Press, New York, N.Y., pp. 569–89.

Lesniak, J., 1977. Histological approach to the sublethal cyanide effects on rainbow trout ovaries. *M.Sc. Thesis*, Concordia University, Montreal, 120 pp.

Lett, P. F., Farmer, G. J., and Beamish, F. W. H., 1976. Effect of copper on some aspects of the bioenergetics of rainbow trout *(Salmo gairdneri). J. Fish. Res. Bd. Canada*, 33, 1335–42.

Lewis, S. D. and Lewis, W. M., 1971. The effect of zinc and copper on the osmolarity of blood serum of the channel catfish, *Ictalurus punctatus*, Rafinesque and golden shiner, *Notemigonus crysoleucas* Mitchill. *Trans. Am. Fish. Soc.*, 100, 639–43.

Lindahl, P. E., Olopson, S., and Schwanbom, E., 1977. Rotary-flow technique for testing fitness of fish. In *Biological Monitoring of Water and Effluent Quality* (Eds. J. Cairns, K. L. Dickson, and G. F. Westlake), ASTM, pp. 75–84.

Lindahl, P. E. and Schwanbom, E., 1971a. A method for the detection and

quantitative estimation of sublethal poisoning in living fish. *Oikos*, **22**(2), 210–4.
Lindahl, P. E. and Schwanbom, E., 1971b. Rotatory-flow technique as a means of detecting sublethal poisoning in fish populations. *Oikos*, **22**(3), 354–7.
Lloyd, R. and Orr, L. D., 1969. The diuretic responses by rainbow trout to sublethal concentrations of ammonia. *Water Res*, **3**, 335–44.
Loomis, T. A., 1974. *Essentials of Toxicology*, 2nd edn., Lea and Febiger, Philadelphia, 223 pp.
Lunn, R. R., Toews, D. P., and Pree, D. J., 1976. Effects of three pesticides on respiration, coughing, and heart rates of rainbow trout (*Salmo gairdneri* Richardson), *Can. J. Zool.*, **54**, 214–9.
Macek, K. J., 1975. Acute toxicity of pesticide mixtures to bluegills. *Bull. Environ. Contam. Toxicol.*, **14**(6), 648–52.
Mawdesley-Thomas, L. E., 1971. Neoplasia in fish: a review. In *Current Topics in Comparative Pathology* (Ed. T. C. Cheng), Vol. 1, pp. 87–170.
McKee, J. E. and Wolf, H. W. (Eds.), 1963. *Water Quality Criteria*, 2nd edn., Publ. 3-A of Resources agency of the California State Water Control Board, Sacramento, California.
McKim, J. M. and Benoit, D. A., 1971. Effects of long-term exposures to copper on survival, growth and reproduction of brook trout, *Salvelinus fontinalis*. *J. Fish. Res. Bd. Canada*, **28**(5), 655–62.
McKim, J. M., Christensen, G. M., and Hunt, E. P., 1970. Changes in the blood of brook trout (*Salvelinus fontinalis*) after short-term and long-term exposure to copper. *J. Fish. Res. Bd. Canada*, **27**, 1883–9.
McLeay, D. J., 1973. Effects of 12 hr and 25 day exposure to kraft pulpmill effluent on the blood and tissue of juvenile Coho salmon. *J. Fish. Res. Bd. Canada*, **30**, 395–400.
Menzie, C. M., 1972. Fate of pesticides in the environment. *Ann. Rev. Ent.*, **17**, 199–223.
Mitrovic, V. V., 1972. Sublethal effects of pollutants on fish. In *Marine Pollution and Sea Life* (Ed. M. Ruivo), FAO, Fishing News (Books) Ltd., London, pp. 252–7.
Mount, D. I. and Stephen, E. C., 1969. Chronic toxicity to the fathead minnow, *Pimephales promelas* in soft water. *J. Fish. Res. Bd. Canada*, **26**(9), 2449–57.
Mulcahy, M. F., 1974. Epizootic neoplasia of northern pike, *Esox lucius* L. in Ireland. *Abs. Symp. Conf. XIth Int. Cancer Congress, Florence*, **218**.
National Academy of Sciences. National Research Council: Food Protection Committee, 1959, *Problems in the Evaluation of Carcinogenic Hazard from Use of Food Additives*, Publ. 749, Washington, D.C.
Neil, J. H., 1957. Some effects of potassium cyanide on speckled trout (*Salvelinus fontinalis*). *Proc. 4th Ontario Ind. Waste Conf. Ontario Water Res. Comm.*, pp. 74–96.
Nigrelli, R. F., 1952. Virus and tumours in fishes. *Ann. N.Y. Acad. Sci.*, **54**, 1076–92.
O'Conner, D. V. and Fromm, P. O., 1975. The effect of methyl mercury on gill metabolism and blood parameters of rainbow trout. *Bull. Environ. Contam. Toxicol*, **13**, 406–11.
Oseid, O. and Smith, L. L. Jr., 1972. Swimming endurance and resistance to copper and malathion of bluegills treated by long-term exposure to sublethal levels of hydrogen sulfide. *Trans. Am. Fish. Soc.*, **101**, 620–5.
Osterman-Golkar, S., 1975. Studies on the reaction kinetics of biologically active

electrophilic reagents as a basis for risk elements. *Doctorate Thesis*, University of Stockholm, Sweden, 55 pp.

Osterman-Golkar, S., Ehrenberg, L., Segerbäck, D., and Hällström, I., 1976. Evaluation of genetic risks of alkylating agents. II. Haemoglobin as a dose monitor. *Mutat. Res.*, **34**, 1–10.

Peterson, R. H., 1974. Influence of fenitrothion on swimming velocities of brook trout (*Salvelinus fontinalis*). *J. Fish. Res. Bd. Canada*, **31**, 1757–62.

Pfuderer, P. and Francis, A. A., 1973. Effects of phthalate esters on heart rate of goldfish. *Fed. Proc.*, **32**, 224.

Plackett, R. L. and Hewlett, P., 1952. Quantal responses to mixtures of poisons. *J. Roy. Stat. Soc.* **B14**, 151–4.

Reinbold, K. A. and Metcalf, R. L., 1976. Effects of the synergist piperonyl butoxide on metabolism of pesticides in green sunfish. *Pestic. Biochem. Phys.*, **6(5)**, 401–12.

Reish, D. V., 1972. The use of marine invertebrates as indicators of varying degrees of marine pollution. In *Marine Pollution and Sea Life* (Ed. M. Ruivo), FAO, Fishing News (Books) Ltd., London, England, pp. 203–7.

Renfro, J. L., Schmidt-Nielsen, B., Miller, D., Benos, D., and Allen, J., 1973. Methyl mercury and inorganic mercury: uptake, distribution and effect on osmoregulatory mechanisms in fish. In *Pollution and Physiology of Marine Organisms* (Eds. F. J. Vernberg and W. B. Vernberg), Academic Press, New York, pp. 59–65.

Roales, R. R. and Perlmutter, A., 1974. Toxicity of zinc and cygon applied singly and jointly to zebra fish embryos. *Bull. Environ. Contam. Toxicol.* **12(4)**, 475–80.

Rosenthal, H. and Alderdice, D. F., 1976. Sublethal effects of environmental stressors, natural and pollutional, on marine fish eggs and larvae. *J. Fish. Res. Bd. Canada*, **33**, 2047–65.

Sanders, H. J., 1969. Chemical mutagens: an expanding roster of suspects. *Bull. Chem. and Engin.*, June, pp. 54–68.

Saunders, N. L. and Sprague, J. B., 1967. Effects of copper-zinc mining pollution on a spawning migration of Atlantic salmon. *Water Res.*, **1**, 419–32.

Schaumberg, F. D., Howard, T. E., and Walden, C. C., 1967. A method of evaluating the effects of water pollutants on fish respiration. *Water Res*, **1**, 731–5.

Scherer, E., 1975. Avoidance of fenitrothion by goldfish, *Carassius auratus*. *Bull. Environ. Contam. Toxicol.*, **13**, 492–6.

Schmid, W., 1973. Chemical mutagen testing on *in vivo* somatic mammalian cells. *Agents and Actions*, **3**, 77–85.

Seba, D. B., 1975. Toxicity index for permits. In *Proc. Symposium on Structure–Activity Correlations in Studies of Toxicity and Bioaccumulation with Aquatic Organisms* (Eds. G. D. Veith and D. E. Konasewich), Publ. of I. J. C. Res. Ad. Bd. Windsor, Ontario, pp. 199–259.

Sellers, C. M. Jr., Heath, A. G., and Bass, M. L., 1975. The effect of sublethal concentrations of copper and zinc on ventilatory activity, blood oxygen and pH in rainbow trout (*Salmo gairdneri*). *Water Res.*, **9**, 401–8.

Shelton, G., 1971. The regulation of breathing. In *Fish Physiology* (Eds. W. S. Hoar and D. J. Randall), Vol. IV, Academic Press, New York, pp. 293–359.

Skidmore, J. F., 1970. Respiration and osmoregulation in rainbow trout with gills damaged by zinc sulfate. *J. Exp. Biol.*, **52**, 481–94.

Sprague, J. B., 1968. Avoidance reactions of rainbow trout to zinc sulphate solutions. *Water Res.*, **2**, 367–72.

Sprague, J. B., 1970. Measurement of pollutant toxicity to fish. II. Utilizing and applying bioassay results. *Water Res.*, **4**, 3–32.
Sprague, J. B., 1971. Measurement of pollutant toxicity to fish. III. Sublethal effects and 'safe' concentrations. *Water Res.*, **5**, 245–66.
Sprague, J. B., 1976. Current status of sublethal tests of pollutants on aquatic organisms. *J. Fish. Res. Bd. Canada*, **33**, 1988–92.
Statham, C. N. and Lech, J. J., 1976. Studies on the mechanisms of potentiation of the acute toxicity of 2-4-D-n-butyl ester and 2'-5-dichloro-4'-nitrosalicylanilide in rainbow trout by carbaryl. *Tox. Appl. Pharm.*, **36**(2), 281–96.
Stich, H. F., Acton, A. B., and Dunn, B. P., 1975. Carcinogens in estuaries, their monitoring and possible hazard to man. *Paper given at Symposium on Environmental Pollution and Carcinogenic Risks*, Lyon, France.
Swift, D. J. and Lloyd, R., 1974. Changes in urine flow rate and haematocrit value of rainbow trout, *Salmo gairdneri* (Richardson) exposed to hypoxia. *J. Fish. Biol.*, **6**, 379–87.
Triolo, A. J. and Coon, J. M., 1966. Toxicologic interactions of chlorinated hydrocarbon and organophosphate insecticides. *J. Agr. Food Chem.*, **14**, 544–55.
Vollenweider, R. A. and Dillon, P. J., 1974. The application of the phosphorus loading concept to eutrophication research. *National Research Council of Canada Report*, #13690, 42 pp.
Wales, J. H. and Sinnhuber, R. O., 1972. Hepatomas induced by aflatoxin in the sockeye salmon *(Oncorhynchus nerka)*. *Nat. Cancer Inst.*, **48**, 1529–30.
Warner, R. E., Peterson, K. K., and Borogman, L., 1966. Behavioural pathology in fish. *J. Applied Ecol.*, **3** (Suppl.), 223–47.
Warren, C. E., 1971. *Biology and Water Pollution Control*, W. B. Saunders, Co., Philadelphia, U.S.A., 434 pp.
Warren, C. E. and Davis, G. E., 1971. Laboratory stream research: objectives, possibilities and constraints. *Am. Rev. Eco. System*, **2**, 111–44.
Webb, P. W. and Brett, L. R., 1973. Effects of sublethal concentrations of sodium pentachlorophenate on growth rate, food conversion efficiency and swimming performance of under yearling sockeye salmon *(Oncorhynchus nerka)*. *J. Fish. Res. Bd. Canada*, **30**, 499–507.
Wedemeyer, G., and Chatterton, K., 1970. Some blood chemistry values for the rainbow trout *(Salmo gairdneri)*. *J. Fish. Res. Bd. Canada*, **27**, 1162–4.
Weisburger, J. H., Hadidian, Z., Frederickson, T. N., and Weisburger, E. K., 1965. Carcinogenesis by simultaneous action of several agents. *Tox. Appl. Pharm.*, **7**, 502.
Westlake, G. F. and Kleerekoper, H., 1974. The locomotor response of goldfish to a steep gradient of copper ions. *Water Resources Res.*, **10**, 103–5.
WHO (World Health Organization), 1964. *Prevention of Cancer Tech. Rept. Ser.*, #276, Geneva, Switzerland.
Wilson, J. G., 1964. Teratogenic interaction of chemical agents in the rat. *J. Pharm. Exp. Therapeut.*, **144**, 429–36.
Wolfe, P. H., 1974. Integumentary neoplasms in oysters. *Abs. Symp. Conf. XIth Int. Cancer Congress, Florence*, **219**.

CHAPTER 10

Terrestrial Plants and Plant Communities

M. TRESHOW

Department of Biology, University of Utah, Salt Lake City, Utah 84112, U.S.A.

10.1	INTRODUCTION	223
10.2	TOXICANT UPTAKE	224
10.3	RESPONSES OF COMMUNITIES TO PERTURBATION	225
	(i) Physiological responses	225
	(ii) Responses of organisms	226
	(iii) Responses of plant communities	227
	(iv) Agricultural systems	231
10.4	DIAGNOSING COMMUNITY DYSFUNCTION	233
	(i) Diagnosing changes	233
	(ii) Predicting changes	235
10.5	CONCLUSIONS	235
10.6	REFERENCES	236

10.1. INTRODUCTION

A terrestrial plant community is best described by its composition and dynamics. It is most important to know for each plant species its numbers, the area covered by it, its rate of reproduction, and its longevity. It is also important to know how soon adult plants are replaced by their young or other species and the frequency with which one species occurs relative to others in the community. All these characteristics used to describe the normal composition and dynamics of the community (sometimes called the baseline condition) are influenced by the qualities of the environment – humidity, temperature, sunlight, soil composition, and the presence or absence of toxicants.

Normally, a plant community is relatively stable and displays an equilibrium condition that changes only slowly. It also tends to be resilient in the sense that short-term perturbations to the system do not persist. But prolonged periods of stress, e.g. due to extremes of climate, may reduce the growth of sensitive species; the more tolerant species then become predominant.

When a toxic agent is introduced into a plant community (often in the air), some species will be affected more adversely than others. These sensitive species may be no longer able to compete successfully for their place in the system, and disappear. High levels of toxic agent or prolonged exposure may destroy large numbers of species with consequent disruption to the plant community. Ultimately the system may survive or recover but the number of species and the amount of ground cover will be greatly reduced.

The response of plants to pollutants is especially important since many of them are more sensitive than are other organisms. Thus they can be the most sensitive indicator, or bioindicator, for the presence of toxic pollutants.

10.2. TOXICANT UPTAKE

Plants may absorb toxicants either directly from the atmosphere, through the leaves, or from the soil or water through the roots.

The usual pathway is through the leaves, as with gaseous chemicals including principally the sulphur and nitrogen oxides, photochemical pollutants, fluoride, chlorine, and ammonia. Other chemicals, present in the atmosphere in particulate forms, may be impacted onto the leaf surface but rarely enter the leaf unless dissolved. Such toxicants include heavy metals such as lead, zinc, cadmium, copper, and nickel.

Lead is known to accumulate on plants growing alongside highways in proportion to the traffic density. The lead content of roadside atmospheres may be 2 to 20 times that of non-roadside atmospheres (Smith, 1976) and the concentration declines with distance from the roads. Sedimentation from the atmosphere contaminates both soil and vegetation. Lead in the top 5 cm of soil may be 30 times that in non-roadside soil. Although concentrations of lead in or on plants may reach high levels, there are no reports of injurious effects on vegetation, nor is lead taken up by plants from the soil.

Particulate materials falling on plant surfaces also include dusts emitted from the kilns of cement plants, soot, and miscellaneous matter from various types of metal processing (Lerman and Darley, 1975). Fall-out of cement kiln dusts may range from 1.5 to 3.8 g per m^2 per day near cement plants equipped with multiple (cyclone emission) control devices. The dusts form a crust in the presence of moisture thus plugging stomata and adversely affecting photosynthesis and growth. Leaf drop is often accelerated. The amount of dust causing such effects has not been established but is somewhat over 1.0 g/m^2/day.

Toxicants present in the soil may or may not move into the plant depending on their solubility in the soil and their absorption by the roots. Photochemical pollutants do not accumulate in the soil, but sulphur and nitrogen oxides are accumulated in both plants and soil. Since nitrogen and sulphur are essential to normal plant growth, they are rarely present in excess, although they may be harmful indirectly.

Sulphur analysis of leaf tissues has been used as an index of exposure to pollution to help diagnose suspected injury, but the sulphur content of leaves is extremely variable and unreliable as a measure of contamination. The normal foliar content is generally in the range of 0.1 to 0.2 per cent of the dry weight, but may range as high as 0.4 per cent in the complete absence of any pollution. Sulphur is extremely mobile in the plant. The sulphur content depends on the tissue analysed, physiology of the species being studied, its metabolic activity, the content and availability of sulphur in the soil, the age of the tissues, and even the time of day.

Fluoride, a normal component of the soil, is taken up by the roots of most plants in only small amounts even when the soil content exceeds several thousand parts per million (Treshow, 1970). There are, however, a few exceptions, as with members of the camelia family, which accumulate high concentrations of fluoride even when the soil content is quite low. Fluoride content of foliage provides a rough index of exposure to atmospheric fluorides, but is related to toxicity only in a general way.

Toxicants such as arsenic accumulate in the plant when soluble forms are present in high quantities (Ratsch, 1974). The exact amount of accumulation depends upon the solubility of the arsenic compounds and the binding properties of the soil. Under certain conditions, sufficient arsenic may be absorbed to injure or kill sensitive plants, thus altering the community structure, or arsenic may enter the leaves and thus be hazardous to any organisms feeding on them.

The solubility of toxicants in the soil is greatly affected by the soil acidity. Thus greater amounts of metals such as zinc and aluminium become soluble and absorbable in acid soils. Should the acidity increase sufficiently, these elements may reach toxic concentrations in the soil solution.

This has raised the question of whether sulphur, deposited in precipitation as sulphuric acid, and later forming an acid reaction, could change the soil acidity to the extent that certain elements such as aluminium, manganese, or zinc become toxic. See Section 14.4 for a discussion of 'acid rain'.

10.3. RESPONSES OF COMMUNITIES TO PERTURBATION

(i) Physiological Responses

The extent of community disturbance depends largely on the sensitivity of individual plants and their physiological processes. The first impact of a toxic chemical usually can be expected at the molecular or physiological level of organization. In other words, it is often a process such as respiration or photosynthesis that is first affected, or still more specifically, it is some enzymatic reaction in that process that is most sensitive to a given toxicant.

The way in which pollutants affect physiological mechanisms in plants has been reviewed extensively (Treshow, 1970; McCune and Weinstein, 1971; Brinckmann, et al., 1971; Mudd and Kozlowski, 1975). The most sensitive enzyme system naturally varies with the pollutant or toxicant. With SO_2, the mechanism of toxicity may be in interfering with sulphate formation and conversion to such essential compounds as cysteine and methionine in peptides and proteins. In most cases, sulphur is required in a reduced state, and this may be impaired by an excess of SO_2 (Mudd and Kozlowski, 1975).

Photochemical pollutants, specifically ozone, affect metabolism in many ways (Heath, 1975). Briefly, one of the more significant, basic effects may be the oxidation of sulphydryl groups which, among other things, impairs protein synthesis and inhibits CO_2 fixation. The unsaturated fatty acid residues of the membrane lipids, on the other hand, are considered by some to be the primary site of ozone injury. Membrane disruption would cause ionic imbalance leading to a multitude of adverse effects including a reported decline in ATP content and increased water loss. The lipid changes reported might, however, arise from sulphydryl oxidation.

Fluoride appears to interfere with a number of different enzyme systems, and many different metabolic processes can be affected (McCune and Weinstein, 1971). Glucose catabolism is especially affected indicating that pathways of respiration, monosaccharide interconversion, or polysaccharide synthesis are the most likely sites of fluoride activity.

Such enzyme inhibition, at first sublethal, next affects the total plant and in turn, the interaction of that organism with others in the community.

(ii) Responses of Organisms

In order to evaluate the impact of a toxicant, it is desirable to be able to detect the earliest response of a system to it, at whatever level of organization it might appear. Classically, the presence of foliar injury visible to the naked eye has been used as the most sensitive criterion. Distinctive types of injury caused by major pollutants have been described extensively and will not be enumerated (Jacobson and Hill, 1970).

Symptoms have been sought principally in the more sensitive species, the bioindicators. For SO_2, plants such as alfalfa or barley are examined; for fluoride, apricot trees, gladioli, St Johns Wort, and sometimes conifer trees are studied; for ozone, tobacco, grape vines, and pines provide good indicators.

Visual observations for the presence of symptoms are made over the area in question, and the severity of injury is recorded. If no symptoms are found, it is presumed that toxic doses of a pollutant have not been reached. The doses of SO_2 causing injury to such forest species as *Pinus strobus* L. are in the range of 0.1 to 1.0 mg/m^3 for a few hours exposure. For exposures of months, injury is reported in the range of 0.05–0.1 mg/m^3 (0.02–0.04 ppm) (Knabe, 1971). Minimum toxic doses of fluoride that might injure the most sensitive species are of the order of 2–5 μg F/m^3 for 24 hours (personal obs.).

The toxic dose of any given pollutant varies tremendously. Not only must the variation in genetic tolerance among species and varieties be considered, but so must the stage of growth, and the age and kind of tissue. The most sensitive stage of growth varies somewhat with the pollutant, but generally, recently matured tissues are most sensitive. At this time the stomata are mature so that the toxicant can enter the leaf, but the individual cells have not yet developed a protective layer of wax.

The time of exposure to a toxicant is particularly important. If the plant is exposed when it is still young and growth is suppressed, more resistant individuals are most likely to invade the area and the sensitive individuals will be crowded out. If the plant is nearing its reproductive stage, exposure to a toxicant may impair its capacity to set blossoms. If the blossoms are just emerging, blossom and fruit drop may be excessive so that production is impaired. If the exposure occurs after the fruit has set, fruit size may be reduced, but the amount of seed produced is not likely to be affected. The plant is approaching senescence, and damage is minimal.

Chemicals may also enter the ground-water or be washed into the soil and affect sensitive root systems. One example of this is where salt, primarily sodium chloride, is applied to reduce icing of highways. A typical New England highway receives about 6 tons/km each winter. Sugar maple trees are both common along these roads and sensitive to salt (Westing, 1969). The frequency and severity of the decline of maples, especially during dry periods, has been observed within 10 to 13 metres of the roads.

While at first it might seem unlikely that the plant community would be affected in the absence of some observable effect on the individual plants, such a possibility should not be disregarded. It is at least conceivable that the earliest visible sign of some change might be to the community itself.

(iii) Responses of Plant Communities

By the time sensitive plant species are injured, more insidious changes may already have been wrought in the plant community. As mentioned earlier, basic processes such as photosynthesis may have been suppressed thereby impairing growth of more sensitive plants. This could alter their ability to compete in the ecosystem so that their density and abundance might be reduced before the appearance of any symptoms of injury in the leaves.

Few studies have been concerned with changes in the plant community that occurred in the absence of leaf necrosis. Often plant injury has also involved the mortality of large numbers of plants. The few studies that have examined the community, demonstrated the way in which a toxicant may indirectly affect the community structure.

The most obvious of these is probably the influence on growth. Growth differences often can be determined in perennial plant species by measuring the width of growth increments, or annual rings, in the main stem. In one such study (Treshow *et al.*, 1967) increment borings were made in trunks of Douglas fir trees (*Pseudotsuga menzieseii* Franc.) growing at increasing distances from a phosphate reduction plant emitting fluorides. Annual growths before, during, and after operation of the plant were measured. Trees on which no injury symptoms had appeared, as well as those with foliar injury were studied. It was found that growth of trees having no visible injury was reduced by 26 per cent. Growth of trees showing needle necrosis was reduced by 40 per cent. Normal growth of trees not killed was resumed after the emissions ceased.

There are many examples of recovery following the cessation of pollution but only a few will be cited. The recovery is especially striking in areas surrounding smelters following the installation of control equipment. In one instance, in a study of plant communities near a smelter in the western United States, Eastmond (1971) found that the communities were recovering from previous damage following installation of SO_2 control equipment. Revegetation was most rapid beyond 3 to 5 km from the smelter where active erosion had decreased. Forbs (herbaceous pasture plants other than grass) began to appear at about 2.5 to 3 km from the smelter; shrub and deciduous tree species at 4 to 5 km; and conifers at about 7 km. Overall, the number of species increased from only 1 to 5 species per 0.005 ha stand within 3 km of the smelter to 20 to 30 per stand at 8 km. Altitude and soil pH also influenced the number of species present. In another case where SO_2 was the principal pollutant near a power plant, normal growth of white pine trees was resumed the year following installation of a 250 m stack that reduced ground-level concentrations (personal observation).

A more indirect mechanism of impact on a plant community is where the activity of some organisms other than the higher plant is affected, and this in turn affects the welfare of the terrestrial plant community. Such may be the case where a toxicant influences the pathogenicity of a fungus or virus and thereby alters the disease interactions. It may also be the case where a toxicant influences the population density of an insect pest species or its parasites thereby raising or lowering the numbers of insects in the system and the damage that might result therefrom.

This interaction recently has been reviewed in detail by Treshow (1975) and Heagle (1973). Most characteristically, toxicants such as ozone or SO_2 presumably weaken the trees in a plant community and render them more sensitive to the insect and disease pathogens in the area. A toxicant may also injure a plant species causing lesions readily invaded by normally weak pathogens such as *Botrytis* or *Lophodermium*. Attacks on weakened trees by insects, such as pine bark beetles (*Dendroctanus* sp.) on Ponderosa pine (Stark *et al.*, 1968) have been reported to increase in the presence of pollutants.

Obvious changes in community structure have been recognized near local pollution sources, notably smelters, for centuries (Treshow, 1968; Miller and Millican, 1973). In more recent years, the changes have been quantified to illustrate the shift from sensitive to tolerant species (Gordon and Gorham, 1963; Eastmond, 1971).

Gordon and Gorham (1963) found a striking decline in the number of species within about 16 km of a sinter plant emitting SO_2. Beyond this distance the number of macrophyte species at each 40 square metre site averaged 43 while at lesser distances it declined steadily to as few as 2 to 4 species per site within 5 km of the source. The SO_2 doses to which these sites were exposed were not reported.

The influence of photochemical toxicants in altering plant community structure has also been established (Miller and Millican, 1973; Taylor, 1973). The most important, dominant members of the pine community in the San Bernadino Mountains rimming the Los Angeles Basin of California also happen to be the most sensitive to oxidants. Thus when Jeffrey (*Pinus jeffreyi* A. Murr.) and Ponderosa pine (*P. ponderosa* Laws) trees were killed, major changes in the community took place similar to those reported in the conifer communities altered by smelter smoke.

Only the most sensitive individuals in the pine population died soon after exposure, but after 20 years, an estimated 1,298,000 trees had been affected. Severe damage in 1969 extended over 19,000 ha. The concentrations of ozone and other photochemical pollutants were not known when the condition was first observed in the 1950s, but ozone concentrations in the affected area regularly exceeded 10 pphm (parts per hundred million) for 10 hours daily in the forest with an average daily maximum of 20 pphm and momentary peaks reaching 52 pphm. The highest values occurred during the growing season from May to July, maximizing the effects.

Quantitative studies have not been made of changes that may have occurred in the plant community involving species other than pines. It is known that only lower concentrations of a pollutant reach the understory, but they still may be sufficient to affect the more sensitive herbaceous plants. Little is yet known of their tolerance.

In one of the few studies of native understory species, Treshow and Stewart (1973) showed that an exposure to 15 pphm of ozone for 2 hours injured leaves of 7 of the 70 species studied. These were *Hedysarum boreale* Nutt., *Bromus tectorum* L., *Populus tremuloides* Michx., *Aster engelmannii* A. Gray, *Gentiana amarella* L., *Geranium richardsonii* Risch and Trautv, and *Senecio serra* Hook. Seventeen other species were injured at concentrations of between 15 and 25 pphm.

Since others have shown that growth and reproductive effects occurred even before visible injury symptoms appear, it is likely that sublethal effects occur with exposures to still lower concentrations. Levels of the above magnitude are common in many urban areas, so it is possible that oxidants are already influencing community structure in such areas.

Anderson (1967) found that fluorides could also influence plant community structure even when plant leaves were not visibly injured. Plant community population shifts are obvious where the dominant species have been killed and the understory exposed to a changing light intensity. But even where there had been no apparent injury to any species, Anderson (1967) found shifts in populations. Oregon grape (*Mahonia repens* Don.) a species highly sensitive to fluorides, decreased in areas near a phosphate reduction plant, as did chickweed (*Stellaria jamesiana* Tor.) and members of the genus *Viola*. Pine grass (*Calamagrostis rubescens* Bukl.) and waterleaf (*Hydrophyllum capitatum* Dougl.) on the other hand, increased in areas of higher fluoride exposures.

Some of the most striking changes in community structure concern lichens. While the demise of lichens may not be as conspicuous as the death of trees in local areas around some smelters, the effect extends over larger areas and is often associated with general community pollution (LeBlanc and DeSloover, 1970). Sulphur dioxide is the principal toxicant responsible, and the impact was observed long before the concentrations in the areas were known. It is only in recent years that air quality data have provided some knowledge of the doses associated with the breakdown of specific lichen communities.

Some of the best-documented descriptions of the effects on lichen communities have been in the area around a smelter in Ontario, Canada. Rao and LeBlanc (1967) related the number of lichen species to the concentrations of soil sulphate. At the normal, background sulphate concentrations of 0.4 mEq per 100 g, there were over 30 epiphytic species at any site. When sulphate concentrations exceeded about 0.4 mEq per 100 g, the number of species declined to 0 at 1.4 mEq per 100 g.

Considering atmospheric concentrations, LeBlanc and Rao (1973) transplanted lichens at different distances from a smelter in the area of Sudbury, Ontario, Canada. The study suggested that exposures to 0.006 to 0.03 ppm for periods longer than 6 months could injure lichens. The number of times maximum concentrations exceeded a much higher value were not considered, however. Similar work reported by LeBlanc and Rao (1975) suggested that annual exposures to 0.006 to 0.03 ppm can be harmful to lichens, but the peak concentrations are dismissed as being of less consequence. Such field observations are, however, contradicted by laboratory studies showing that doses exceeding 4.0 ppm of SO_2 are necessary to produce injury (Showman, 1972).

In another study, Anderson (1976) exposed lichen communities on Lodgepole pine to doses of 0.5, 2, or 5 ppm of SO_2 for 2 and 6 hours. When community structure was recorded before, and 3 years after, the exposure, it was found that the populations of foliose lichens had decreased significantly on the fumigated limbs. The amount of bare bark had increased significantly as had the cover of crustose lichens.

The mechanisms by which lichens are injured by toxicants are not well defined. It is known that SO_2 impairs the photosynthetic mechanisms at higher concentrations, but this does not explain the extreme sensitivity of this group of organisms. One postulate is that increased acidity of the substrate may be unsuitable for the lichen (Anderson, 1976).

The disruption of the lichen community could have secondary effects on the forest if the blue—green algae component of the lichen symbiont were important to the nitrogen fixation and nitrogen budget of the forest ecosystem.

Patterns of change in terrestrial plant communities have been well illustrated following exposure to ionizing radiation (Woodwell, 1970). The effects of chronic irradiation of a late successional oak-pine forest have been closely studied at Brookhaven National Laboratory in New York. Following 6 months' exposure from a ^{137}Cs source, five distinct zones of vegetation changes were evident. Each became more pronounced over 7 more years of irradiation.

There were striking differences in the effects on the different layers of vegetation, but this may well have been a response of the greater exposures of larger plants rather than their greater sensitivity. Similar responses, where the taller species are the most affected, are seen near other pollutant sources where the overstory species provide a pollution filter for the understory species. Fumigation of understory species by themselves often reveals that some are more sensitive than the trees that provide the canopy.

In the central area where exposure rates exceeded 200 R/day, no higher plants survived. It is interesting to note that some lichens and mosses survived total exposures of over 1,000 R while these same plant types are the most sensitive to other pollutants, notably SO_2.

A second zone of sedge (*Carex pennsylvanica*) survived over 150 R/day around the central area.

A shrub zone surrounding this consisted largely of two species of *Vaccinium* and one of *Gaylossacia*. Exposure rates here exceeded 40 R/day.

A fourth zone of oak survived where exposure rates were over 16 R/day.

Finally, the normal oak-pine forest zone received less than 2 R/day. Here there was no obvious change in number or kinds of species.

(iv) Agricultural Systems

The plant communities in most danger from toxicants are those modified by man. These include artificial systems simplified to consist of as few as one species in which food, horticultural or forest crops are raised. When the crop species or variety is sensitive to a toxicant, the entire crop may be destroyed, or if exposed to lower doses, yields may be reduced in proportion to the dose.

There are many instances of losses especially in the years before the extreme sensitivity of certain varieties was known. In the Los Angeles, California area, production of such ornamental cut-flower crops as orchids was severely impaired as early as the 1940s, several years before the cause was known. Later it was learned that photochemical pollution was responsible for the losses. The dose of toxicants that caused the early damage can only be postulated. As with all toxicants, the dose first causing injury cannot be established with precision because of the tremendous influence of such environmental variables as nutrition, moisture, temperature, light, and inherent and predispositional sensitivity of the plant.

PAN (Peroxyacyl nitrates) cause silvering, glazing, or bronzing symptoms on plants exposed to toxic doses. More significantly, PAN also adversely affects plant growth, development and production (Taylor, 1968; Thompson and Kats, 1975; Marx, 1975).

Perhaps even more than with other pollutants, the toxicity of PAN is strongly influenced by a number of environmental variables, including humidity, age of leaf tissues, duration of exposure, light intensity, photoperiod, and air temperature. Under controlled fumigation conditions, such sensitive plants as Romaine lettuce, pinto beans, and Clintland oats were injured with 0.5 pphm when plants were exposed for 7 hours in midday. In the field, injury has been observed following 6 to 8 hours exposure to PAN concentrations not exceeding 1 pphm of PAN.

When ozone was found to be the principal toxicant of photochemical pollution, it was gradually established that ozone had caused significant production losses in a number of crops. Thompson and Taylor (1969) studied the effect of ambient oxidants on commercial production of lemon and navel orange trees. Leaf drop from trees grown in the ambient atmospheres of the Los Angeles basin averaged 66 per cent, compared with 22 per cent drop from trees grown in filtered air. Yields of trees in ambient atmospheres were up to 50 per cent less than in the control group.

During this study the oxidant concentration exceeded 10 pphm for 5 to 30 hours per month in November, December, January, and February. From March to October, the hours of over 10 pphm ranged from 40 to 280. Maximum concentrations exceeded 30 pphm with peaks reaching 50 pphm. The air quality standard is 8 pphm for 1 hour.

In addition to citrus fruits, yields of grapes and other important crops have been greatly reduced. Studies were conducted in Yonkers, New York in which 60 to 70 per cent of the ambient photochemical oxidants were excluded (MacLean and Schneider, 1976). Comparison of these treatments revealed that yields of tomatoes and beans in untreated air were reduced 33 and 24 per cent, respectively. The average daily oxidant (principally ozone) concentrations during the 43-day exposures of bean plants were 1.2 and 4.1 pphm in the filtered and unfiltered chambers, respectively. For the 99-day exposures of tomato plants, the values were 1.5 and 3.5 pphm. Hourly oxidant peaks reached 20 pphm. In other areas, certain varieties of potatoes and onions have been found to be extremely sensitive to ozone, and yields have been substantially reduced even when concentrations rarely exceeded 15 pphm.

Sulphur dioxide has long been known to cause crop injury including yield reductions of sensitive species raised near major sources of the toxicant (Thomas, 1961; Guderian and van Haut, 1970). The extent of these losses has been calculated and monetary compensation awarded to farmers whose crops were affected. Losses were estimated according to the amount of necrosis and chlorosis (yellowing) of the total leaf area.

The impact of various SO_2 doses on plants has generally been found to be a function of the amount of leaf injury produced; that is, reduction of a crop yield is proportional to the percentage of leaf area destroyed (O'Gara, 1922). The yield lost could be calculated from the equation: $y = 100 - bx$, where y is the yield expressed as the percentage of full yield in the absence of the pollutant, x is the percentage of leaf area destroyed estimated on the leaves present at the time of fumigation and b is the slope of the line obtained by plotting yield against leaf destruction. This equation was determined for each species at different stages of growth. The values varied with the sensitivity of the species. The sensitivity was based on a 'threshold dose' determined from fumigation studies and incorporated into a generalized equation developed by O'Gara (1922).

The generalized equation used to describe the damage from SO_2 was $(C - C_R)t = K$; where C = the concentration of SO_2; C_R = the threshold concentration below which no injury developed; t = the time in hours required to initiate damage; and K = the effective exposure (see Chapter 5). The equation has been modified a number of times since O'Gara developed it in 1922, but there are so many variables affecting the toxic exposure that it seems unwarranted to attempt refining the equation further.

Agricultural systems are especially vulnerable to the wide array of pesticide formulations applied to maximize growth of the crop species. Residues of various chemicals applied to control weed or insect pests often accumulate in agricultural soils more rapidly than they are degraded. Since insecticides are designed to kill insects they should not, ideally, be toxic to plants, but herbicides are as likely to kill desirable species as the intended target. The impact may not be limited directly to the plant community or agricultural crop but may act indirectly through effects on such soil microorganisms as nitrogen-fixing bacteria. Growth and production of higher plants are then impaired. The pesticides have the further potential of being dispersed over neighbouring natural terrestrial communities. In one instance, herbicides applied to control non-forage plants drifted beyond, and killed nearby forest communities (Treshow, 1970). The pesticides also may leach or wash into aquatic systems causing further damage.

The impact of chemical manufacturing wastes (phenols, acids, metals, etc.) applied to soil can also be considerable, ranging from selection toward tolerant species to complete destruction of vegetation and soil sterility.

10.4. DIAGNOSING COMMUNITY DYSFUNCTION

(i) Diagnosing Changes

How can we assess or measure the degree of dysfunction in a plant community or even detect it in the early stages? In areas close to major sources of toxicants, whether cities or industries, damage has often been obvious, as where thousands of hectares of woodland have been killed. Even where the trees have not been killed, the 'burning' or necrosis of leaves caused by a toxicant may be evident.

However, the impact may not always be so obvious. Perhaps growth of only a few species in the community has been affected causing subtle and inconspicuous alterations of community structure.

It is under such conditions, especially, that newer techniques of community analysis must be considered, and it therefore becomes more critical than ever to understand the normal, or 'baseline' community dynamics.

By the time that obviously visible symptoms or changes in either individual plants, or the community, take place, damage may have already been incurred. The normal variation that occurs in a community from year to year may be great. Old plants die, new ones are 'born', and other plants may establish themselves in the system as the existing populations modify the environment. The frequency with which each species occurs, its density, the portion of the total system which it covers, and its productivity vary from one year to the next in response to the ever-changing environment. In other words, the community structure varies each year around an average condition and even this average may change gradually over the years. With such natural variation, how can we detect changes that might be imposed by some abnormal component of the environment, some toxicant?

Community responses are best determined by establishing the normal condition prior to the installation of a pollutant source. One of the most common techniques by which this is done is to select a dozen or more study sites for each vegetation type at increasing distances, and in different directions, from the source. Such sites, or plots, should comprise about 0.04 hectares each, depending on the community type. Within each plot, a number of subplots, often 10 to 25 depending on the variability and diversity of the species, are sampled. Sampling consists of determining the frequency with which each plant species occurs, that is, the percentage of its occurrence, and the percentage of cover it contributes, i.e. the area it occupies in the community. The density, the average number of individuals per unit area, is also determined. The productivity can also be found by sampling and weighing the biomass of vegetation in other representative plots.

A more refined method of studying community structure is that of cluster analysis. Data collected by the above or similar methods are analysed to learn the degree to which the different species present are associated with each other.

These methods are laborious, and yet in order to know the normal annual fluctuations, the study must be conducted, and even repeated several times, over a period of years. The procedure may be somewhat simplified by evaluating only the frequency and cover data, but these must still be determined over a period of years to detect any abnormal changes.

Another method, still in its infancy, is detection of changes using high-altitude, infrared photography. Such methods can reveal the general character and health of the dominant vegetation in an area, but can tell little about the associated understory vegetation if masked beneath a forest canopy. To interpret the results of infrared aerial photography it must be supplemented by on-ground observations.

Estimation of community productivity and biomass also requires substantial work and replication. Plant material from the species being studied over a given area must be collected and weighed and the process must be repeated sufficiently to establish the normal range of variation.

The ability of a terrestrial plant community to tolerate a toxicant depends largely on the inherent resistance of the individual members of a population as well as of the species as a whole. The stability of a community exposed to a stress, such as the presence of toxicants with which it has not evolved, depends also on the capacity of the community to detoxify, or neutralize the toxicants imposed.

There are a few examples of detoxification, although certain pollutants such as SO_2 enter the normal biogeochemical cycle and are soon assimilated along the standard pathways. For instance, sulphur dioxide entering a plant in small quantities soon enters the metabolic processes and the sulphur is incorporated into the cellular proteins.

When more sulphur enters the cellular system than can be converted, the tissues are often killed, but, at least one species, *Muhlenbergia asperifolia* (Ness and May) Parodi, is capable of transporting the excess sulphur to, and out of, the roots and leaves (Campbell, 1972). The sulphur then enters the normal soil pathways in slightly elevated amounts.

(ii) Predicting Changes

Methods for predicting the impact of toxicants on a plant community over a period of a decade or more are still almost non-existent. There are, however, possible approaches to making such predictions. Basically, two attributes of the normal population must be known: mortality and replacement (growth and reproduction) rates of each existing plant species. These can only be learned through detailed studies of community dynamics as discussed earlier. The dose of the toxicant or combination of toxicants required to influence individual plant species and the normal community structure, must also be known.

Theoretically, this information could be combined and an equation formulated to describe the community structure. This model could be used to determine the ultimate structure of the system over a period of years of exposure to various doses of toxicant.

A second approach could involve subjecting whole plant communities to known doses of toxicants and studying the subsequent community structure. This is being done by Lewis *et al.* in Montana (1976) where the baseline dynamics of a grassland community have been established. Known, different concentrations of SO_2 are released over one-half hectare plots in this community and the community structure is being studied following these fumigations. It is hoped that this information can be used to predict more long-term effects as well as the effects on similar plant communities near power plants under construction.

A third approach would be similar but involve only a few species in a 'microcosm' system. Guderian (1966) tested the response of such a system consisting of clover and forage crops to sulphur dioxide. The more sensitive clover was visibly injured and gradually disappeared to be replaced by the more tolerant grasses.

10.5. CONCLUSIONS

For the most part, research on the ecotoxicology of terrestrial plant communities has been directed toward the molecular and organismal levels of organization. It is clear that any toxic influence on one subsystem, or level, of organization will influence all others. We know the doses of many toxicants that are injurious to certain individual plants. We understand some of the physiological mechanisms whereby some major pollutants often cause this injury. However, specifically, we should learn more regarding:

(1) how the total community responds to pollutants, especially when no visible injury occurs to the plant foliage;
(2) the baseline, or normal, community dynamics. We must establish the baseline dynamics of major plant community types. Only then can we understand and predict the long-term responses of communities to doses of a given toxicant.

10.6. REFERENCES

Anderson, F. K., 1967. Air pollution damage to vegetation in Georgetown Canyon, Idaho. University of Utah, *M. S. Thesis*, 102 pp.

Anderson, F. K., 1976. Phytosociology of lichen communities subjected to controlled sulphur dioxide fumigation. University of Utah, *Ph.D. Thesis*.

Brinckmann, E., Lüttge, U., and Fisher, K., 1971. The action of SO_2 and bisulphite compounds on photosynthesis, ion transport and the regulation of stomatal openings in leaves of higher plants (in German). *Ber. Deut. Bot. Ges.*, 84, 523–4.

Campbell, D. E. Jr., 1972. A comparison of SO_2 tolerance between two populations of *Muhlenbergia asperifolia*. University of Utah, *M. S. Thesis*, 75 pp.

Eastmond, R. J., 1971. Response of four plant communities to eroded soils and smelter smoke in northern Utah. University of Utah, *Ph.D. Thesis*, 83 pp.

Gordon, A. G. and Gorham, E., 1963. Ecological aspects of air pollution from an iron-sintering plant at Wawa Ontario. *Canad. J. Bot.*, 41, 1063–78.

Guderian, R., 1966. Reactions of plant communities of forage crops to sulphur dioxide effects (in German). *Schriftenreihe der Landesanstalt für Immissions- und Bodenmetzungsschutz der Landes NW,* No. 4, Verlag W. Girardet, Essen, pp. 80–100.

Guderian, R. and van Haut, H., 1970. Detection of SO_2-effects upon plants (in German). *Staub-Reinhalb. Luft,* 30, 22–35.

Heagle, A. S., 1973. Interactions between air pollutants and plant parasites. *Rev. Phytopathol.*, 11, 365–388.

Heath, R. L., 1975. Ozone. In *Responses of Plants to Air Pollutants* (Eds. J. B. Mudd and T. T. Kozlowksi), Academic Press, New York.

Jacobson, J. S. and Hill, A. C., 1970. Recognition of air pollution injury to vegetation. *Air Pollution Control Assoc. Inform. Rept.*, No. 1.

Knabe, W., 1971. Air quality criteria and their importance for forests. *Mitt. Forstl. Bundes-Versuchsanst. Wien,* 92, 129–50.

LeBlanc, F. and DeSloover, J., 1970. Relation between industrialization and the distribution and growth of epiphytic lichens and mosses in Montreal. *Canad. J. Bot.*, 48, 1485–96.

LeBlanc, F. and Rao, D. N., 1973. Effects of sulfur dioxide on lichen and moss transplants. *Ecology*, 54, 612–7.

LeBlanc, F. and Rao, D. N., 1975. Effects of air pollutants on lichens and bryophytes. In *Responses of Plants to Air Pollutants* (Eds. J. B. Mudd and T. T. Kozlowski), Academic Press, New York, pp. 237–72.

Lerman, S. L. and Darley, E. F., 1975. Particulates. In *Responses of Plants to Air Pollutants* (Eds. J. B. Mudd and T. T. Kozlowski), Academic Press, New York.

Lewis, R. A., Glass, N. R., and Lefohn, A. S. (Eds.), 1976. The bioenvironmental impact of a coal-fired power plant. *2nd Interim Report, Colstrip Mont. U.S. Environmental Protection Agency Env. Res. Lab.*, Corvallis, Oreg.

MacLean, D. C. and Schneider, R. E., 1976. Photochemical oxidants in Yonkers, New York: Effects on yield of bean and tomato. *J. Environ. Qual.*, **5**, 75–8.

Marx, J. L., 1975. Air pollution: Effects on plants. *Science*, **187**, 731–3.

McCune, D. C. and Weinstein, L. H., 1971. Metabolic effects of atmospheric fluorides on plants. *Environ. Pollut.*, **1**, 169–74.

Miller, P. R. and Millican, A. A., 1973. Extent of oxidant injury to some pines and other conifers in California. *Plant Dis. Rept.*, **55**, 555–9.

Mudd, J. B. and Kozlowski, T. T. (Eds.), 1975. Responses of Plants to Air Pollutants, Academic Press, New York, 383 pp.

O'Gara, P. J., 1922. Sulfur dioxide and fume problems and their solutions. *Ind. Eng. Chem.*, **14**, 744.

Rao, D. N. and Leblanc, F., 1967. Influence of an iron sintering plant on corticolous epiphytes in Wawa, Ontario. *Bryologist*, **70**, 141–57.

Ratsch, H. C., 1974. Heavy metal accumulation in soil and vegetation from smelter emissions. *National Environmental Research Center, EPA Ecological Res. Ser. EPA* 660/3–74–012.

Showman, R. E., 1972. Residual effects of sulfur dioxide on the net photosynthetic and respiratory rates of lichen thalli and cultured lichen symbionts. *Bryologist*, **75**, 335–41.

Smith, W. H., 1976. Lead contamination of the roadside ecosystem. *J. Air Polln. Control Assn.*, **26**, 753–66.

Stark, R. W., Miller, P. R., Cobb, F. W. Jr., Wood, D. L., and Parmeter, J. R. Jr., 1968. Photochemical oxidant injury and bark beetle (*Coleoptera:Scolytoidea*) infestation of Ponderosa pine. I. Incidence of bark beetle infestation in injured trees. *Hilgardia*, **39**, 121–6.

Taylor, O. C., 1968. Effects of oxidant air pollutants. *J. Occup. Med.*, **10**, 485–96.

Taylor, O. C., 1973. Oxidant air pollutant effects in a western coniferous forest ecosystem. *Task C Report*, Univ. Calif. Statewide Air Pollution Research Center, Riverside, Calif.

Thomas, M. O., 1961. Effects of air pollution on plants. In *Air Pollution, World Health Organization Monograph*, Columbia University Press, New York, 442 pp.

Thompson, C. R. and Kats, G., 1975. Effects of ambient concentrations of peroxyacetyl nitrate on navel orange trees. *Environ. Sci. Technol.*, **9**, 35–8.

Thompson, C. R. and Taylor, O. C., 1969. Effects of air pollutants on growth, leaf drop, fruit drop, and yield of citrus trees. *Environ. Sci. Technol.*, **3**, 934–40.

Treshow, M., 1968. Impact of air pollutants on plant populations. *Phytopathology*, **58**, 1103–13.

Treshow, M., 1970. Environment and Plant Response, McGraw-Hill Book Co. Inc., New York, 422 pp.

Treshow, M., 1975. Interaction of air pollutants and plant disease. In *Responses of Plants to Air Pollutants* (Eds. J. B. Mudd and T. T. Kozlowski), Academic Press, New York, 383 pp.

Treshow, M., Anderson, F. K., and Harner, F., 1967. Responses of Douglas fir to elevated atmospheric fluorides. *For. Sci.*, **13**, 114–20.

Treshow, M. and Stewart, D., 1973. Ozone sensitivity of plants in natural communities. *Biol. Conserv.*, **5**, 209–14.

Westing, A. H., 1969. Plants and salt in the roadside environment. *Phytopathology*, **59**, 1177–81.

Woodwell, G. M., 1970. Effects of pollution on the structure and physiology of ecosystems. *Science*, **168**, 429–33.

CHAPTER 11

Ecotoxicology of Aquatic Plant Communities

C. HUNDING

National Agency of Environmental Protection, The Freshwater Laboratory, 52 Lysbrogade, DK-8600 Silkeborg, Denmark

R. LANGE

The Norwegian Marine Pollution Research and Monitoring Programme, 29, Munthes Gate, N-Oslo 2, Norway

11.1	INTRODUCTION	239
11.2	PRINCIPLES OF TOXICITY TESTS	240
11.3	THE FATE AND SIGNIFICANCE OF THE POLLUTANT IN TEST SOLUTIONS AND IN NATURAL WATERS	241
11.4	THE UPTAKE AND ACCUMULATION OF TOXICANTS IN PLANT TISSUE	244
11.5	TEST SYSTEMS COMPARED WITH NATURAL SYSTEMS. INTERACTIONS BETWEEN TOXICANTS AND BETWEEN PLANT SPECIES	247
11.6	SOME LONG-TERM EFFECTS OF TOXIC SUBSTANCES ON AQUATIC PLANT COMMUNITIES	248
11.7	CONCLUDING REMARKS	252
11.8	REFERENCES	253

11.1. INTRODUCTION

The aim of the present paper is to review and discuss present knowledge of the principles of ecotoxicology of aquatic plant communities. More than 70% of the earth's surface is covered by the oceans and the major primary producers in these areas are small planktonic algae. Most earlier toxicological studies have been concerned with these algae, and other aquatic plants have been studied only rarely. Possibly the main reasons for selecting planktonic algae as routine test organisms are the relative ease with which they can be handled in the laboratory and, perhaps more important still, their short generation periods. However, the shallow-water areas of aquatic environments are more liable to be polluted than the oceans. These areas are often dominated by plant communities other than the phytoplankton and though the latter still play an important part, it would seem more relevant to include the whole plant community in the investigation.

11.2. PRINCIPLES OF TOXICITY TESTS

Below will be found some remarks on general principles related to tests on the toxicity of pollutants on aquatic plants. For further details, reference is made to the report of a group of experts appointed by ACMRR/IABO of FAO, who have recently completed their task 'to review and critically evaluate methods of toxicity tests on aquatic organisms' (FAO, in press).

The toxicity of a particular element or compound may be tested at different levels of biological organization. For monocultures of unicellular algae the action of the toxicant may be tested by using any of the following properties: growth rate (increase in cell number or increase in the content of organic matter in the cell population), motility, gamete production, cell division, cytoplasmatic streaming, protein synthesis, permeability of the cytoplasmatic membrane, blockage of uptake sites for a particular nutrient etc., CO_2 fixation or production, O_2 uptake or production.

When considering the influence of a toxicant on growth, true cause—effect relationships may be difficult to assess, since different processes are involved, for example:

(1) gross photosynthesis of the plant;

(2) respiration of the plant;

(3) cell division processes.

Thus a diminished growth rate may be the result of a reduction in the rate of photosynthesis, increased respiration rate, inhibition of cell division mechanisms, or combinations of these factors. It is well known, for example, that respiration is affected by changes in temperature to a greater extent than photosynthesis, as photochemical processes are almost independent of temperature. Thus with low illumination and increased temperature, decreased growth rates may be recorded, since respiration is increased but photosynthesis is unaltered (e.g. Cairns *et al.*, 1975). In addition, the primary cause of a decrease in growth rate may be due to the action of a toxic substance on particular processes, such as the uptake mechanisms for various essential plant nutrients. Most earlier studies, however, have considered the influence of toxicants on growth rate (i.e. increase in cell number or algal dry weight) or the rate of photosynthesis. The last has been measured preferentially by means of the ^{14}C-method, originally described for measurements of phytoplankton primary production (Steemann-Nielsen, 1952). What is actually measured by the ^{14}C-method is still a matter for discussion (e.g. Hobson *et al.*, 1976), in particular when certain environmental factors act as stressors. Although information obtained by the ^{14}C-method may serve as a first approximation, it is the influence of the toxicant on net production which is most important. Net production is the basis for all other forms of life in the community, and a reduced net production is associated with a lower input of energy to the system.

Attempts have been made to measure the effects of pollutants on the community as a whole using the following indicators: community structure, species diversity, community nutrient cycling including turnover rates of vital biological elements. A community may be changed if exposed to compounds or elements toxic to individual cells or individual organisms within the community.

Many factors may modify the toxic effect of a substance on the growth of aquatic plants; considered to be of particular importance in this context are:

(1) types and composition of species;

(2) developmental stage(s) of the test species, or the species present in the community;

(3) the physiological state of the individual species or of various species of the community;

(4) the number or biomass of the plants;

(5) composition of test culture medium or of abiotic factors in the environment;

(6) procedure of the experiments; applications of water-insoluble toxicants to test systems, loss of toxicant from the test system by evaporation etc.;

(7) presence of other toxicants in the test culture medium or in the environment: the question of additive, synergistic or antagonistic effects;

(8) short-term versus long-term incubations including the question of adaptation of individual species to the particular toxicant, and short-term 'shock' effects;

(9) small-scale experiments versus large-scale experiments.

Evidently both vertebrates and invertebrates, as well as microalgae and other aquatic plants, show great variation in their response to various stress factors depending on the developmental stage of the organism. This aspect has been best illustrated in unicellular plants in synchronous culture. The physiology of the plant, as well as its response to toxicants, changes through successive stages of the plant's development. Many current test systems consist of a mixture of different developmental stages, for instance, cultures of microalgae under continuous illumination. In these systems the (mode of) action of the toxicant may not be detected and the results of such tests should be interpreted with caution.

11.3. THE FATE AND SIGNIFICANCE OF THE POLLUTANT IN TEST SOLUTIONS AND IN NATURAL WATERS

To assess the result of a toxicity test it is necessary to know the concentration of the pollutant in the test medium, as well as the fate of the pollutant. Is the observed effect a direct one, i.e. caused by the added pollutant? Or are we dealing with an indirect effect of the pollutant? These questions are not easy to answer for various reasons, some of which are given below. One problem is the lack of appropriate analytical chemical methods, and another that of possible interactions between the pollutant and the constituents of the test solution or the natural receiving water.

The study of dissolved organic compounds in the sea is very recent, a probable reason being that work in this field had to await the development of new analytical techniques. The major difficulty in isolating these compounds is the great preponderance of inorganic salts in the sea water, as shown by the following example: Duursma (1960) found the dissolved organic carbon and dissolved organic nitrogen in the Norwegian Sea coastal water to be 1.0 mg/l and 0.1 mg/l, respectively. The ratio of inorganic salt to dissolved organic carbon is thus approximately 35,000 to 1, and the ratio of inorganic salt to dissolved organic nitrogen 350,000 to 1. Such facts make it easy to understand that most of the work up to now has been methodological.

Dissolved organic compounds enter the sea from many different sources. It is possible that the largest fraction originates from the marine organisms themselves, in particular from zoo- and phytoplankton. Another fraction originates in the terrestrial environment and is brought to the sea by rivers or from the atmosphere. The organic pollutants are to be found in the pool of dissolved organic substances in sea water, and it is those which are not found naturally that seem to play a particularly important role in relation to the pollution of the aquatic environment. It is not easy to demonstrate, however, either the role of these substances or that of those naturally present. It has proved difficult, for instance, to differentiate between dissolved and particulate organic compounds because the distinction between the two fractions depends on the filtration or centrifugation technique employed. It is thus impossible to decide whether the dissolved organic substances of natural sea water are in solution or if they exist as colloidal matter. Organic compounds are easily adsorbed on to the hydroxides of iron or aluminium. For this reason, it is assumed that at least some of the dissolved organic compounds of sea water are adsorbed on colloidal or larger particles. It is known that river water may contain high concentrations of colloidal organic matter and when it reaches the ocean, the colloids present coagulate rapidly.

The response to a given concentration of toxicant varies not only with the receptor plant community but also with the waters in which a community resides. For example Glooschenko (1971) found differences in the effects of DDT and dieldrin *in situ* on natural phytoplankton depending on whether they were tested in Lake Erie or Lake Ontario. Another example of the dependence of effects on the nature of the test waters is provided by experiments of Skaar *et al.* (1974) which demonstrated a low nickel-binding capacity of phosphate-starved cells of the diatom *Phaeodactylum tricornutum*; addition of phosphate to the culture medium increased the binding capacity. However, when nickel ions were added before or together with the phosphate, almost no increase in nickel-binding capacity of phosphate-starved cells was found. Similarly, the influence of copper on $^{14}CO_2$ uptake of natural phytoplankton was found to vary in different lakes (Steemann-Nielsen and Bruun-Laursen, 1976). Whereas addition of copper to the water generally gave lower uptake rates, the opposite was found after addition of small doses of copper to phytoplankton sampled in a lake with a high humus content. It is assumed that the copper is bound to humus under natural conditions and as such is unavailable to the algae. When sufficient copper is added, the binding capacity of the humus is exceeded and thus copper is made available to the plants. A low concentration of copper in the state at which the plants can utilize it, is necessary for growth, but at higher copper levels toxic effects occur. This example shows the possible dual role of a pollutant and the importance of the concentration in this respect.

Copper is found in natural waters in both particulate and dissolved forms (Table 11.1) (Kamp-Nielsen, 1972) and unfortunately, little is known about the equilibrium constants between the different copper species (Sylva, 1976). This aspect is important for toxicological studies, since only some (or one of the) copper species present in solution may be taken up by the plants and hence exhibit toxicity (e.g. Mandelli, 1969; Gächter et al., 1973). Moreover, more accurate estimates are needed of the rates at which, for instance, copper is mobilized from various copper compounds under different environmental conditions. The rate at which the most toxic species of, for example, copper, is supplied to the plants is of more importance than the concentration of total copper in the water. The discrepancies between the results of different workers with regard to measurements of copper toxicity on plants may be due to the varying states of the metal in different test media or in natural waters. Similar considerations apply to most other heavy metals.

Heavy metals seem to have a more severe toxic effect when present in the ionic state than when in the form of various complexes. Deep ocean water usually contains these metals in the unchelated state, whereas copper ions, for example, are present in only small quantities in surface waters. This is ascribed to the presence of

Table 11.1 Chemical States of Copper in Natural Waters. (Reproduced from Kamp-Nielsen, 1972, with permission of Pergamon Press, Ltd.

Copper fractions	Examples
Dissolved, ionic	Cu^+, Cu^{2+}
Dissolved, inorganic complexed	$CuCO_3$, $Cu(CN)^-$, $CuCl_4^{2-}$
Dissolved, organic complexed	Cu-peptides, Cu-humates, Cu-porphyrins
Adsorbed to particles	Cu^{2+} adsorbed to $Fe(OH)_3$-micelles, seston, algae
Particulate, inorganic	CuS, CuO, $Cu_2(OH)_2CO_3$
Particulate, organic	Copper in algae, animals, seston

a relatively high level of dissolved organic matter in the upper water layers where these substances form complexes with the copper ions. Steemann-Nielsen and Wium-Andersen (1970) suggested that the low primary production observed in the water at the centre of an upwelling area, compared with that occurring in the peripheral waters is due to the different states at which copper is present in the two types of water. The occurrence of relatively high levels of dissolved organic matter in surface layers of natural fresh and marine waters is ascribed to the high metabolic activity exhibited by the plankton, which excrete dissolved organic compounds. This excretion seems to increase when the organisms are under stress (e.g. when competing for nutrients or light), and thus also when the organisms are exposed to toxic substances (Steemann-Nielsen and Wium-Andersen, 1970). It may be argued, therefore, that the dissolved organic molecules act as external detoxicants. Davies (1976) similarly found internal detoxication of mercury by binding of the metal to cell constituents, predominantly proteins, in *Dunaliella tertiolecta*.

11.4. THE UPTAKE AND ACCUMULATION OF TOXICANTS IN PLANT TISSUE

In general, the exposure of natural and laboratory populations of aquatic plants to high concentrations of toxic compounds of heavy metals and organohalogens may cause a reduction in rates of photosynthesis or growth. At low concentrations there are, however, species differences. Thus for example, the concentration of a substance may inhibit growth or photosynthesis of one species while another is unaffected or even stimulated (e.g. Fisher and Wurster, 1973). A consistent view regarding the general effects of toxicants cannot be given (e.g. Leland et al., 1976), firstly, because aquatic plants differ in their inherent response to toxicants and, secondly, because the effects of the toxicants are also strongly influenced by the chemical composition of the surrounding medium. Similarly, the concentrations of toxicant residues reported for different aquatic plants from various water types (e.g. Reish and Kauwling 1976) may be of limited significance. Lorch et al. (1976) found that the accumulation factors for the green alga *Pediastrum tetras* varied from 4 to 1,710. They were able to relate this variation to the culture conditions and, in particular, to the chemical composition of the culture medium. Although the results of these earlier toxicological studies on aquatic plants are sometimes conflicting and difficult to interpret, they have illustrated some important principles of ecotoxicology. These principles are discussed briefly below, and special attention is paid both to the studies on the effects of toxicants on natural aquatic plant communities and to laboratory studies offering ecologically meaningful results.

The uptake of both toxic heavy metals and organohalogens by aquatic plants seems to be initiated most often by a phase of rapid and passive adsorption to the cell wall. For example, Fujita and Hashizume (1975) found that about 20% of the total amount of mercury found in dividing cells of the diatom *Synedra ulna* was taken up by adsorption and could be eliminated by washing with distilled water. A further 50% was apparently accumulated in the inner part of the cells and, the remaining 30% seemed relatively firmly physically or chemically adsorbed to the cell membrane. The adsorption isotherm of mercury uptake by *Synedra* was:

$$\log C' = 2.603 + \frac{\log C}{1.3}$$

where C' and C are concentrations (molar) in cells and culture medium respectively. However, it could not be concluded from the experiments whether the uptake of mercury by *Synedra* was due solely to adsorption in a monolayer on the cell surface or if it was the combination of biological uptake and passive adsorption which obeyed the isotherm. Davies (1973) suggested that the uptake of zinc in the diatom *Phaeodactylum tricornutum* starts as a rapid adsorption onto the cell membrane followed by diffusion-controlled uptake rate and then by binding to proteins within the cell. This binding of zinc to protein may control the concentration in the cell. During the growth cycle the concentration of zinc reached a maximum and then decreased as the amount of protein in the cell declined. Similar uptake patterns have been described for other metals, in e.g. Skaar *et al.* (1974) who studied nickel uptake in the same diatom species.

The initial uptake of organohalogen compounds by unicellular algae also appears to be a rapid process. The uptake of DDT, for example, was found to be completed in less than 15 seconds after addition of ^{14}C-labelled DDT to a culture of *Chlorella* sp. (Södergren, 1968). The uptake was predominantly an adsorption to the cell membrane and no difference was found between the uptake rates of living and dead cells. The adsorption capacity of *Chlorella* cells was at least 6.3×10^{-6} μg DDT per cell, and thus equivalent to 0.32 μg DDT per mg algal dry weight. The accumulation of DDT in continuous-flow cultures of the algae followed the equation:

$$C_c(t) = C_0(1 - e^{-(Qt/v)})$$

where

C_c = concentration in the cells
C_0 = concentration in the influent
Q = the volume of liquid added per day to the system
t = time in days
v = culture volume.

Salonen and Vaajakorpi (1976) demonstrated the same initial rapid uptake of DDT in a small pond in Finland. DDT was added to the water and its concentration was followed for 60 days in the abiotic and biotic compartments of the system (Figure 11.1). It was found that small suspended particles, whether alive or dead, took up DDT more rapidly than larger particles.

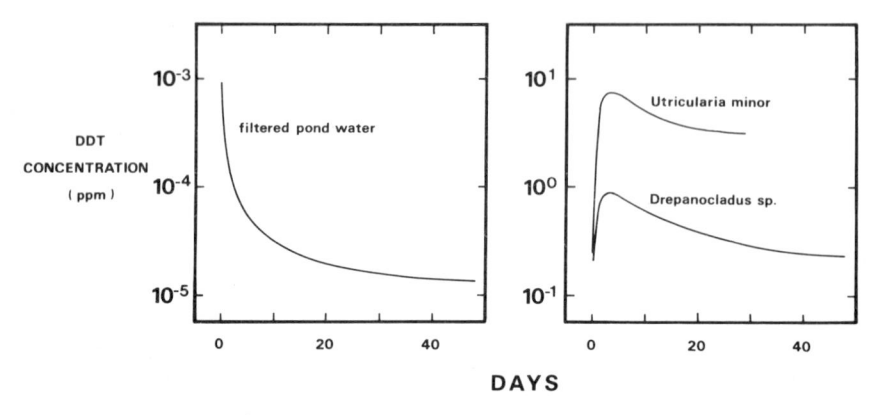

Figure 11.1 The DDT concentrations in filtered pond water and in some plants as a function of time after labelling the pond. (Reproduced by permission of Georg Thieme Verlag, from Salonen and Vaajakorpi, 1976)

There are only a few studies on higher aquatic plants which allow the calculation of reliable concentration (accumulation) factors for toxicants in plant tissues. Mortimer and Kudo (1975) maintained the mercury at constant concentrations in the water of experimental containers by controlling the input of mercury into the test system. In addition, they continuously measured the actual concentrations of mercury chloride and methylmercury chloride in plants, bed sediments, and on the glass walls of the containers. The aquatic macrophyte *Elodea densa* was used as the test plant and its growth was determined for the 17 days' experiment. Uptake was directly proportional to time and to water concentration. The relationship of uptake to the mercury concentration in the flowing water was expressed as the concentration factor:

$$\frac{\mu g \text{ Hg/g plant}}{\mu g \text{ Hg/g water}}$$

The daily increment of mercury uptake was related to water concentration, and to the weight of the plant tissue, and was estimated from the specific activity values determined at days 3, 7, 10, 14, and 17. The transfer coefficients were calculated as:

$$\frac{\mu g \text{ Hg/g plant/day}}{\mu g \text{ Hg/cm}^3 \text{ water}} = \text{cm}^3 \text{ water/g plant/day (day}^{-1})$$

which is equivalent to the volume of water cleared of mercury per gram of plant per day. The transfer coefficients revealed a reasonably consistent grouping around 5,000 day^{-1} for both mercury compounds at concentrations up to about 10 ppb of mercury.

11.5. TEST SYSTEMS COMPARED WITH NATURAL SYSTEMS. INTERACTIONS BETWEEN TOXICANTS AND BETWEEN PLANT SPECIES

It is usual for only one organism and only one toxic substance to be used at one time in practical laboratory tests. In the natural environment, however, a number of pollutants are usually present simultaneously along with several species of aquatic plants. This raises the question of combined action. Do different toxic substances antagonize each other, or are their harmful effects additive or synergistic? Furthermore, does the coexistence of plants influence toxicity and accumulation rates?

Hutchinson and Czyrska (1975) exposed the floating aquatic weeds *Salvinia* and *Lemna* to various concentrations of heavy metals. The growth of *Salvinia* was better at 0.01 and 0.05 ppm of cadmium when in competition with *Lemna* than when grown alone. In the absence of competition, *Salvinia* failed to survive at 0.05 ppm of cadmium. In contrast, the growth of *Lemna* was significantly reduced when it was grown together with *Salvinia*. Thus in the laboratory, the effect of the metals on the growth of only one of the test plants was changed when the two species were grown together. The rate at which the metals were accumulated in the plants was also altered when the plants were grown together. In this case the cadmium level in *Lemna* tissue increased while the cadmium concentration in *Salvinia* was correspondingly reduced. It should be mentioned, however, that the concentrations of the toxicants given above are calculated on the basis of the amount of toxicant added initially and the volume of culture medium. If one of the plant species had a more efficient uptake mechanism(s) for the toxicant (e.g. due to a larger surface area) the concentration at which the other plant is exposed in the competition experiment is lower than calculated. Thus the experiments demonstrate the differences in the initial uptake capacities for toxicants in different plant species. It seems, however, that there are also other mechanisms involved. It was shown, for instance, that at the same initial toxicant concentration, the accumulation of cadmium in plant tissue of *Lemna* is higher when grown together with *Salvinia* than when grown alone.

The metals tested in the study by Hutchinson and Czyrska (1975) apparently acted synergistically in *Lemna*. When nickel and copper were present at the same time, the uptake rates of both metals were increased. Similarly, the presence of zinc appeared to result in a higher cadmium uptake. The accumulation of both substances was higher in *Lemna* tissue when the plant was grown together with *Salvinia* than when grown alone. On the other hand, cadmium and zinc levels in *Salvinia* were the same whether it was grown alone or together with *Lemna*.

Interactions of other toxicants such as PCB's, DDT and DDE have also been demonstrated in laboratory toxicity tests on microalgae (Figure 11.2). Mosser *et al.* (1974) observed substantial growth inhibition when the marine diatom *Thalassiosira pseudonana* (strain 3H) was treated with 10 ppb PCBs and 100 ppb DDE simultaneously, while growth of the diatom was only slightly reduced when

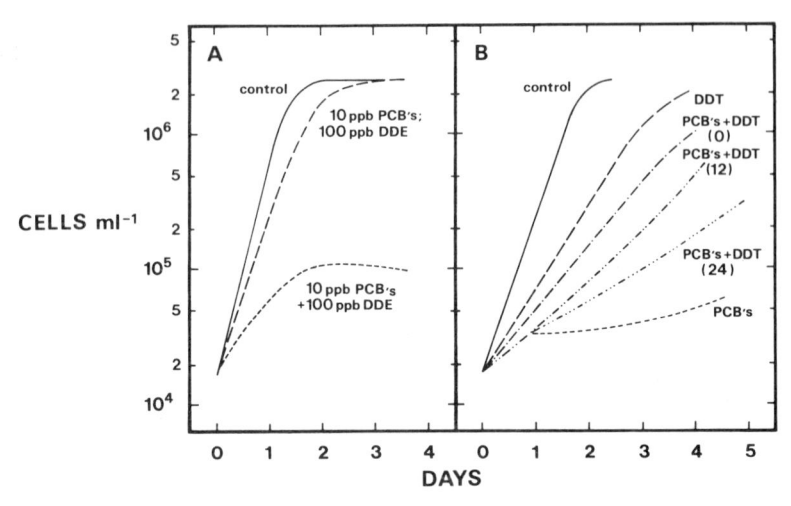

Figure 11.2 Interactions among PCBs, DDT, and DDE in *Thalassiosira pseudonana*. The concentrations of PCBs and DDT in B were 50 ppb and 500 ppb, respectively. The number of hours after which DDT is added to the PCBs treated cultures of the alga is indicated in brackets. (Reproduced by permission of Springer-Verlag, New York, Inc., from Mosser *et al.*, 1974)

treated with just one of these substances at the same concentrations (Figure 11.2A). So in this case, PCBs and DDE acted synergistically. In contrast, the interaction of PCBs and DDT was antagonistic. When the diatom was treated with 50 ppb PCB's, growth was almost stopped. The simultaneous addition of 500 ppb DDT restored growth rate to about two thirds of that in control cultures. The addition of DDT after 12–24 h inoculation with PCBs also reversed the inhibition caused by the latter (Figure 11.2B). The mechanism appeared to be an intracellular interaction rather than a physical process (coprecipitation etc.) because the process was reversible. Almost no interaction was demonstrated between DDT and DDE. Hence the effect of simultaneous exposure to these substances was additive.

The examples given above clearly demonstrate that toxicity test experiments conducted on only one plant species and with one toxic substance only will not always allow prediction of the effects of toxicants on natural plant communities.

11.6. SOME LONG-TERM EFFECTS OF TOXIC SUBSTANCES ON AQUATIC PLANT COMMUNITIES

In recent years, attempts have been made to obtain more reliable data on the influence of toxic substances on entire natural communities and ecosystems. The aim of the CEPEX programme (Reeve *et al.*, 1976) is to identify sensitive, short-term indicator reactions of long-term sublethal pollutants. This may be accomplished by correlating metabolic, behavioural or other quickly measurable variables with subtle effects on community and population structure and function. Current ecotoxicological studies can be separated into two groups. In the CEPEX programme and related studies, laboratory experiments, basically derived from classical toxicological tests, are intended to be used in monitoring the 'state of the environment'. For the same purpose, others have aimed at identifying indicator species for the various types of pollution.

The following discussion concerns some aspects of recent long-term and large-scale studies in ecotoxicology and includes investigations on indicator species, adaptation of aquatic plant populations and communities to toxic substances, and changes in community structure and function as a consequence of exposure of the systems to toxicants.

Whitton (1970) attempted to demonstrate correlations between the abundance of different plants and the discharge of a number of pollutants in various concentrations in British rivers. He showed, for instance, that the filamentous green algae *Microspora* and *Ulothrix* tended to be resistant to copper, lead, and zinc, whereas *Oedogonium* was sensitive. In contrast, the genera *Mougeotia*, *Spirogyra* and *Zygnema* (Zygnematales), showed a wide range of behaviour with respect to zinc. However, it was concluded that the existing information is inadequate to enable decisions to be made as to species or genera of higher benthic algae suitable to serve as indicators of metal pollution. Littler and Murray (1975) found that brown algal species and surf grass were replaced by blue—green algae, sea lettuce (*Ulva lactuca*) and some finely branched red algae in the vicinity of a small domestic outfall at San Clemento Island. The last were characterized as having higher productivities and shorter life histories. Obviously, the experience gained from such natural long-term exposures of plant communities to pollutants offers a more realistic basis for predictions about the effects of polluting substances on natural plant communities than the results of acute toxicity tests.

In a recent paper, Stockner and Antia (1976) warned against using the conclusions based on observations made by the usual acute toxicity tests because the physiological mechanism of adaptation to long-term exposure is neglected in most of these tests. Stockner and Costella (1976) studied the effect of pulpmill effluent on the growth of the marine planktonic algae *Skeletonema costatum* and *Amphidinium carteri*. They interpreted the prolonged lag phases of algal growth to mean that time was required for physiological adaptation prior to commencement of exponential growth. Generally, these algae required from several days to as long as 3 weeks to adapt to the higher concentrations of kraft effluent from the pulpmill (Figure 11.3). No loss of toxicity of the kraft effluent was observed on long-term storage (90 days). During the initial phases of exposure to pollutants, the algae tolerated only low concentrations of the pollutants, whereas much higher levels were accepted by repeated exposure to gradually increasing concentrations (Stockner and Antia, 1976). As pointed out earlier, however, such long-term experiments may be difficult to interpret in terms of the exact cause—effect

250   *Principles of Ecotoxicology*

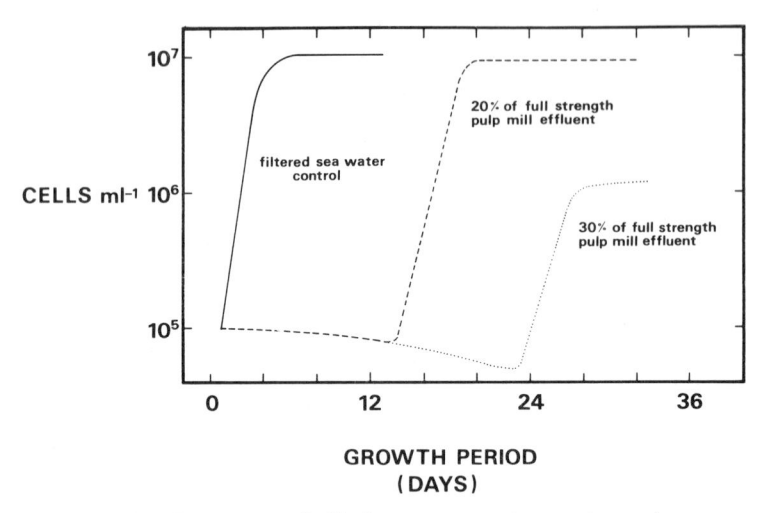

[Figure: Graph showing CELLS ml⁻¹ (y-axis, 10⁵ to 10⁷) vs GROWTH PERIOD (DAYS) (x-axis, 0 to 36). Curves labeled "filtered sea water control", "20% of full strength pulp mill effluent", and "30% of full strength pulp mill effluent".]

Figure 11.3 Response of *Skeletonema costatum* to various concentrations of pulpmill effluent and to filtered sea-water controls. Only the response to the kraft fraction of the pulpmill effluent is shown. (Reproduced by permission of Minister of Supply and Services, Canada, from Stockner and Costella, 1976)

relationships because physicochemical factors may obscure the possible physiological adaptation of the plants. In addition, it is difficult to assess whether the observed adaptation is a rapid, reversible physiological response or an alteration of the genetic material of the cells. A distinction has also to be made between the adaptation of the species and that of the community. Populations of most algal species show great variations in their sensitivity to various toxicants (Lazaroff and Moore, 1966; Reeve *et al.*, 1976). In natural plant communities, some species may become dominant after long-term exposure to pollutants, while others are eliminated. Several laboratory experiments with mixed cultures of algae have demonstrated similar changes in the species composition upon exposure to toxicants (e.g. Mosser *et al.*, 1972; Jensen *et al.*, 1975; Reeve *et al.*, 1976).

Our conception of the ecosystems has changed frequently in recent years (e.g. Kerr and Neal, 1976). Current theories concerning natural communities and ecosystems have moved the focus from the classical characteristics (indicator species diversity etc.) towards other distinguishing features. A promising theory may be the type which emphasizes the analysis of the size distribution of organisms in aquatic communities (e.g. Sheldon *et al.*, 1972; Kerr and Neal, 1976). This ecological pattern is likely to be useful also in the context of ecotoxicology.

When semi-long-term (of up to 280 h of exposure, e.g. Jensen *et al.*, 1975) or long-term and large-scale experiments (e.g. the CEPEX programme, Reeve *et al.*, 1976) on the effects of toxicants are conducted on phytoplankton communities, small algal species tend to dominate. If these observations prove to be the general rule of community response to stress factors, including toxic substances, significant changes in plant community structure and function occur. It is well known that heavy doses of plant nutrients to lakes over prolonged periods result in a high eutrophication status. The increase in eutrophication level is usually correlated with an alteration of the species composition and the small planktonic algae (the nannoplankton) become dominant (e.g. Pavoni, 1963). Dominance of small phytoplankton cells may be favourable to the higher trophic levels of the grazer food chains of the aquatic system, since the herbivorous zooplankton preferably feed on the small suspended particles in the water. Predominance of the small algal cells, however, may also result when the herbivorous zooplankton is more susceptible to the toxicant than the other organisms of the planktonic community. Grazing of the zooplankton on the small algae is thereby reduced, and these algae will dominate over other algal species due to their relatively high growth rates. Some of the results obtained in the CEPEX programme also indicated that the herbivorous zooplankton were more susceptible to copper than other planktonic organisms. These findings demonstrate the need to analyse entire communities and ecosystems when studying the effects of pollutants because investigations of only one functional group within the system (e.g. the phytoplankton) can offer only inadequate information on the fate and effects of the pollutant.

When the aquatic environments are dominated by small size classes and the total biomass is maintained at the same level or increased, a higher uptake and accumulation capacity for toxic substances results. The effect of toxicants on the size distribution of plant communities may be self-promoting by analogy with the self-promoting eutrophication processes observed in aquatic systems receiving large amounts of plant nutrients. An important aim of future ecotoxicological studies is to identify the critical maximum dose of toxicants which the various types of aquatic plant communities can tolerate. Maki and Johnson (1976) studied the effect of a toxicant, the lampricide TFM (3-trifluoromethyl-4-nitrophenol) on the metabolism of benthic communities in artificial streams. During exposure to 9.0 mg/l TFM, gross primary production was suppressed by 20–25%, while respiration was increased by 3–50%. The stream community, however, demonstrated a capacity to adjust to the influence of the toxicant as shown by the rapid return of metabolic rates to pretreatment levels after the exposure period. It was not demonstrated in the study whether an alteration in species composition had occurred or if all the organisms present were able to survive the treatment.

After an initial adaptation period, the planktonic communities dominated by the small size classes of cells may be assumed also to possess a higher metabolic activity than the unpolluted communities, at least at relatively low toxicant concentrations. Under these conditions a different status of element cycling results with increased turnover rates in the aquatic systems. Indications of accelerated turnover rates of elements in natural ecosystems have in fact been demonstrated as a result of the increased discharge of many elements into the aquatic ecosystems (e.g. Wollast et al., 1975; SCOPE, 1976).

The biotic and abiotic compartments of the biosphere interact, and we may assume that present-day life forms have been adapted to the environmental conditions prevailing during the last few centuries. Unfortunately, we do not at present understand the consequences of this increased rate of cycling of the biologically essential elements.

11.7. CONCLUDING REMARKS

1. Our present knowledge of the behaviour of toxicants, their effects on natural aquatic plant communities, and the consequences of possible alterations of these communities for the structure and function of aquatic ecosystems is extremely poor. There seem to be indications of alterations of communities of both macrophytes and plankton. In macrophytic communities, plants with higher metabolic rates and shorter life histories replace the original vegetation. In the planktonic plant communities, the small algal species and cells tend to become dominant when the communities are exposed to relatively low doses of toxic substances. This may result in an increase in plant community metabolism and also in ecosystem metabolism, since the herbivorous zooplankton feed predominantly on the small suspended particles in the water. The turnover rates of elements within the system, and the uptake and accumulation rates of toxicants will also be increased. The two processes act in antagonism, since the higher uptake rates of toxicants are usually accompanied by a suppressed metabolic activity of the individual cells and organisms. The ecosystem response is thus dependent on the internal 'buffer' capacity or adaptability of the system towards the toxicants. This hypothesis needs to be confirmed, however, through more detailed studies of the effects of long-term exposures to various toxicants or mixtures of toxicants on large-scale test communities and ecosystems.

2. The present lack of understanding of the effects of toxicants on natural systems is also due to inadequate information on the baseline situations of the systems. The difficulties of such assessments are ascribed mainly to the natural spatial and temporal variations in both abiotic and biotic factors in the systems. The planktonic communities are dominated by small organisms, thus creating the problem of reproducibility of ecosystem analysis because even minor natural changes in the environment may result in rapid changes in the species composition, and in the type and level of the community metabolism.

3. It is apparent from the above discussion of the current acute toxicity tests that no consistent view can be expressed with regard to the general effects of toxicants in laboratory experiments and in nature. For future acute toxicity test systems, the biological effects of one toxicant, or preferably that combination of toxicants most relevant to the individual case should be tested. It would be best if the toxicity test were carried out on the natural plant community including the recipient water, or with a standard test medium to which the plant community has been adapted. Alternatively, it is desirable that a well-known 'standard' test plant

(or 'standard mixture' of plant species) should be cultivated in the recipient water. The former procedure is to be preferred because the latter offers more information about the test algae and its physiology than about the toxicity of the test substance(s) to the plants present in the recipient water. It should, however, be strongly emphasized that these screening tests should be as simple as possible to perform (though including the precautions in Section 11.2). Even so, the results of these tests should be interpreted with caution. Extrapolation to natural systems is prevented mainly by their higher complexity due to their abiotic and biotic interactions, and because the characteristics of populations are fundamentally different from those of communities and ecosystems.

11.8. REFERENCES

Cairns, J. Jr., Heath, A. G., and Parker, B. C., 1975. The effects of temperature upon the toxicity of chemicals to aquatic organisms. *Hydrobiologia*, **47**, 135–71.

Davies, A. G., 1973. The kinetics of and a preliminary model for the uptake of radio-zinc by *Phaeodactylum tricornutum* in culture. In *Radio-active Contamination of the Marine Environment, Proceedings of a Symposium*, Seattle, 1972, pp. 403–20. STI/PUB/313, International Atomic Energy Agency, Vienna.

Davies, A. G., 1976. An assessment of the basis of mercury tolerance in *Dunaliella tertiolecta*. *J. Mar. Biol. Assoc. U.K.*, **56**, 39–57.

Duursma, E. K., 1960. Dissolved organic carbon, nitrogen and phosphorus in the sea. *Neth. J. Sea Res.*, **1**, 1–148.

FAO, in press. *Report of the ACMRR/IABO Working Party on Biological Effects of Pollutants*, Rome, Italy.

Fisher, N. S. and Wurster, C. F., 1973. Individual and combined effects of temperature and polychlorinated biphenyls on the growth of three species of phytoplankton. *Environ. Pollut.*, **5**, 205–12.

Fujita, M. and Hashizume, K., 1975. Status of uptake of mercury by the fresh water diatom *Synedra ulna*. *Water Res.*, **9**, 889–94.

Gächter, R., Lum-Shue-Chan, K., and Chau, Y. K., 1973. Complexing capacity of the nutrient medium and its relation to inhibition of algal photosynthesis by copper. *Schweiz. Z. Hydrol.*, **35**, 252–61.

Glooschenko, W. A., 1971. The effect of DDT and dieldrin upon ^{14}C uptake by *in situ* phytoplankton in lakes Erie and Ontario. *Proc. 14th Conf. Great Lakes Res.*, 219–33. International Assoc. Great Lakes Research, CCIW, Burlington, Ont., Canada.

Hobson, L. A., Morris, W. J., and Pirquet, K. T., 1976. Theoretical and experimental analysis of the ^{14}C technique and its use in studies of primary production. *J. Fish. Res. Board Canada*, **33**, 1715–21.

Hutchinson, T. C. and Czyrska, H., 1975. Heavy metal toxicity and synergism to floating aquatic weeds. *Verh. Internat. Verein Theor. Angew. Limnol.*, **19**, 2102–11.

Jensen, S., Lange, R., Berge, G., Palmork, K. H., and Renberg, L., 1975. On the chemistry of EDC-tar and its biological significance in the sea. *Proc. Roy. Soc. London*, **B189**, 333–46.

Kamp-Nielsen, L., 1972. Some comments on the determination of copper fractions in natural waters. *Deep Sea Res.*, **19**, 899–902.

Kerr, S. R. and Neal, M. W., 1976. Analysis of large-scale ecological systems. *J. Fish. Res. Board Canada*, 33, 2083–89.
Lazaroff, N. and Moore, R. B., 1966. Selective effects of chlorinated insecticides on algal populations. *J. Phycol.*, 2 (Suppl.), 7–8.
Leland, H. V., Wilkes, D. J., and Copenhaver, E. D., 1976. Heavy metals and related trace elements. *J. Water Pollut. Control Fed.*, 48, 1459–86.
Littler, M. M. and Murray, S. N., 1975. Impact of sewage on the distribution, abundance and community structure of rocky intertidal macro-organisms. *Mar. Biol.*, 30, 277–92.
Lorch, D. W., Melkonian, M., Weber, A., and Wettern, M., 1976. Accumulations of lead by green algae. *Proc. Intern. Symp. Exp. Use Algae Cultures in Limnology*, Sandefjord, Norway, Oct. 26–28, 1976.
Maki, A. W. and Johnson, H. E., 1976. Evaluation of a toxicant on the metabolisms of model stream communities. *J. Fish. Res. Board Canada*, 33, 2740–6.
Mandelli, E. F., 1969. The inhibitory effects of copper on marine phytoplankton. *Contrib. Mar. Sci.*, 14, 47–57.
Mortimer, D. C. and Kudo, A., 1975. Interaction between aquatic plants and bed sediments in mercury uptake from flowing water. *J. Environ. Qual.*, 4, 491–5.
Mosser, J. L., Fisher, N. S., and Wurster, C. F., 1972. Polychlorinated biphenyls and DDT alter species composition in mixed cultures of algae. *Science*, 176, 533–5.
Mosser, J. L., Teng, T.-C., Walther, W. G., and Wurster, C. F., 1974. Interactions of PCBs, DDT and DDE in a marine diatom. *Bull. Environ. Contam. Toxicol.*, 12, 665–8.
Pavoni, M., 1963. The importance of nannoplankton in comparison with net-plankton (in German). *Schweiz. Z. Hydrol.*, 25, 219–341.
Reeve, M. R., Grice, G. D., Gibson, V. R., Walter, M. A., Darcy, K., and Ikeda, T., 1976. A controlled environmental pollution experiment (CEPEX) and its usefulness in the study of larger zooplankton under toxic stress. In *Effects of Pollutants on Aquatic Organisms* (Ed. A. P. M. Lockwood), Cambridge University Press, pp. 145–62.
Reish, D. J. and Kauwling, T. J., 1976. Marine and estuarine pollution. *J. Water Pollut. Control Fed.*, 48, 1439–59.
Salonen, L. and Vaajakorpi, H. A., 1976. Bioaccumulation of ^{14}C-DDT in a small pond. *Environ. Qual. Saf.*, 5, 130–40.
SCOPE, 1976. *Nitrogen, Phosphorus and Sulphur – Global Cycles* (Eds. B. M. Svensson and R. Söderlund), *Ecological Bulletin No. 22, SCOPE Report 7*, Stockholm, 170 pp.
Sheldon, R. W., Prakash, A., and Sutcliffe Jr., W. H., 1972. The size distribution of particles in the ocean. *Limnol. Oceanogr.*, 17, 327–40.
Skaar, H., Rystad, B., and Jensen, A., 1974. The uptake of ^{63}Ni by the diatom *Phaeodactylum tricornutum*. *Physiol. Plant*, 32, 353–8.
Södergren, A., 1968. Uptake and accumulation of ^{14}C-DDT by *Chlorella* sp. (Chlorophyceae). *Oikos*, 19, 126–38.
Steemann-Nielsen, E., 1952. The use of radio-active carbon (^{14}C) for measuring organic production in the sea. *J. Cons. Perm. Int. Explor. Mer.*, 18, 117–40.
Steemann-Nielsen, E. and Bruun-Laursen, H., 1976. Effect of $CuSO_4$ on the photosynthetic rate of phytoplankton in four Danish lakes. *Oikos*, 27, 239–42.
Steemann-Nielsen, E. and Wium-Andersen, S., 1970. Copper ions as poison in the sea and in freshwater. *Mar. Biol.*, 6, 93–7.
Stockner, J. G. and Antia, N. J., 1976. Phytoplankton adaptation to environmental stresses from toxicants, nutrients and pollutants – a warning. *J. Fish. Res. Board Canada*, 33, 2089–96.

Stockner, J. G. and Costella, A. C., 1976. Marine phytoplankton growth in high concentrations of pulpmill effluent. *J. Fish. Res. Board Canada*, **33**, 2758–65.

Sylva, R. N., 1976. The environmental chemistry of copper (II) in aquatic systems. *Water Res.*, **10**, 789–92.

Whitton, B. A., 1970. Toxicity of heavy metals to freshwater algae: A review. *Phykos.*, **9**, 116–25.

Wollast, R., Billen, G., and Mackenzie, F. T., 1975. Behaviour of mercury in natural systems and its global cycle. In *Ecological Toxicology Research. Effects of Heavy Metals and Organohalogen Compounds* (Eds. A. D. McIntyre and C. F. Mills), Plenum Press, New York and London, pp. 145–66.

CHAPTER 12

Toxic Effects of Pollutants on Plankton

G. E. WALSH

Environmental Research Laboratory, United States Environmental Protection Agency, Gulf Breeze, Florida 32561, U.S.A.

12.1 INTRODUCTION 257
12.2 PHYTOPLANKTON 259
12.3 ZOOPLANKTON 267
12.4 RECOMMENDATIONS 269
12.5 REFERENCES 270

12.1. INTRODUCTION

There are four main sources of aquatic pollution: industrial wastes, municipal wastes, agricultural run-off, and accidental spillage. Non-point sources, such as automobile exhausts, add appreciable amounts of pollutants to air that may enter aquatic systems in rainfall or dry fall-out. These sources add pesticides, heavy metals, oil, petroleum products, and a large number of organic and inorganic compounds to water. Lakes and oceans serve as sinks for many pollutants. Plankton comprise a large portion of the living matter in natural waters and function in biogeochemical cycles. They are affected by pollutants, transfer them to sediments and other organisms, and function in their biological transformation.

In natural waters, such as oceans, lakes, rivers, and swamps the greatest amount of biological production is done by the smallest organisms, the plankton. These microscopic plants and animals comprise communities that drift aimlessly with tides and currents, yet they incorporate and cycle large amounts of energy that they pass on to higher trophic levels. Thus communities of plankton, as distinct as those of swamp, forest, or grassland, support other communities of aquatic species and man.

In this chapter, pollution is considered as it affects plankton communities and species. Plankton (Gr. 'wandering') is a general term for those organisms that drift or swim feebly in the surface water of ponds, lakes, streams, rivers, estuaries, and oceans. It is composed of organisms with chlorophyll (phytoplankton) and animals (zooplankton). Phytoplankton is the primary producer community and consists mainly of algae such as diatoms, dinoflagellates, and a variety of forms from other divisions of the plant kingdom. The zooplankton contains consumer species from

Table 12.1 Size Classes of Plankton. (Reproduced by permission of the International Council for the Exploration of the Sea, from Cushing et al., 1958)

Name	Linear dimensions	Organisms
Megaloplankton	1 cm	Squids, salps
Macroplankton	1 mm–1 cm	Large zooplankton
Mesoplankton	0.5–1.0 mm	Small zooplankton, large diatoms
Microplankton	0.06–0.5 mm	Most phytoplankton
Nannoplankton	0.005–0.06	
Ultraplankton	0.0005–0.005 mm	Bacteria, flagellates

all major groups of animals except sponges, bryozoans, brachiopods, ascidians, and mammals (Johnson, 1957).

Plankton is composed of algae and animals in various stages of development. All life stages of holoplanktonic species are completed in the plankton, whereas only a portion of the life cycles of meroplanktonic species occurs in plankton. Planktonic forms are relatively small in size and often lack locomotory organs. Size classes are given in Table 12.1. These classes are often referred to in the scientific literature.

Plankton shares space with swimming animals known collectively as 'nekton' (Gr. 'swimming'). Together, plankton and nekton form the pelagic community, that community associated with open water and not with shore or bottom.

Since many pollutants occur in surface waters, plankton functions are likely to be affected by them. Plankton communities generally exhibit cyclical stability, that is, they vary in composition in relation to changes in light, temperature, and availability of nutrients. Relationships between producers and consumers are determined by:

(1) numbers of each;
(2) efficiency with which energy is incorporated by algae;
(3) the rate of renewal of dominant populations;
(4) the ability of producers to renew consumed production;
(5) the relationship between energy required for maintenance and that available for production in the dominant species (Schwarz, 1975).

All these involve relationships between populations within the plankton community and regulate, to a large extent, those functions that determine the biological characteristics of natural aquatic systems.

At the ecosystem level, natural water bodies may be described by the properties given in Table 12.2. These properties are associated with rates of energy utilization, nutrient cycling, predator–prey relationships, and size of the energy reservoir within the system. A pollutant that affects any one of them can affect the others in relation to the resiliency of the system.

Table 12.2 Metabolic Properties that can be Used to Characterize Ecosystems. (Reprinted, with permission from Reichle, 1975, *Bioscience*, published by the American Institute of Biological Science)

Property	Symbol
Gross Primary Production	GPP
Autotrophic Respiration	R_A
Net Primary Production	NPP
Heterotrophic Respiration	R_H
Net Ecosystem Production	NEP
Ecosystem Respiration	R_E
Production Efficiency	R_A/GPP
Effective Production	NPP/GPP
Maintenance Efficiency	R_A/NPP
Respiration Allocation	R_H/R_A
Ecosystem Productivity	NEP/GPP

Some systems can restore themselves quickly after being perturbed by adjusting community structure or population function. Others are less resilient to environmental change and disappear, leading to a series of developmental changes that results in a stable community under the new conditions. Although composition of the community changes, species diversity may or may not change because stability is a function of community history and is not related to diversity. Therefore, species diversity, by itself, is not necessarily a good indicator of pollution effects on plankton.

There is a large scientific literature dealing with the effects of pollutants on planktonic species of fresh and marine waters. Much less has been reported about effects on communities and ecosystems. Mathematical models can be used to predict the effects of pollution on plankton communities, and a few simple models will be used here to suggest possible effects of selected pollutants.

12.2. PHYTOPLANKTON

Many studies on algal species and specific pollutants have been published. Most have described effects upon population growth or photosynthesis and indicate that, generally, algae are as sensitive to pollutants as animals. Growth and photosynthesis are closely related, each being a function of the utilization of light and nutrients. Dugdale (1975) described the growth of an algal population as being proportional to the effect of light on photosynthesis (Ryther, 1956; Yentsch, 1974), the concentration of nutrients, and the maximum specific growth rate. Pollutants can affect the relation between growth rate and each of these variables. For example, if an industrial effluent is coloured or contains suspended solids, light may be filtered or absorbed by it, resulting in a reduced growth rate. MacIsaac and Dugdale (1976) demonstrated that reduction of light caused reduction in rate of uptake of ammonia and nitrate by marine phytoplankton.

Some chemicals interfere with the Hill reaction of photosynthesis. The Hill reaction is a light-dependent transfer of electrons from cell water to nicotinamide adenine dinucleotide phosphate (NADP) and is inhibited by such compounds as triazines, ureas, carbamates, and acylanilides. Thus chemical pollutants can also block the effect of light on the photosynthetic mechanism and inhibit growth. Walsh (1972) demonstrated that, in four species of marine algae, the EC_{50}'s (concentrations that reduced rate of growth or rate of oxygen evolution by 50%) for Hill reaction inhibitors such as ametryne, atrazine, and diuron were approximately the same for growth and rate of photosynthesis. There was no such correlation between compounds such as silvex, diquat, and trifluralin that had other modes of action (Table 12.3). Overnell (1975) showed that light-induced oxygen evolution from the freshwater species *Chlamydomonas reinhardii* was very sensitive to cadmium, methylmercury, and lead. Moore (1973) found that organochlorine compounds reduce utilization of bicarbonate by estuarine phytoplankton. See the review of Whitacre *et al.* (1972) for effects of many chlorinated hydrocarbons on carbon fixation by phytoplankton.

Pollutants can, therefore, affect photosynthesis and other aspects of energy utilization and incorporation and, thus cause changes in population growth rates. Such changes are most easily seen in systems polluted by algal and plant nutrients. Although there does not seem to be any consistent relationship between nutrient concentrations and biomass of phytoplankton, the rate of autotrophic production of a system may be regulated to a great extent by pollutants (Table 12.4). Goldman (1974), in an extensive study of the eutrophication of Lake Tahoe, found that nitrogen, iron, and phosphorus had a great effect on the rate of primary productivity. He also demonstrated that the rate of primary productivity increased markedly between 1959 and 1971 (Table 12.5). Measurements of primary productivity together with measurements of heterotrophy were the most sensitive indicators of eutrophication in the lake. Measurement of change in the rate of photosynthesis over a number of years may be a sensitive method for the detection of pollution by nutrients.

Pollutants may also affect species composition of the plankton community. Eutrophic systems commonly contain mainly blue–green algae (Cyanophyta), especially in summer (Walsh, 1975), although they do not seem to be more sensitive to pollutants than other algae (Fitzgerald *et al.*, 1952; Whitton, 1970; Venkataraman and Rajyalakshmi, 1971; Voight and Lynch, 1974; DaSilva *et al.*, 1975). Walsh (1975) showed that, in four coastal ponds, cyanophytes comprised nearly 100% of the total algal number, and that the ratio of photosynthesis to respiration was as high as 2.7, indicating a very high productivity. The reason for blooms of cyanophytes is not known, but they often occur after blooms of green algae (Chlorophyta). When cyanophytes begin to grow, the water often contains

Table 12.3 EC$_{50}$ Values (parts per million) for Growth and Photosynthesis (rate of oxygen evolution) by Four Genera of Marine Unicellular Algae when Exposed to Hill Reaction Inhibitors (ametryne, atrazine, and diuron) and Non-Hill Reaction Inhibitors (silvex, diquat, and trifluralin). (Reproduced by permission of the Editor, *Hyacinth Contr. J.* from Walsh, 1972).

Herbicide	Chlorococcum growth	Chlorococcum photo.	Dunaliella growth	Dunaliella photo.	Isochrysis growth	Isochrysis photo.	Phaeodactylum growth	Phaeodactylum photo.
Ametryne	0.01	0.02	0.04	0.04	0.01	0.01	0.02	0.01
Atrazine	0.10	0.10	0.30	0.30	0.10	0.10	0.20	0.10
Diuron	0.01	0.02	0.02	0.01	0.01	0.02	0.01	0.02
Silvex	25	250	25	200	5.0	250	5.0	300
Diquat	200	5,000	30	5,000	15	5,000	15	5,000
Trifluralin	2.5	500	5.0	500	2.5	500	2.5	500

Table 12.4 Rates of Phytoplankton Production in Oligotrophic and Eutrophic Lakes. (Reproduced by permission of the National Academy of Sciences from Rodhe, 1969)

Type of lake	Rate of production (g C/m^2/yr)
Oligotrophic	7–25
Natural eutrophic	75–250
Polluted eutrophic	350–700

only small amounts of nitrogen and phosphorus. Fogg *et al.* (1973) described the ability of cyanophytes to fix nitrogen, and this could give them a competitive advantage when nitrogen is in very low concentration. Also, blue–green algae require sodium and potassium, the optimum being 2.5 to 5.0 ppm in water (Emerson and Lewis, 1942). These elements are often found in the required concentrations in polluted waters, a fact that led Provasoli (1958) to suggest that their presence is a factor that selects cyanophytes over chlorophytes.

Lund (1969) suggested that, since cyanophytes are seldom grazed upon by herbivores, they have a competitive advantage over chlorophytes which are grazed upon. Talling (1965) suggested that blue–green algae are not harmed by high temperature and intense illumination in summer and that they may be selected because of this adaptation.

Davies (1976) measured the effect of Hg^{2+} on the specific growth rate of *Dunaliella tertiolecta*, a marine alga, in culture. The specific growth rates of control and treated cultures with up to 2.03 micromolar Hg^{2+} varied between 0.65 and 0.70. At 5.02 micromolar Hg^{2+} it was 0.43, and at 10.0 micromolar Hg^{2+} it was 0.11. At 10.0 micromolar Hg^{2+}, growth was uncoupled from cell division and giant cells were produced. Davies suggested that growth without division was due to

Table 12.5 Total Primary Productivity of Phytoplankton per Year in Lake Tahoe. (Reproduced by permission of the United States Environmental Protection Agency from Goldman, 1974)

Year	Productivity (mg·C/m^2/yr)	Percentage increase from 1959
1959	38,958	
1968	46,685	20
1969	50,525	30
1970	52,467	35
1971	58,655	51

inhibition of methionine production by mercury. It is important, therefore, when testing for effects of some pollutants, that total volume or weight of living matter be measured, and not only cell numbers.

Algal growth may be inhibited or stimulated by pollutants that affect nutrient relationships. The mechanisms whereby they affect the chemical state of a nutrient or its uptake and utilization vary according to nutrient, physical and chemical conditions of the water, and algal species. Meijer (1972) showed that a decrease in the phosphate concentration in the growth medium caused an increase in toxicity of copper to *Chlamydomonas* sp. as measured by oxygen production and an increase in cell numbers. There was a strong effect of 2 mg PO_4/l at the very high concentration of 2 mg Cu/l. Steeman-Nielsen *et al.* (1969) reported similar results and suggested that the toxic effect of copper is due to binding to the cytoplasmic membrane, thus arresting cell division. Zarnowski (1972) demonstrated this to be true with iron and *Chlorella pyrenoidosa* and stated that phosphorus compounds cause oxidation and precipitation of iron from water. Sheih and Barber (1973) showed that mercury in low concentration (0.03 mM $HgCl_2$) as mercuric chloride stimulated the rate of potassium turnover in *C. pyrenoidosa* but did not affect the concentration of internal potassium. At higher concentrations (0.5 mM $HgCl_2$) mercury caused changes in permeability of the cell membrane that resulted in a net efflux of potassium.

The stimulatory effect of low concentrations of heavy metals on algae is well known. DeFilippis and Pallaghy (1976) hypothesized that 1 mM $ZnCl_2$, 1 μM $HgCl_2$, or 0.1 μM phenylmercuric acetate stimulated growth by preventing loss of carbon compounds from *Chlorella* sp. They suggested that heavy metals either interfere with production of glycollate or inhibit its secretory mechanism.

Such actions by heavy metals are often related to ionic state. In sediments at low redox potential, Hg^{2+} is reduced to Hg^0 with the formation of dimethylmercury by methyl radical addition. Organic mercurial compounds are sometimes more toxic than inorganic compounds. Harriss *et al.* (1970) showed that 0.1 ppb (μg/l) of methylmercury reduced photosynthesis (as ^{14}C uptake) and growth of the marine diatom *Nitzschia delicatissima*. However, dimethylmercury was less toxic; 1 ppb was required for an effect. The authors stated that the alga was sensitive to much lower concentrations of mercurial compounds than fishes. Davies (1976) suggested that some algae detoxify mercury by reduction of Hg^{2+} to Hg^+ in the outer layers of the cell. Zingmark and Miller (1975) noted that net photosynthesis by *Amphidinium carteri* was decreased 92% by 100 ppb mercury. However, Hg^{2+} was reduced to Hg^0 during the test.

A very important aspect of uptake and toxicity of heavy metals to phytoplankton is the role of chelating agents. Plum and Lee (1973) showed that up to 20% of the iron in a lake was associated with organic matter. Andren (1973) found that mercury in the water of three estuaries was associated with dissolved organic matter. Dissolved mercury was associated with dissolved organic matter of the humic and fulvic types. In a Norwegian lake, Beneš *et al.* (1976), showed that sodium, calcium, aluminium, chromium, iron, zinc, manganese, mercury, and other elements were bound strongly to humus. Slowey *et al*. (1967) indicated that up to 50% of the total copper present in sea water from the Gulf of Mexico was complexed with phospholipid, amino lipid, or porphyrin fractions. The samples were taken from the sea surface, and the authors speculated that the organic compounds were present because of rupture of fragile planktonic organisms.

Perdue *et al*. (1976) described competition between metals for binding sites on dissolved organic matter; thus iron, for example, may be present in water as (1) free colloidal particles of iron hydroxide, possibly associated with colloidal organic matter, and (2) dissolved complexes of iron with naturally occurring organic substances. The authors gave evidence that aluminium competes with iron for sites on molecules of dissolved organic matter. Hendrickson *et al*. (1974) demonstrated the high affinity of cadmium for dissolved organic matter from a lake in Wisconsin.

Formation of chelates implies that the amount of dissolved matter may, in some cases, regulate toxicity of metals because chelation decreases toxicity to a very great extent. Lomonosov (1969) showed that copper dichloride pyridinate inhibited growth of *Scenedesmus quadricauda* at 0.4 mg Cu/l. Copper chelates were much less toxic. Walsh *et al*. (unpublished) have found that when the marine unicellular alga *Dunaliella tertiolecta* is exposed to nickel in the presence of the chelator ethylenediaminetetraacetic acid (EDTA) its EC_{50} was 9.1 ppm. When EDTA was excluded from growth media the EC_{50} was 2.4 ppm.

They have also found (unpublished) that the herbicide 2,4-D inhibits the toxicity of nickel and aluminium. One type of interaction between pollutants can be formulated by the expression (Colby, 1967)

$$E = \frac{XY}{100} \tag{12.1}$$

where

E = expected population density of algae as a percentage of the control

X = population density as a percentage of the control when exposed to chemical X

Y = population density as a percentage of the control when exposed to chemical Y

When the observed value of E is less than the expected value, the combination acts synergistically; when equal, the combination acts additively; when greater, the interaction is antagonistic. Table 12.6 shows that nickel and 2,4-D react antagonistically; therefore, in evaluating effects of pollutants on algae, it is necessary to consider the chemical nature of dissolved materials, including other pollutants.

In other types of interactions, Tsay *et al*. (1976) demonstrated that copper ions (2 ppm) inhibited toxicity of the insecticide paraquat (1,1-dimethyl-4,4'-bipyridinum ion, 2 ppm) to *C. pyrenoidosa*, whereas cyanide ion (16 ppm)

Table 12.6 Calculated and Observed Values of E (equation 12.1) for Three Genera of Marine Unicellular Algae Exposed to Nickel (as $NiCl_2$) and the Technical Acid of 2,4-D (Walsh, unpublished)

Concentration	Chlorococcum calc.	Chlorococcum obs.	Dunaliella calc.	Dunaliella obs.	Thalassiosira calc.	Thalassiosira obs.
1 ppm Ni + 50 ppm 2,4-D	84.0	84.5	79.6	98.9	30.2	35.8
2 ppm Ni + 50 ppm 2,4-D	55.9	75.9	64.3	96.0	29.7	37.6
4 ppm Ni + 50 ppm 2,4-D	34.9	63.8	53.6	88.6	13.9	36.4

enhanced herbicide activity. The toxic effects of phenylurea herbicides to *Chlorella vulgaris* were shown by Kruglov and Kvyatkovskaya (1975) to be inversely related to the humus content of the soil.

It is clear, therefore, that many pollutants exert their effects in relation to the uptake dynamics of nutrients, heavy metals, carbon, etc. The dynamics of such uptake are explained by the expressions of Volterra (1926):

$$\frac{dN_1}{dt} = N_1[\epsilon_1 - r_1(h_1N_1 + h_2N_2)] \quad (12.2)$$

$$\frac{dN_2}{dt} = N_2[\epsilon_2 - r_2(h_1N_2 + h_2N_2)] \quad (12.3)$$

where

N_1, N_2 = numbers of two competing species
ϵ_1, ϵ_2 = their coefficients of increase with unlimited nutrient
r_1, r_2 = constants of susceptibility to nutrient shortage
h_1, h_2 = constants of nutrient consumption.

The equations are based on the assumptions that

(1) rate of depletion of a nutrient pool is directly proportional to size of the population;
(2) depletion of the pool reduces the growth rate;
(3) the coefficients of increase, which are related to efficiency of utilization of nutrients, are different for each species.

The equation predicts dominance of one species over others; i.e. the species that is most susceptible to the influence of pollution will be eliminated in competition with a species of lesser susceptibility.

Dominance is determined by many factors, and, for any given species, may be related to temperature, salinity, irradiation, pollution, etc. Mandelli (1969) reported that the growth of selected species of dinoflagellates and diatoms was inhibited by copper between 0.055 and 0.256 $\mu g/ml$. However, there was a positive correlation between the log of the ratio of copper uptake, biomass of algae and temperature, but a negative correlation between that ratio and salinity. Walsh and Grow (1971) demonstrated a direct correlation between carbohydrate concentration in algal cells and concentration of urea herbicides in the growth medium.

If it is assumed that the coefficients of increase of different species are affected in different ways by pollutants in relation to environmental conditions, then changes in species composition of a community are to be expected. Such differences were shown by Hollister and Walsh (1973) who reported the responses of 18 species of marine unicellular algae to four compounds that inhibit the Hill reaction. There were large differences in the coefficients of increase among the species. Bowes (1972) reported that 80 μg/l of DDT had no effect on growth of the marine alga *Dunaliella tertiolecta*, had slight effect on growth of *Cyclotella nana*, *Thalassiosira fluviatilis*, *Amphidinium carteri*, *Coccolithus huxleyi*, and *Porphyridium* sp., but caused a lag of nine days in growth of *Skeletonema costatum*. Fisher et al. (1973) showed that estuarine clones of the algae *Thalassiosira pseudonana*, *Fragilaria pinnata*, and *Bellerochia* sp. were more resistant to PCBs (polychlorinated biphenyls) than were oceanic clones. Dunstan et al. (1975) demonstrated effects of oil on marine phytoplankton and concluded that a significant effect of oil could be the stimulation of some species by aromatic compounds of low molecular weight, resulting in the alteration of natural phytoplankton community structure and its trophic relationships. These works suggest that, in mixed culture, the competitive exclusion principle of Gause (1934) would operate, causing species best adapted to the polluted condition to dominate.

Changes in structure of a marine algal community under stress from PCBs and DDT were demonstrated by Mosser et al. (1972) and Fisher et al. (1974). In untreated mixed cultures, *T. pseudonana* was dominant in number over *D. tertiolecta*. When treated with 1 ppb of PCBs or 10 ppb of DDT, the competitive success of *T. pseudonana* decreased while that of *D. tertiolecta* increased, even though the pollutant concentration had no effect on *T. pseudonana* in pure culture.

Another important aspect of pollution is the fact that algae accumulate many compounds and heavy metals by absorption or by adsorption to the cell wall. In general, pollutants such as pesticides are not degraded to a large extent by algae (Butler et al., 1975) and are passed on to herbivores when eaten. Hollister et al. (1975) showed that four species of marine phytoplankton, when exposed to only 10 pptr (parts per trillion) of the insecticide mirex, concentrated it to over 100 ppb (parts per billion) on cells. Numerous other publications report uptake of pollutants by phytoplankton (Falchuk et al., 1975; Stokes, 1975; DeFilippis and Pallaghy, 1976; Glooschenko and Lott, 1976).

Uptake and accumulation of substances may prove to be the most important aspect of phytoplankton pollution dynamics. Since they are grazed upon by many animals, the phytoplankton may be a vital link in the transfer of pollutants from water to higher trophic levels, including man.

12.3. ZOOPLANKTON

Zooplankton comprise a very dynamic portion of the total plankton. There is greater species diversity among the zooplankton than phytoplankton, and composition varies strongly with seasonal production of meroplankton such as medusae, eggs, larvae, and juveniles of the benthos and nekton. Functionally, the zooplankton includes detritivores, herbivores, carnivores, and omnivores, all of which excrete dissolved and particulate organic and inorganic materials that can serve as nutrients for saprovores, phytoplankton, and coprovores. Organic and inorganic materials are lost from the water column when organisms die and settle to the bottom, where they enter geochemical cycles involving sediments.

Zooplankton population production is a function of the intrinsic rate of increase, natural mortality, mortality due to pollution, and rate of predation:

$$\begin{array}{c}\text{Rate of increase}\\\text{of population}\end{array} = \begin{array}{c}\text{Rate of change in the absence}\\\text{of predators and pollution}\end{array} - \begin{array}{c}\text{Rate of}\\\text{predation}\end{array} - \begin{array}{c}\text{Rate of loss}\\\text{due to effects}\\\text{of pollution}\end{array}$$

It is difficult to measure an effect of pollution on the rate of predation, but such an effect is theoretically probable. For example, if a compound such as chlorinated hydrocarbon insecticide were present, it could affect the efficiency of a prey arthropod to avoid a predatory fish. Since arthropods are generally more sensitive to the direct influence of such pollutants than fishes, their rates of population increase could be expected to decline because of increased predation and direct mortality due to insecticide.

Hansen (1974) studied the effect of 'Aroclor' 1254, a polychlorinated biphenyl (PCB), on settling of planktonic larvae (meroplankton). He used the composition of estuarine animal communities that developed while meroplankton were subjected to three concentrations of the pollutant for four months. Arthropods constituted over 75% of the organisms in untreated aquaria and aquaria that received 0.1 μg/l of PCB. Two other aquaria that received 1.0 and 10 μg/l were dominated by chordates, primarily tunicates. Just as important, however, is the fact that species diversity, as measured by the Shannon-Weaver index (Shannon and Weaver, 1963), was not affected, indicating that such an index cannot always measure pollution effects on planktonic larvae.

Conversely, temporal changes in zooplankton community composition may be modified if benthic and nektonic contributions to the meroplankton are affected by pollution. Numerous reports show that zooplanktonic species and meroplanktonic larvae are often much more sensitive to pollutants than adults. Armstrong and Millemann (1974) showed that, for the mussel, *Mytilus edulis*, the developmental stage most sensitive to the insecticide 'Sevin' occurred shortly after fertilization, at the time of appearance of the first polar body. Sensitivity decreased in each succeeding stage. The EC_{50} of 'Sevin' at the first polar body stage was 5.3

ppm. At the 32-hr stage, the EC_{50} was 24.0 ppm. Affected eggs were characterized by dysjunction of blastomeres, reduction in rate of development, and asynchronous and unaligned cleavages. Bookhout and Costlow (1976) demonstrated that certain zoeal stages of the crabs *Callinectes sapidus* and *Rhitropanopeus harrisii* were the most sensitive to the insecticides mirex, methoxychlor, and malathion, and that the total time from hatching to the first crab stage was prolonged by methoxychlor and malathion. The total time of larval development of the crabs, *Menippe mercinaria* and *R. harrisii*, was lengthened by 0.01 to 10.0 ppb of mirex. Crabs that had been reared in 0.1 ppb of mirex concentrated it 2,400 times in their bodies (Bookhout *et al.*, 1972). Rosenberg (1972) found reduction in the swimming activity of stage II nauplii of the barnacle, *Balanus balanoides*, by 2 ppm of chlorinated aliphatic hydrocarbons. Stages V and VI were affected by 4 ppm. Young juvenile mullet, *Mugil cephalus*, were more susceptible to mirex exposure than older juveniles or adults (Lee *et al.*, 1975), and cadmium reduced the incubation time (50% hatching time) of herring (*Clupea harengus*) eggs. Conversely, Bills (1974) reported that formalin and malachite green were more toxic to rainbow trout, *Salmo gairdneri*, fingerlings than to the eyed egg stage.

Generally, greater susceptibility of young to pollutants could have a strong effect upon all factors that determine the rates of increase of prey and predator species. Also, change in the carrying capacity of herbivorous zooplankton could affect the carrying capacity for phytoplankton.

Reduced stocks of young that survive to maturity could greatly affect the size of a reproductive population. For example, temporal extension of larval stages or weakening of young would make them vulnerable to excessive predation and ultimately reduce the size of the reproducing population during the next production year. Nimmo *et al.* (in press), using the mysid *Mysidopsis bahia*, found a 24-hour delay in the formation of brood pouches and reduced production of young by females exposed to 6.4 µg Cd/l.

Adult zooplankton are affected by pollutants and accumulate relatively large amounts. Uptake can occur either through ingestion of contaminated food or directly from water. When Burnett (1973) fed a marine copepod, *Tigriopus* sp., phytoplankton exposed to DDE, growth rate and egg production were reduced. Maki and Johnson (1975) demonstrated reproductive inhibition in the freshwater species, *Daphnia magna*, by DDT and a commercial formulation of PCBs. Reproduction was inhibited by 0.30 ppb of DDT and 10.0 ppb of PCB. In static toxicity tests that used survival as the criterion of effect, the LC_{50}s were 0.67 ppb for DDT and 24.0 for PCB. Thus the EC_{50} based on the number of young produced was approximately one-half of the EC_{50} based on survival. Baudouin and Scoppa (1974) compared the responses of several species of freshwater zooplankton to twelve heavy metals and concluded that members of the genus *Daphnia* were considerably more sensitive than were copepods. They recommended that this genus be used in toxicity tests with metals. Khan and Khan (1974) showed that *Daphnia* accumulated larger amounts of photodieldrin than other aquatic invertebrates. Its magnification ratio (concentration in animal/concentration in water) was 63,000 when exposed to 6 ppb. Body residues were greatest when contaminated algae were fed to *Daphnia pulex* (Khan *et al*., 1975) but up to 70% of the pollutant was depurated in 2 days. Dieldrin and photodieldrin were not metabolized (Neudorf and Khan, 1975).

Varying sensitivities to pollutants among zooplankton species could cause changes in community structure by affecting variables such as rate of increase, rate of predation, mortality, and population density. This situation was demonstrated by Sprules (1975) who showed that major changes in composition of crustacean zooplankton communities of acid-stressed lakes were related to pH. Industrial acidification caused decreases in the number of species and changes in species dominance, both of which were affected as pH fell from 7.0 to 3.8.

In general, zooplankton do not degrade pesticides to a great degree. Sameoto *et al*. (1975) reported 68 to 1,757 ppb DDT and 24 to 937 ppb DDE in the lipids of three species of euphausiids in Canada. No other metabolites were found. Darrow and Harding (1975) were unable to demonstrate metabolic products of DDT in marine copepods, but Burnett (1973) stated that DDT was converted to DDE in the marine copepod, *Tigriopus*. In fresh water, DDT was converted to DDE by *Daphnia pulex* (Neudorf and Khan, 1975).

Lee (1975) reported that marine copepods, euphausiids, crab zoea, ctenophores, and jelly-fish rapidly took up benzpyrene, methyl cholanthrene, and naphthalene from sea-water solution. They were metabolized to various hydroxylated and more polar metabolites by the crustaceans but not by ctenophores or jelly-fish. Four species of copepods depurated the hydrocarbons, but depuration was never complete.

12.4. RECOMMENDATIONS

There are many models for explaining and predicting the effects of pollutants on the ecosystem dominated by plankton, and these models may serve as hypotheses for planning research. There is a lack of data for verification of the present models and for generation of new, more accurate models.

At the species level, data are needed with regard to effects of pollution on species composition of plankton communities, and this implies understanding of effects on single species. At present, it is impossible to predict effects on communities from data on single species because little is known about the requirements under conditions of competition, for temperature, salinity, maximal energy utilization, and other physiological and ecological considerations.

We need to know more about lethal and sublethal effects of pollutants on geographical variants of single species. Oceanic clones tend to be more susceptible than estuarine clones. The reason for this is not known, though it may rest in genetic differences. Pollutants can affect nucleic acids within cells, and it would be beneficial to know the genetic basis of tolerance.

At the ecosystem level, data on pollution and the basic functions listed in Table 12.2 are needed. Studies on nutrient cycling may lead to the development of sensitive methods for detection of sublethal effects. Possible uses of artificial ecosystems should be investigated in this regard.

Mathematical and conceptual models must be developed to explain the effects of pollution on structure and function of populations and communities and their physical, chemical, and biological environments.

Basic marine microbial ecology requires study. Information on microbial population structure, species successions, and interactions must be obtained. Very few data are presently available.

The interactive and sequential events occurring in microbial degradation of oil and other pollutants must be clearly understood. For example, at present, the succession of groups of microorganisms in mixed cultures degrading petroleum hydrocarbons is only partially understood (see Chapter 13). The mechanisms of biodegradation *in situ*, such as co-metabolism, have not been sufficiently studied so that the biodegradative events that follow microbial seeding of spills cannot be predicted with accuracy.

The action of microorganisms in the production and accumulation of carcinogenic compounds in the marine environment needs to be elucidated.

The role of microorganisms, i.e. the bacteria, yeasts, and fungi, in the marine food chains must be elucidated, since amplification of carcinogenic compounds may occur and, thereby, pose more serious environmental problems.

12.5. REFERENCES

Andren, A. W., 1973. The geochemistry of mercury in three estuaries from the Gulf of Mexico. *Ph.D. Dissertation*, University of Florida, Gainesville, 140 pp.

Armstrong, D. A. and Millemann, R. E., 1974. Effects of the insecticide 'Sevin' and its first hydrolytic product, 1-naphthol, on some early developmental stages of the bay mussel *Mytilus edulis. Mar. Biol. (Berl.)*, 28, 11–5.

Baudouin, M. F. and Scoppa, P. S., 1974. Acute toxicity of various metals to freshwater zooplankton. *Bull. Environ. Contam. Toxicol.*, 12, 745–51.

Beneš, P., Gjessing, E. T., and Steinnes, E., 1976. Interactions between humus and trace elements in freshwater. *Water Res.*, 10, 711–6.

Bills, T. D., 1974. Toxicity of formalin, malachite green, and the mixture to four life stages of rainbow trout. *M.Sc. Thesis*, University of Wisconsin, La Crosse, 41 pp.

Bookhout, C. G. and Costlow, J. D., 1976. Effects of mirex, methoxychlor, and malathion on development of crabs. *U.S. Environmental Protection Agency*, Pub. No. EPA-600/3–76–007, Gulf Breeze, Florida, 85 pp.

Bookhout, C. G., Wilson, A. J. Jr., Duke, T. W., and Lowe, J. I., 1972. Effects of mirex on the larval development of two crabs. *Water, Air, Soil Pollut.*, 1, 165–80.

Bowes, G. W., 1972. Uptake and metabolism of 2,2-bis-(*p*-chlorophenyl)-1,1,1-trichloroethane (DDT) by marine phytoplankton and its effect on growth and electron transport. *Plant Physiol*, 49, 172–6.

Burnett, R. D., 1973. DDT in marine phytoplankton and crustacea. *Diss. Abstr. Int. B. Sci. Eng.*, **34**, 533.
Butler, G. L., Deason, T. R., and O'Kelley, J. C., 1975. Loss of five pesticides from cultures of twenty-one planktonic algae. *Bull. Environ. Contam. Toxicol.*, **13**, 149–52.
Colby, R. S., 1967. Calculating synergistic and antagonistic responses of herbicide combinations. *Weeds*, **15**, 20–2.
Cushing, D. H., Humphrey, G. F., Banse, K., and Laevastu, T., 1958. Report of the committee on terms and equivalents. In *Measurements of Primary Production in the Sea*, Coun. Perm. Internat. Explor. Mer, Copenhagen, pp. 15–16.
Darrow, D. C. and Harding, G. C. H., 1975. Accumulation and apparent absence of DDT metabolism by marine copepods, *Calanus* sp., in culture. *J. Fish. Res. Bd. Canada*, **32**, 1845–9.
DaSilva, E. J., Henriksson, L. E., and Henriksson, E., 1975. Effect of pesticides on blue–green algae and nitrogen-fixation. *Arch. Environ. Contam. Toxicol.*, **3**, 193–204.
Davies, A. G., 1976. An assessment of the basis of mercury tolerance in *Dunaliella tertiolecta*. *J. Mar. Biol. Assoc. U.K.*, **56**, 39–57.
DeFilippis, L. F. and Pallaghy, C. K., 1976. The effect of sublethal concentrations of mercury and zinc on *Chlorella*. I. Growth characteristics and uptake of metals. *Z. Pflanzenphysiol.*, **78**, 197–207.
Dugdale, R. C., 1975. Biological modelling I. In *Modelling of Marine Systems* (Ed. J. C. J. Nihoul), Elsevier Pub. Co., New York, pp. 187–205.
Dunstan, W. M., Atkinson, L. P., and Natoli, J., 1975. Stimulation and inhibition of phytoplankton growth by low molecular weight hydrocarbons. *Mar. Biol.*, **31**, 305–10.
Emerson, R. and Lewis, C. M., 1942. The photosynthetic efficiency of phycocyanin in *Chroococcus* and the problem of carotenoid participation in photosynthesis. *J. Gen. Physiol.*, **25**, 579–95.
Falchuk, K. H., Fawcett, D. W., and Vallee, B. L., 1975. Role of zinc in cell division of *Euglena gracilis*. *J. Cell. Sci.*, **17**, 57–8.
Fisher, N. S., Carpenter, E. J., Remsen, C. C., and Wurster, C. F., 1974. Effects of PCB on interspecific competition in natural and gnotobiotic phytoplankton communities in continuous and batch cultures. *Microbiol. Ecol.*, **1**, 39–50.
Fisher, N. S., Graham, L. B., Carpenter, E. J., and Wurster, C. F., 1973. Geographic differences in phytoplankton sensitivity to PCBs. *Nature*, **241**, 548–9.
Fitzgerald, G. P., Gerloff, G. C., and Skoog, F., 1952. Studies on chemicals with selective toxicity to blue–green algae. *Sewage Indust. Wastes*, **24**, 888–96.
Fogg, G. E., Stewart, W. D. P., Fay, P., and Walsby, A. E., 1973. *The Blue-Green Algae*, Academic Press, New York, vii, 459 pp.
Gause, G. F., 1934. *The Struggle for Existence*, Williams and Wilkins, Baltimore, 163 pp.
Glooschenko, V. and Lott, J. N. A., 1976. Effects of chlordane on the green alga *Scenedesmus quadricauda*. *Abst. Thirty-ninth Ann. Meeting, Am. Soc. Limnol. Oceanogr.*, Savannah, Georgia.
Goldman, C. R., 1974. Eutrophication of Lake Tahoe emphasizing water quality. *U.S. Environmental Protection Agency, Ecological Res. Ser.*, EPA-660/3–74–034, Corvallis, Oregon, xvii, 408 pp.
Hansen, D. J., 1974. 'Aroclor' 1254. Effect on composition of developing estuarine animal communities in the laboratory. *Contrib. Mar. Sci.*, **18**, 19–33.

Harriss, R. C., White, D. B., and Macfarlane, R. B., 1970. Mercury compounds reduce photosynthesis by plankton. *Science*, 170, 736–7.

Hendrickson, D. W., Armstrong, D. E., Veith, G. D., and Glass, G. E., 1974. Nature of organic derivatives of selected toxic metals in natural waters. *Univ. Wisconsin, Water Resources Center, Tech. Rep.*, WIS WRC 74–07, iii, 23 pp.

Hollister, T. A. and Walsh, G. E., 1973. Differential responses of marine phytoplankton to herbicides: oxygen evolution. *Bull. Environ. Contam. Toxicol.*, 9, 291–5.

Hollister, T. A., Walsh, G. E., and Forester, J., 1975. Mirex and marine unicellular algae: accumulation, population growth and oxygen evolution. *Bull. Environ. Contam. Toxicol.*, 14, 753–9.

Johnson, M. W., 1957. Plankton. In *Treatise on Marine Ecology and Paleoecology*, Vol. I, *Ecology* (Ed. J. W. Hedgepeth), pp. 443–59, Geol. Soc. Am. Mem. No. 67, vii, 1296 pp.

Khan, H. M. and Khan, M. A. Q., 1974. Biological magnification of photodieldrin by food chain organisms. *Arch. Environ. Contam. Toxicol.*, 2, 289–301.

Khan, H. M., Neudorf, S., and Khan, M. A. Q., 1975. Absorption and elimination of photodieldrin by *Daphnia* and goldfish. *Bull. Environ. Contam. Toxicol.*, 13, 582–7.

Kruglov, Y. V. and Kvyatkovskaya, L. B., 1975. Algae as indicators of herbicide contamination of soil. *Rocz. Glebozn.*, 26, 145–9.

Lee, J. H., Nash, C. E., and Sylvester, J. R., 1975. Effects of mirex and methoxychlor on striped mullet, *Mugil cephalus* L. *U.S. Environmental Protection Agency, Pub.*, No. EPA-660/3–75–015, Gulf Breeze, Florida, 18 pp.

Lee, R. F., 1975. Fate of petroleum hydrocarbons in marine zooplankton. *Proc. 1975 Conf. Prevent. Cont. Oil Poll.*, Am. Pet. Inst., Washington, D.C. pp. 549–53.

Lomonosov, M. V., 1969. The toxicity of copper complexes towards *Scenedesmus quadricauda. Mikrobiologiya*, 38, 729–31.

Lund, J. W. G., 1969. Phytoplankton. In *Eutrophication: Causes, Consequences, Correctives*, pp. 306–30. U.S. Nat. Acad. Sci. Washington, D.C., vii, 661 pp.

MacIsaac, J. J. and Dugdale, R. C., 1976. Inorganic nitrogen uptake by marine phytoplankton under *in situ* and simulated *in situ* incubation conditions: results from the northwest African upwelling region. *Limnol. Oceanogr.*, 21, 149–52.

Maki, A. W. and Johnson, H. E., 1975. Effects of PCB ('Aroclor' 1254) and *p,p'*-DDT on production and survival of *Daphnia magna* Strauss. *Bull. Environ. Contam. Toxicol.*, 13, 412–6.

Mandelli, E. E., 1969. The inhibitory effects of copper on marine phytoplankton. *Contrib. Mar. Sci.*, 14, 47–57.

Meijer, C. L. C., 1972. The effect of phosphate on the toxicity of copper for an alga (*Chlamydomonas* sp.) (in Dutch). *TNO-nieuws*, 27, 468–73.

Moore, S. A., Jr., 1973. Impact of pesticides on phytoplankton in Everglades estuaries. *South Florida Environmental Project, Ecological Rep.*, No. DI-SFEP-74–15, 100 pp.

Mosser, J. L., Fisher, N. S., and Wurster, C. F., 1972. Polychlorinated biphenyls and DDT alter species composition in mixed cultures of algae. *Science*, 176, 533–5.

Neudorf, S. and Khan, M. A. Q., 1975. Pick-up and metabolism of DDT, dieldren and photodieldrin by a freshwater alga (*Ankistrodesmus amalloides*) and a microcrustacean (*Daphnia pulex*). *Bull. Environ. Contam. Toxicol.*, 13, 443–50.

Nimmo, D. R., Rigby, R. A., Bahner, L. H., and Sheppard, J. M. The acute and chronic effects of cadmium on the estuarine mysid *Mysidopsis bahia. Bull. Environ. Contam. Toxicol.*, 19(1), 80–85.

Overnell, J., 1975. The effect of some heavy metal ions on photosynthesis in a freshwater alga. *Pestic. Biochem. Physiol.*, 5, 19–26.

Perdue, E. M., Beck, K. C., and Reuter, J. H., 1976. Organic complexes of iron and aluminum in natural waters. *Nature*, 260, 418–20.

Plum, R. H., Jr. and Lee, G. F., 1973. A note on the iron-organic relationship in natural water. *Water Res.*, 7, 581–5.

Provasoli, L., 1958. Nutrition and ecology of protozoa and algae. *Ann. Rev. Microbiol.*, 12, 279–308.

Reichle, D. E., 1975. Advances in ecosystem analysis. *BioScience*, 25, 257–64.

Rodhe, W., 1969. Crystallization of eutrophication concepts in northern Europe. In *Eutrophication: Causes, Consequences, Correctives*, pp. 50–64, U.S. Nat. Acad. Sci. Washington, D.C., vii, 661 pp.

Rosenberg, R., 1972. Effects of chlorinated aliphatic hydrocarbons on larval and juvenile *Balanus balanoides. Environ. Pollut.*, 3, 313–8.

Ryther, J. H., 1956. The measurement of primary production. *Limnol. Oceanogr.*, 1, 72–84.

Sameoto, D. D., Darrow, D. C., and Guildford, S., 1975. DDT residues in euphausiids in the upper estuary of the Gulf of St. Lawrence. *J. Fish. Res. Bd. Canada*, 32, 310–4.

Schwarz, S. S., 1975. The flow of energy and matter between trophic levels (with special reference to the higher levels). In *Unifying Concepts in Ecology* (Eds. W. H. van Dobben and R. H. Lowe-McConnell), W. Junk, Publishers, Centre for Agricultural Publishing and Documentation, Wageningen, The Netherlands, pp. 50–60.

Shannon, C. E. and Weaver, W., 1963. *The Mathematical Theory of Communication*, University of Illinois Press, Urbana.

Sheih, Y. J. and Barber, J., 1973. Uptake of mercury by *Chlorella* and its effect on potassium regulation. *Planta*, 109, 49–60.

Slowey, J. F., Jeffrey, L. M., and Hood, D. W., 1967. Evidence for organic complexed copper in sea water. *Nature*, 214, 377–8.

Sprules, W. G., 1975. Midsummer crustacean zooplankton communities in acid-stressed lakes. *J. Fish. Res. Bd. Canada*, 32, 389–95.

Steeman-Nielsen, E., Kamp-Nielsen, L., and Wium-Anderson, S., 1969. The effect of deleterious concentrations of copper on the photosynthesis of *Chlorella pyrenoidosa. Physiol. Plant.*, 22, 1121–33.

Stokes, P., 1975. Uptake and accumulation of copper and nickel by metal-tolerant strains of *Scenedesmus. Verh. Internat. Verein. Limnol.*, 19, 2128–37.

Talling, J. F., 1965. The photosynthetic activity of phytoplankton in East African lakes. *Int. Rev. Ges. Hydrobiol. Hydrograph*, 50, 1–32.

Tsay, S.-F., Lee, J.-M., and Lynd, J. Q., 1976. The interactions of Cu^{++} and CN^- with paraquat phytotoxicity to a *Chlorella. Weed Sci.*, 18, 596–8.

Venkataraman, G. S. and Rajyalakshmi, B., 1971. Tolerance of blue–green algae to pesticides. *Cur. Sci. (Bangalore)*, 6, 143–4.

Voight, R. A. and Lynch, D. L., 1974. Effects of 2,4-D and DMSO on procaryotic and eucaryotic cells. *Bull. Environ. Contam. Toxicol.*, 12, 400–5.

Volterra, V., 1926. Variations and fluctuations in the number of individual species of animal communities (in Italian). *Mem. Acad. Lincei, Ser. 6*, 2, 31–113.

Walsh, G. E., 1972. Effects of herbicides on photosynthesis and growth of marine unicellular algae. *Hyacinth Cont. J.*, 10, 45–8.

Walsh, G. E., 1975. Utilization of energy by primary producers in four ponds in northwestern Florida. In *Proceedings: Biostimulation and Nutrient Assessment*

Workshop, pp. 249—74, U.S. Environmental Protection Agency, Ecological Research Series, EPA-660/3—75—034, iv, 319 pp.

Walsh, G. E. and Grow, T. E., 1971. Depression of carbohydrate in marine algae by urea herbicides. *Weed Sci.*, **19**, 568—70.

Whitacre, D. M., Roan, C. C., and Ware, G. W., 1972. Pesticides and aquatic microorganisms. *Search*, **3**, 150—7.

Whitton, B. A., 1970. Toxicity of heavy metals to freshwater algae: A review. *Phykos*, **9**, 116—25.

Yentsch, C. S., 1974. Some aspects of the environmental physiology of marine phytoplankton: a second look. *Oceanogr. Mar. Biol. Ann. Rev.*, **12**, 41—75.

Zarnowski, J., 1972. The effect of ethylenediamine-tetraacetic acid on the growth of *Chlorella pyrenoidosa* and its role in the dynamics of metabolism and accessibility of iron and calcium. *Acta Hydrobiol.*, **14**, 353—73.

Zingmark, R. G. and Miller, T. G., 1975. The effects of mercury on the photosynthesis and growth of estuarine and oceanic phytoplankton. *Belle W. Baruch Libr. Mar. Sci.*, **3**, 45—57.

CHAPTER 13

Toxic Effects of Pollutants on Microorganisms

R. R. COLWELL

Department of Microbiology, University of Maryland, College Park, Maryland 20742, U.S.A.

13.1 INTRODUCTION 275
13.2 MICROBIAL DEGRADATION OF POLLUTANTS 276
13.3 MICROBIAL ECOTOXICOLOGICAL EFFECTS OF PETROLEUM 278
13.4 CARCINOGENICITY OF POLLUTANTS ASSOCIATED WITH
     MICROBIAL ACTIVITY 284
13.5 MICROBIAL EFFECTS ASSOCIATED WITH POLYCHLORINATED
     BIPHENYLS 285
13.6 EFFECTS OF COMBINATIONS OF POLLUTANTS 286
13.7 TRANSFORMATIONS AND MOBILIZATION OF POLLUTANTS
     BY MICROBIAL ACTION 286
13.8 HEAVY METAL RESISTANCE, ANTIBIOTIC RESISTANCE, AND
     TRANSFER OF RESISTANCE FACTORS AMONG MICROORGANISMS . . 287
13.9 MICROORGANISMS AS INDICATORS OF ENVIRONMENTAL POLLUTION . 288
13.10 REFERENCES 289

13.1. INTRODUCTION

Bacteria and fungi play a fundamental role in the biogeochemical cycles in nature. These microorganisms remineralize organic matter to carbon dioxide, water, and various inorganic salts. Because bacteria are ubiquitous, and capable of rapid growth when provided with nutrients and conditions favourable for metabolism and cell division, they are involved in catalysis and synthesis of organic matter in the aquatic and terrestrial environments. Many substances, such as lignin, cellulose, chitin, pectin, agar, hydrocarbons, phenols, and other organic chemicals, are degraded by microbial action. The rate of decomposition of organic compounds depends upon their chemical structure and complexity and upon environmental conditions.

The nitrogen cycle, including fixation of molecular nitrogen and denitrification, is mediated by microorganisms in the natural environment. Other biogeochemical cycles, including the sulphur, phosphorus, iron, and manganese cycles also depend primarily upon microbial activity. Transformation and mobilization of heavy metals, degradation of pesticides, herbicides, and other man-made, allochthonous materials are left, ultimately to the microorganisms, for recycling. The toxic effects of pollutants on the autochthonous microbial populations, therefore, become of major significance in ecotoxicology.

13.2. MICROBIAL DEGRADATION OF POLLUTANTS

A wide variety of synthetic organic compounds contaminate the environment from chemical and industrial processes. In many instances, organic loads entering receiving waters add to the existing organic pools and cause perturbations in the natural degradation processes of the aquatic microbial community. Many chemicals employed in industrial processes are both refractory and toxic, and removal of these pollutants from the aquatic environment occurs primarily by microbial activities. Microbial degradation is dependent upon physical and chemical environmental variables, as well as on the toxicity of the chemical.

A list of the entire spectrum of industrial compounds entering the aquatic environment, either through chemical processes, accidental contamination, or as waste by-products, would fill volumes of books and catalogues. In a recent National Science Foundation report *ca.* 200 compounds were identified as being of national concern, because of their relative abundance as environmental contaminants or their relative toxicity. The most extensively studied of the chemical compounds are those representing the greatest threat to environmental health (Nelson and Van Duuren, 1975).

Physical and chemical factors may render a given compound more or less susceptible to microbial degradation. For example, irradiation in the visible and ultraviolet ranges can aid in the degradation of polymerized plastics and dechlorination of halogenated substrates and, perhaps, in the cleavage of alkylated biphenyls and fused aromatic ring systems. Photodegradation has also been implicated in the potential formation of chlorinated dibenzofurans from chlorinated biphenyls producing more toxic compounds of unknown biodegradative potential (Crosby and Moilanen, 1973).

Interaction of hydrophobic aquatic contaminants with dissolved organic substances and particulate matter can result in physical partitioning of the compound from the water column, bringing the susceptible substrate into closer association with those microorganisms capable of degrading the compounds. Such partitioning can also cause a concentration of the contaminant to toxic levels, thereby suppressing or retarding biodegradation or affecting the biological components of the ecosystem. Thus solubilization or partitioning of pollutants into dissolved organic phases can stimulate biodegradation, through availability of co-metabolizable substrates or inhibition of normal decomposition activity. Many biological compounds, such as lipids, proteins, nucleic acids, and amino acids concentrate or increase the solubility of chlorinated biphenyls and many polycyclic aromatic hydrocarbons.

Of the large variety of existing synthetic industrial chemicals, a number of classes of compounds have been studied (Table 13.1).

Table 13.1 Chemical Contaminants found in Aquatic Environments. (Reproduced by permission of John Wiley & Sons, Inc., from Colwell and Sayler, 1977)

Chemical class	Model compounds	Examples of degrading organisms	Reference
Phenolics	Phenol	*Pseudomonas putida*	Der Yang and Humphrey (1975)
Phenolics-halogenated	Pentachlorophenol	Soil bacteria	Kirsch and Etzel (1973)
Aromatic substrates monocyclic	Xylene	*Pseudomonas putida*	Gibson et al. (1974)
Monocyclic-halogenated	Fluorobenzoate	*Pseudomonas* spp.	Goldman (1972)
Bicyclic	Biphenyl	Soil bacteria	Lunt and Evans (1970)
Bicyclic-halogenated	PCB ('Aroclors')	*Pseudomonas* spp.	Sayler et al. (1977a)
Polycyclic	Benzo(a)pyrene	*Pseudomonas* spp.	Barnsley (1975)
Alkylated	Dibutylphthalate	*Brevibacterium* spp.	Engelhardt et al. (1975)
Chlorinated aliphatics	Trichloroethane	Marine bacteria	Jensen and Rosenberg (1975)
Glycols	Ethylene glycol	Aquatic bacteria	Evans and David (1974)
Petroleum	Hexadecane	Aquatic bacteria	Walker and Colwell (1974a)
	Mixed hydrocarbon substrate		

Petroleum offers a good example of the effects of pollutants on microorganisms since it is increasingly a problem in the world oceans, with toxic effects noted for some marine organisms. Since the earliest recorded oil spill in 1907 (Bourne, 1968), numerous reports have been published on oil pollution. These include 1,100 scientific manuscripts (Moulder and Varley, 1971), as well as popular publications (Marx, 1971; Nelson-Smith, 1973) and a number of symposia directed at the problem of oil pollution (Anonymous, 1969, 1970, 1971, 1973, 1975; Carthy and Arthur, 1968; Colwell, 1977; Hepple, 1971; Holmes and Dewitt, 1970; Hoult, 1969; D'Emidio, 1972; Ahearn and Meyers, 1973).

13.3. MICROBIAL ECOTOXICOLOGICAL EFFECTS OF PETROLEUM

The role of microorganisms in oil pollution has been emphasized by Davis (1967) and Sharpley (1966) in two textbooks on petroleum microbiology and several reviews on the subject have also been published (ZoBell, 1969; Friede et al., 1972; Atlas and Bartha, 1973a; Crow et al., 1974).

The community structure of the total viable microbial populations, the population shifts occurring as seasonal events or upon introduction of new metabolites, and the metabolic capabilities of the natural microbial flora are only beginning to be understood. The full extent of the effect of petroleum on both micro- and macrobiological communities remains to be clarified. Although it has been shown that components of petroleum persist in the marine environment (Hartung and Klingler, 1968; Holcomb, 1969; Horn et al., 1970; Blumer, 1971; Blumer et al., 1971, 1972; Blumer and Sass, 1972a,b), it is relatively easy to demonstrate in the laboratory that many marine bacteria, under optimal conditions, can remove selected fractions of oil, usually the n-alkanes, in a matter of days or weeks so that a certain percentage of oil by weight will disappear in a given period of time. However, it cannot be said that all the oil will disappear in a proportionate time, since the remaining fractions may be more refractory to microbial attack. In nature, conditions rarely are favourable for maximum biodegradation; hence, the rate of degradation can be slow and oil may persist in the marine environment for a long time. Floodgate (1972) surmises that some components of the 10 million tons of oil that found its way into the sea in World War II are still in the oceans. The significant questions of ecotoxicology of petroleum, then, are whether degradation demonstrated to occur in the laboratory will occur in the oceans; whether oil reaches depths of 1,000 m and, if so, the rates and degrees of oil degradation at depths greater than 1,000 m. It is not clearly understood whether concentrations of oil below some threshold level are degraded, since Jannasch (1970) noted that when the concentration of carbon in sea water was below a threshold level, bacterial growth was limited.

The qualitative and quantitative differences in hydrocarbon content of petroleum, influence its susceptibility to degradation, a major consideration in determining ecotoxicological effects of petroleum. Two crude oils and two refined oils were examined for susceptibility to degradation by bacteria from an oil-contaminated environment, Baltimore Harbour in Chesapeake Bay, located on the southeastern Atlantic Ocean (Walker *et al.*, 1975a,b). The oils studied included South Louisiana crude oil, which contains a high percentage of saturates (56%), Kuwait crude oil, which contains 34% saturates and a significant amount (18%) of resins (pyridines, quinolines, carbazoles, sulphoxides, and amides), Fuel oil No. 2, a refined product composed of only saturates (61%) and aromatics (39%), and Fuel oil No. 6 (Bunker C) which can contain up to 45% asphaltenes (phenols, fatty acids, ketones, esters, and porphyrins) and resins and forms 'tar balls' which may sink, if not dispersed. Degradation of the oils by bacteria was compared and it was found that after seven weeks, the quantity of the four oils remaining after exposure to bacteria from oil-contaminated sediment was approximately the same, indicating that, quantitatively, degradation of the four oils proceeded similarly. However, a comparison of the degradation of selected components of the four oils illustrated that there were, in fact, significant differences in susceptibility to degradation of fractions of the four oils, as was shown by column chromatography and computerized mass spectrometry (Walker *et al.*, 1975c).

Other investigators have recorded similar observations. Mulkins-Phillips and Stewart (1974b) compared the degradation of Venezuelan and Arabian crude oil by a *Nocardia* sp. The isolate degraded 77% and 13% of the *n*-alkanes in an unresolved fraction of Venezuelan crude oil, compared to 94% and 35%, respectively, of an Arabian crude oil. Atlas and Bartha (1973a) demonstrated that two paraffinic crudes, Sweden and Louisiana, were degraded similarly (70%) by mixed bacterial cultures grown in sea water. Jobson *et al.* (1972) compared growth of microbes on an inferior and high-grade crude oil. Microbes isolated on the inferior crude oil were more capable of using it than microbes isolated on high-grade crude oil. Westlake *et al.* (1974) reported that crude-oil composition affected the rate and extent of biodegradation, specifically of the *n*-saturate fraction, resins, and asphaltenes.

Microorganisms present in water and sediment samples collected from two areas in the same geographical location in Chesapeake Bay, an oil-polluted environment (Baltimore Harbour) and an oil-free environment (Eastern Bay), were compared for degradative capability on South Louisiana crude oil. Bacteria previously exposed to oil were found to be the most effective in degrading the crude oil.

Microorganisms present in samples collected from oil-polluted harbours located in different geographical areas were also compared for ability to degrade a South Louisiana crude oil. In this case, the cultures from Baltimore Harbour were found to be more effective in degrading the oil than cultures present in San Juan Harbour sediment. An increase in resins and asphaltenes occurring in the Baltimore Harbour cultures was noted, compared with one San Juan culture; these fractions may have been accumulated 'refractory' components or may have derived from microbial synthesis (Walker and Colwell, 1975a). Atlas and Bartha (1973b) reported that the accumulation of such components as fatty acids in flask cultures may prevent complete biodegradation of oil. Whether this occurs *in situ* is not known. It is of interest that deep-ocean sediment samples yielded greater petroleum-degrading potential than those from Eastern Bay, an unpolluted site in Chesapeake Bay (Walker et al., 1976b). However, Baltimore Harbour, the polluted site, yielded the greatest degradative potential.

The neritic and deep-ocean environments differ considerably, especially with respect to such variables as temperature and pressure, which can have an effect on hydrocarbon degradation. Experiments using hydrocarbons have been done (Schwarz et al., 1974a,b, 1975) using ^{14}C-labelled hydrocarbons and a mixed hydrocarbon substrate which revealed that at one atm pressure and 4°, the stationary phase for the n-hexadecane culture was reached within 4 weeks, but at 500 atm, the stationary phase was reached only after 32 weeks. Utilization of 94% of the hexadecane was accomplished after 8 weeks at 1 atm, but only after incubation for 40 weeks at 400 atm. Thus low temperatures and high pressures restrict the utilization of hexadecane by microorganisms and contribute to the persistence of these compounds in nature (Schwarz et al., 1974a,b, 1975).

Limitation of nutrients, especially nitrates and phosphates, affects the degradation of petroleum occurring in the environment, contributing to the persistence of selected fractions of petroleum. Petroleum degradation is severely limited in the natural environment, unless nitrate and phosphate are added as a nutrient supplement (Atlas and Bartha, 1972b, 1973c).

Climatological conditions affect microbial degradation of petroleum and seasonal effects on the rate and extent of degradation have been observed. Two stations in Raritan Bay were monitored at five intervals between July 1971 and May 1972, and it was discovered that the highest counts of petroleum degraders were observed in July (Atlas and Bartha, 1973d). Increased numbers of petroleum degraders, however, were found at two time periods, December–February and June–July, in Chesapeake Bay (Colwell et al., 1974). At each season larger numbers of petroleum degraders were observed in the oil-polluted environments of Chesapeake Bay than in the unpolluted environments. To determine if enrichment with psychrophilic petroleum degraders occurred during winter months, sediment bacteria from a polluted environment in Chesapeake Bay were sampled during January and September and these samples were tested for petroleum degradation at 0°. Differences were observed in the amount of oil degraded by microorganisms in those samples collected in January compared with samples collected in September, but they were not significant, suggesting that enrichment with psychrotrophic or psychrophilic petroleum degraders during winter months is not important.

The environmental variables involved in microbial degradation occurring in temperate and arctic zones are similar, except that in temperate or arctic zones, the specific conditions are established with much less variability. For example, oil polluting a permanently cold environment would be subjected to degradation by the pre-existing, pyschrophilic or psychrotrophic microbial populations. The principal effects of low temperatures that must be considered are decreased volatilization and increased water solubility of the volatile hydrocarbons in petroleum (Atlas and Bartha, 1972a; Walker and Colwell, 1974b; Colwell et al., 1976b). Hence, the pollutant, oil in this case, may be more toxic in colder environments.

Bacteria (*Flavobacterium, Brevibacterium* and *Arthrobacter* spp.) have been reported to utilize 35–70% of the paraffinic crude oils, whereas fungi (*Penicillium* and *Cunninghamella* spp.) utilized 85–92%. Comparative studies by Walker et al. (1977a,b) demonstrated that yeasts (*Candida*) and fungi (*Penicillium*) degraded South Louisiana crude oil more extensively than bacteria (*Pseudomonas* and *Coryneforms*). Growth of bacteria in flask cultures on mixed hydrocarbon substrate was observed to precede growth of yeasts and fungi (Walker and Colwell, 1974a, 1975b). Also, oscillations were noted in the occurrence of several bacterial genera in mixed cultures growing on a mixed hydrocarbon substrate. Combinations of bacteria, yeasts, and filamentous fungi provided approximately twice as much degradation of the mixed hydrocarbon substrate, compared with each organism grown individually (Walker and Colwell, 1975b). Thus the mixed populations found in nature are more effective in degradation and detoxification than would be expected, based on pure-culture studies performed in the laboratory.

Reports by Atlas and Bartha (1973e) and Mulkins-Phillips and Stewart (1974a) describe effects of dispersants on the biodegradation of oil. Atlas and Bartha (1973e) found that dispersants increased the rate of mineralization, but did not affect the extent of biodegradation. Mulkins-Phillips and Stewart (1974a) examined four dispersants and found only one, 'Sugee' 2, that was associated with increased degradation of n-alkanes in crude oil, compared with crude oil alone.

ZoBell (1969), in summarizing work accomplished prior to 1969, reported rates of degradation of crude oils, lubricating oils, cutting oils, and oil wastes, ranging from 0.02 to 2.0 g/m^2/day at 24° to 30°. Rates of oil biodegradation have been measured as g/m^2/day, g/m^3/yr, mg/day/bacterial cell or per cent oil removed after a known number of days. Thus it is often difficult to make comparisons of results based on the few studies that have been reported. Differences in experimental conditions provide additional problems when comparing the results of studies in which biodegradation rates of petroleum were calculated (Kinney et al., 1969; Johnston, 1970; Kator et al., 1971; Bridie and Bos, 1971; Robertson et al., 1973). Although rates of biodegradation of petroleums in the environment have been estimated, often the experiments involved measuring the amount of oil at the beginning and termination of an experiment, assuming the rate to be linear. Furthermore, there are no studies of the fate of the petroleum components during biodegradation. In a recent study of the biodegradation of Louisiana crude oil (Walker et al., 1976a), maximum degradation of the total residue was observed to occur during the logarithmic phase of bacterial multiplication (1.42 mg/day), with a levelling off at the stationary phase. Asphaltenes, resins, and aromatics increased after the stationary phase was reached, but saturates were degraded throughout the seven-week growth phase. Thus biodegradation of petroleum in the marine environment is influenced by a complex of ecological factors and it is an oversimplification to cite a rate of biodegradation occurring in the ocean or other environments.

In general, microorganisms from an industrially polluted environment or from an oil-polluted harbour will carry out a more extensive degradation of a crude oil than microbes from an unpolluted environment. However, this is not an absolute rule, since deep-ocean sediment has been shown to contain microorganisms carrying out more extensive degradation of crude oil than, for example, sediment inocula collected in San Juan Harbour. As stated above, comparison of neritic and deep-sea microorganisms for ability to degrade petroleum under simulated *in situ* conditions demonstrates that, at least for pure hydrocarbons, deep-sea conditions clearly are associated with significantly slowed down degradation.

The fate of oil, after extensive exposure under various environmental conditions, has been examined by several investigators (Blumer *et al.*, 1972; Rashid, 1974). Paraffinic oils were found to persist as oils, not asphaltenes, for a 16-month period of monitoring. Changes in the oil were found to be due to weathering and microbial degradation. However, it is important to relate alterations in the structure of microbial populations with concomitant degradation of the oil substrate. This has been, in part, accomplished in the laboratory, but requires further study, particularly *in situ*.

From simulated *in situ* experiments, using flow-through sea-water tanks, designed to study biodegradation, it was discovered that oil-soluble phosphate and nitrate was required to enhance biodegradation of oil in sea water (Atlas and Bartha, 1973a). Thus sea water may not contain sufficient amounts of nitrate and phosphate to support significant degradation of oil, with the result that the oil, or fractions of the oil, will persist in the environment for long periods of time. In general, warmer temperatures appear to promote volatilization of low-boiling hydrocarbons, hence, faster utilization of metabolizable hydrocarbons. However, selection for psychrophilic petroleum degraders in colder climates may occur, but this point has not yet been resolved.

Crude oils and fuel oils have been found to have relatively little effect on heterotrophic microorganisms, in general, except in the case of yeasts, which tend to be slightly inhibited by fuel oil (Walker and Colwell, 1975a). However, crude oil and fuel oil have been found to limit the growth of specific groups of microorganisms, such as the lipolytic, proteolytic, and chitinolytic bacteria, with fuel oil observed to be more toxic than crude oil (Walker *et al.*, 1974b), despite the fact that both fuel oil and crude oil can support the growth of certain individual species of heterotrophic and cellulolytic bacteria (Walker *et al.*, 1975e).

Addition of petroleum hydrocarbons to water samples collected from an oil-free environment has been shown to limit the growth of the bacteria normally present in the water (Walker and Colwell, 1975a, 1977). However, addition of petroleum hydrocarbons to water collected from an oil-polluted environment promoted growth of bacteria already present in the water (Walker and Colwell, 1975a), suggesting that selection for oil-resistant species had already occurred in the oil-polluted environment.

The positive chemotactic responses of motile marine bacteria have been shown to be reversed by the addition of sublethal concentrations of hydrocarbons, yet another ecotoxicological effect of oil in that it can interfere with microbial chemotaxis (Young and Mitchell, 1973). This is an important ecological phenomenon since microorganisms depend on chemotaxis for attachment to substrates for growth and nutrition.

With respect to the effects of petroleum, the available data reveal that there is a predictable pattern of events in the biodegradation of petroleum, irrespective of geographical location, with variations in time of occurrence in the steps in the degradative process and in sequential changes in the mixed microbial population structure, according to climate, season, etc. Furthermore, it is clear that no single microbial species will degrade any given oil completely. Bacteria are selective and many different bacterial species in mixed cultures are required for significant degradation. Furthermore, bacterial oxidation of hydrocarbons produces many intermediates which may be more toxic than the original hydrocarbon components. It is obvious that it is microorganisms that must be relied upon to carry out degradation of hydrocarbon decomposition products in nature. Unfortunately, the fraction of crude oil most readily and completely subject to attack by bacteria is the least toxic, e.g. the normal paraffins. Toxic aromatic hydrocarbons, especially the carcinogenic polynuclear aromatics, are degraded only very slowly. It must be added, also, that evidence has been presented showing synthesis of carcinogenic hydrocarbons by various species of bacteria (ZoBell, 1971).

An ecotoxicological effect of petroleum degradation is the severe requirement for oxygen during bacterial oil degradation (ZoBell, 1969). The complete oxidation of one gallon of crude oil requires all the dissolved oxygen in 320,000 gallons of air-saturated sea water. It can be concluded that where the oxygen content has been lowered by other kinds of pollution, either bacterial degradation of oil may cause additional ecological damage via oxygen depletion or not occur at all because of insufficient oxygen.

An aspect of the oil pollution problem which is only just beginning to be understood is the enrichment of heavy metals and organic compounds in the petroleum hydrocarbon pollutant (Duce *et al.*, 1972; Walker and Colwell, 1974c,d, 1976b).

No body of data exists concerning the effect of petroleum on autotrophic microbes, i.e. on those microorganisms upon which the biogeochemical cycles occurring in the sea depend. For example, the sulphur bacteria and nitrogen-fixing bacteria have not yet been examined to determine the effects of releases of petroleum hydrocarbons, both chronic and accidental, on these microorganisms.

Thus from all studies accomplished to date, it can be concluded that normal alkanes are the saturated hydrocarbons most readily susceptible to degradation.

Biodegradation of normal alkanes, however, always results in a residual base 'envelope', which can be detected using gas—liquid chromatography. These saturated hydrocarbons are 1- to 6-ring alkanes, as determined by using computerized low-resolution mass spectrometry (Walker *et al.*, 1975c). Also, microorganisms can produce polar *n*-pentane-insoluble components (asphaltenes) (Jobson *et al.*, 1972; Zajic *et al.*, 1974) and long-chain *n*-alkanes (Walker and Colwell, 1976a). Clearly, crude and fuel oils are degraded differently by microorganisms, with Bunker C fuel oil being degraded to a much lesser extent than South Louisiana crude oils. In fact, the aromatic fraction of Bunker C fuel oil is degraded only very slightly (Walker *et al.*, 1976b). For example, profiles of hydrocarbons in Baltimore Harbour sediment cores, i.e. presence in sediment samples according to depth, revealed that the more refractory compounds, including cycloparaffins, aromatics, and polynuclear aromatics are found in the deeper sediments (Walker *et al.*, 1974a, 1975c,d). Thus it is obvious that, although microorganisms are capable of degrading petroleum and petroleum by-products, there is a great deal of variability in the extent of degradation of the petroleum components, with the more 'refractory' components, such as aromatics and polynuclear aromatics, accumulating in the aquatic ecosystem.

13.4. CARCINOGENICITY OF POLLUTANTS ASSOCIATED WITH MICROBIAL ACTIVITY

Little is known about the long-term hazards of oil contamination on aquatic ecosystems. However, one class of compounds in petroleum that is relatively resistant to microbial degradation is the polycyclic aromatic hydrocarbons, members of which are known to be carcinogenic (Miller and Miller, 1971). By-products of oil degradation, arising either from weathering or incomplete microbial mineralization, may also demonstrate carcinogenic activity. Several bacterial systems have been developed to detect compounds which are reactive with DNA and may cause mutation. Since most carcinogenic chemicals are also mutagens, these systems provide a quick and inexpensive means of screening large numbers of compounds or samples for potential carcinogenicity. The most widely used system is that developed by Ames and his colleagues (Ames *et al.*, 1975). About 85% of known carcinogens which have been tested by the Ames system have reacted positively as mutagens (McCann *et al.*, 1975). Many of these are promutagens, i.e. chemicals which are not mutagenic in themselves but which become so when acted upon by mammalian enzyme systems. They are detected in the bacteria system by inclusion in the assay of a mammalian liver microsome fraction. The mixed-function oxidases are implicated as the responsible enzymes in chemical conversion to carcinogenicity (Heidelberger, 1975). Since potent mutagens are reactive in the Ames systems in microgram quantities, the system is suitable for assaying for potential carcinogens in environmental samples.

To determine whether mutagenic metabolites were produced during biodegradation of oil by indigenous Chesapeake Bay organisms, Colgate Creek water and sediment samples were used to inoculate an inorganic Chesapeake salts' solution medium supplemented with oil as sole carbon and energy source (motor-oil medium, pH 7.0) (Voll et al., 1977). Positive results in the Ames system were obtained with some samples of Colgate Creek water and sediment upon initial assay. Work is still in progress to test the Chesapeake Bay samples by the direct incorporation technique of Ames (Ames et al., 1975) which is more sensitive in detecting low levels of mutagens or mutagens insoluble in water.

Although results to date are inconclusive, such systems do appear suitable for detection of potential carcinogens in samples collected for ecological analysis, but environmental samples present problems and difficulties not generally encountered with pure chemicals. Mutagenic substances may be present at concentrations below the level of detection of the assay, but significant concentration of test samples prior to assay can magnify levels of potentially toxic components. False negatives can arise with promutagens, if other components in the sample inhibit the liver homogenate activation system. With appropriate processing of environmental samples, these sources of error can be detected and eliminated. The most appropriate methods for processing and assaying samples from the environment and using bacterial assay systems will, no doubt, vary with the nature and source of the sample.

Degradation phenomena reported in laboratory studies of pure culture of microorganisms reflect only potential degradation that may occur in the natural environment. Physical—chemical properties of any chemical, environmental variables, and the concentration of the chemical, as well as the concentration and diversity of the microbial flora of a specific habitat, all are factors in the biodegradation process.

13.5. MICROBIAL EFFECTS ASSOCIATED WITH POLYCHLORINATED BIPHENYLS

Biodegradation of polychlorinated biphenyls has been studied by a number of investigators. Ahmed and Focht (1973a,b) demonstrated decomposition of mono and dichlorobiphenyls by *Achromobacter* spp. isolated from sewage which resulted in the production of *p*-chlorobenzoic acid. In later studies (Alexander, 1975), respirometric data were presented that indicated degradation of PCBs to pentachlorobiphenyl, with the rate of degradation decreasing in direct proportion to increasing chlorine content of the PCB.

Kaiser and Wong (1974) investigated the pure culture degradation of 'Aroclor' 1254, a PCB mixture of biphenyl and other components, including a compound with seven chlorine residues per molecule. Results showed that significant degradation of the PCB could take place under appropriate environmental conditions. The bacterium was isolated by enrichment techniques from aquatic samples. Wong and Kaiser (1975) were able to demonstrate the degradation of

'Aroclor' 1221 (21% chlorine by weight) by mixed bacterial cultures in batch degradation studies.

Previously reported effects of PCB on microbial activity and growth responses have ranged from no effect on lipid biosynthesis in *E. coli* and *B. fragilis* (Greer *et al.*, 1974), to slight stimulation in cell growth in *E. coli* (Keil and Sandifer, 1972), to inhibition in the growth of selected estuarine bacteria (Bourquin, 1975).

Organisms capable of PCB metabolism have been recovered from estuarine and marine environments (Sayler *et al.*, 1977b); both PCB and PCB-degrading bacteria were found at the sites tested. However, no correlation between PCB levels and numbers of microorganisms degrading PCB were noted. High concentrations of 'Aroclor' 1254 were found to have no significant effect on the respiration of an estuarine *Pseudomonas* spp. capable of PCB degradation. However, stimulation of O_2 uptake was observed when 'Aroclor' 1254 was coated on diatomaceous earth and added to a pure culture at a concentration of 200 mg/l. Results of experiments carried out by several investigators indicate a significant potential for the removal of highly chlorinated biphenyls from estuarine water through biodegradation processes of a typical estuarine bacterial component.

The composite results from all the studies to date show that polychlorinated biphenyls are subject to microbial degradation, with the specific rate of degradation inversely related to the average chlorination of the PCB mixture and directly dependent upon the concentration of PCB. Furthermore, a wide variety of bacterial genera have demonstrated PCB degradation potential, although only preliminary results are available on the naturally occurring PCB degradation processes. There is no doubt that PCBs can induce changes in microbial population composition and activity.

13.6. EFFECTS OF COMBINATIONS OF POLLUTANTS

The biodegradative aspects of ecotoxicology at the microbial level indicate that many, if not most, of the pollutants entering the ecosystem can be degraded. However, there are effects of pollutants in combination, for example, mercury and oil; when these are combined, there is a suppressed degradation of oil and an enhanced toxicity because of the partitioning of the mercury in the oil phase (Walker and Colwell, 1976a). The toxic effect need not always be an enhancement since the toxic effect may be reduced if chelation occurs to a pollutant in a mixture. Clearly, given bodies of water and various mixtures of pollutants will yield an array of ecotoxicological effects.

13.7. TRANSFORMATIONS AND MOBILIZATION OF POLLUTANTS BY MICROBIAL ACTION

Implied in the biodegradative aspects of ecotoxicology described above are the transformations of pollutants into either more toxic forms or less toxic substances.

For example, under anaerobic conditions, sedimentary bacteria will methylate mercury, whereas under aerobic conditions, the microorganisms will produce the elemental form of mercury. The microorganisms act as mobilizing agents, in both cases, with mercury transported from sediment to water via microbial activity or mobilized through the food chain if the microorganisms concentrate mercury and are, in turn, fed upon by higher forms of the food chain (Sayler et al., 1975; Sayler and Colwell, 1976; Colwell et al., 1976a). Methylation of metals such as mercury, cadmium, and tin, by microorganisms results in either primary or secondary ecotoxicological effects, depending upon whether the pollutant is allochthonous to the environment or the transformation via microbial action yields the polluting substance.

Bioamplification of mercury levels in the oyster *Crassostrea virginica* has been shown to be mediated by bacteria (Sayler et al., 1975), indicating the fundamental role of bacteria in mercury mobilization and accumulation of mercury at higher levels in food webs. Where there is a significant mercury-resistant bacterial population actively metabolizing as well as accumulating mercury compounds, their involvement in the bioaccumulation of mercury will be significant. This aspect of ecotoxicology at the microbial level, demonstrated for mercury and, no doubt, similar for other heavy metals, has yet to be fully appreciated.

Effects of metal ions on the microbial populations of estuaries has also been reported (Mills and Colwell, 1977), with significant effects on photosynthesis noted.

13.8. HEAVY METAL RESISTANCE, ANTIBIOTIC RESISTANCE, AND TRANSFER OF RESISTANCE FACTORS AMONG MICROORGANISMS

The severe impoverishment of the normal fauna by sewage sludge dumping in selected ocean sites is an effect that has been amply documented. Relatively unrecognized, however, is the effect on the microbial populations. In sewage sludges, relatively high concentrations of heavy metals are common. In such toxic environments, bacterial populations are selected by virtue of the rapid dissemination of resistance transfer factors (R plasmids). Antibiotic-resistant strains of *Enterobacteriaceae* reveal R plasmids which are highly transmissible between non-pathogenic and pathogenic donors and recipients by conjugation or transduction. The plasmids confer resistance to a wide spectrum of antibiotics and other antimicrobials, including heavy metals (Smith, 1967; Novick, 1969; Davies and Rownd, 1972; Summers and Silver, 1972; Koditschek and Guyre, 1974).

A large body of evidence has been accumulated showing a high incidence of R factor coliforms in both raw and treated sewage, river water, salt water, and in the New York Bight of the Atlantic Ocean (Feary et al., 1972; Koditschek and Guyre, 1974).

The incidence of antibiotic-resistant bacteria in Chesapeake Bay was routinely monitored at selected stations over a 12-month period (Allen et al., 1977). It was found that in polluted areas of Chesapeake Bay, there was a significantly larger number of antibiotic-resistant bacteria that were, in addition, resistant to heavy metals. The data suggest that, indeed, in environments receiving a large influx of heavy metals, selection of heavy metal and antibiotic-resistant bacteria occurs. Transfer of plasmids appears to be significant in the transfer of resistance from the allochthonous to the autochthonous forms (Sizemore and Colwell, 1977; Guerry and Colwell, 1977).

13.9. MICROORGANISMS AS INDICATORS OF ENVIRONMENTAL POLLUTION

Bacteria have been employed as indicators of abnormal or atypical environmental conditions for many years. The most widely recognized indicators are the coliform group, used to monitor the presence and quantity of faecal pollution (Kabler *et al.*, 1964). Walker and Colwell (1973) showed that the number of petroleum-degrading microorganisms in water and sediments of Chesapeake Bay are related to the concentration of oil present. The concept of hydrocarbonoclastic microorganisms as indicators of hydrocarbons has been supported by several workers. A high correlation noted between ratios of hydrocarbonoclastic and total aerobic heterotrophic bacteria and levels of hydrocarbons in oil-rich salt-marsh sediments was reported by Hood *et al.* (1975), who also noted that the presence of hydrocarbons alters the relative abundance of the most predominant groups of the normal aerobic heterotrophic bacteria.

Similar relationships between mercury-resistant bacteria and concentration of mercury in the waters and sediments of Chesapeake Bay have been reported (Nelson and Colwell, 1975). Natural aquatic environments have measurable, reasonably definable microbial biota. Thus fluctuations in the environment result in a change in the delicate balance of the microbial community structure. In general, a single indicator group as the faecal coliforms, has been used to detect changes arising from pollution. If several indicator groups were used in combination, the effects of environmental changes could be more accurately measured. For example, the presence of crude oil has been shown to decrease the relative concentrations of cellulolytic bacteria in salt-marsh ecosystems (Crow *et al.*, 1975) and in Chesapeake Bay (Walker *et al.*, 1975e). Especially attractive is the potential for early warning of environmental change since microbiological responses are rapid and can be detected within hours or days. The microbial potential, perhaps measured as a community structure index, or other mathematical formulation, should be more fully investigated as an ecotoxicological yardstick of health. Clearly, the microbial aspects of ecotoxicology should be explored since here lies, indeed, a fertile ground for discovery and application in environmental pollution.

13.10. REFERENCES

Ahearn, D. G. and Meyers, S. P. (Eds.), 1973. *The Microbial Degradation of Oil Pollutants*, Louisiana State University Publ. No. LSU-SG-73–01, Baton Rouge, La.

Ahmed, M. and Focht, D. D., 1973a. Oxidation of polychlorinated biphenyls by *Achromobacter* sp. CB *Bull. Environ. Contam. Toxicol.*, **10**, 70–2.

Ahmed, M. and Focht, D. D., 1973b. Degradation of polychlorinated biphenyls by two species of *Achromobacter*. *Can. J. Microbiol.*, **19**, 47–52.

Alexander, M., 1975. Environmental and microbiological problems arising from recalcitrant molecules. *Microbial Ecol.*, **2**, 17–27.

Allen, D. A., Austin, B. and Colwell, R. R., 1977. Antibiotic resistance patterns of metal-tolerant bacteria isolated from an estuary. *Antimicrobial Agents and Chemotherapy*, **12**, 545–7.

Ames, B. N., McCann, J., and Yamasaki, E., 1975. Methods for detecting carcinogens and mutagens with the *Salmonella*/mammalian-microsome mutagenicity test. *Mutat. Res.*, **31**, 347–64.

Anonymous, 1969. *Proceedings, Joint Conference on Prevention and Control of Oil Spills*, American Petroleum Institute and Federal Water Pollution Control Administration, American Petroleum Institute, Washington, D.C.

Anonymous, 1970. Legal, economic and technical aspects of liability and financial responsibility as related to oil pollution. *George Washington University Report*, Natl. Technical Information Service, PB-198–776, Washington, D.C.

Anonymous, 1971. *Proceedings, Joint Conference on Prevention and Control of Oil Spills*, American Petroleum Institute, Environmental Protection Agency and U.S. Coast Guard, American Petroleum Institute, Washington, D.C.

Anonymous, 1973. Background papers for a workshop on inputs, rates and effects of petroleum in the marine environment. National Academy of Sciences, Washington, D.C.

Anonymous, 1975. *Petroleum in the Marine Environment*, National Academy of Sciences, Washington, D.C.

Atlas, R. M. and Bartha, R., 1972a. Biodegradation of petroleum in sea water at low temperatures. *Can. J. Microbiol.*, **18**, 1851–5.

Atlas, R. M. and Bartha, R., 1972b. Degradation and mineralization of petroleum in sea water: limitation by nitrogen and phosphorus. *Biotechnol. Bioeng.*, **14**, 309–18.

Atlas, R. M. and Bartha, R., 1973a. Fate and effects of polluting petroleum in the marine environment. *Residue Rev.*, **49**, 40–85.

Atlas, R. M and Bartha, R., 1973b. Inhibition by fatty acids of the biodegradation of petroleum. Antonie Van Leeuwenhock, *J. Microbiol. Serol.*, **39**, 257–71.

Atlas, R. M. and Bartha, R., 1973c. Stimulated biodegradation of oil slicks using oleophilic fertilizers. *Environ. Sci. Technol.*, **7**, 538–41.

Atlas, R. M. and Bartha, R. 1973d. Abundance distribution and oil biodegradation potential of microorganisms in Raritan Bay. *Environ. Pollut.*, **4**, 291–300.

Atlas, R. M. and Bartha, R., 1973e. Effects of some commercial oil herders, dispersants and bacterial inocula on biodegradation in seawater. In *The Microbial Degradation of Oil Pollutants* (Eds. D. G. Ahearn and S. P. Meyers), Louisiana State University Publ. No. LSU-SG-73–01, Baton Rouge, La.

Barnsley, E. A., 1975. The bacterial degradation of fluoranthene and benzo-(a)pyrene. *Can. J. Microbiol.*, **213**, 1004–111.

Blumer, M., 1971. Scientific aspect of the oil spill problem. *Environ. Affairs*, **1**, 54–73.

Blumer, M., Ehrhardt, M., and Jones, J. H., 1972. The environmental fate of stranded crude oil. *Deep-Sea Res.*, **20**, 239–59.
Blumer, H., Sanders, H. L., Grassle, J. F., and Hampson, G. R., 1971. A small oil spill. *Environment*, **13**, 2–12.
Blumer, M. and Sass, J., 1972a. Indigenous and petroleum-derived hydrocarbons in a polluted sediment. *Mar. Pollut. Bull.*, **3**, 92–4.
Blumer, M. and Sass, J., 1972b. Oil pollution: persistence and degradation of spilled fuel oil. *Science*, **176**, 1120–2.
Bourne, W. R. P., 1968. Effects of oil pollution on bird populations. *Field Studies*, **2** (Suppl.), 99–121.
Bourquin, A. W., 1975. Effect of polychlorinated biphenyl formulations on the growth of estuarine bacteria. *Appl. Microbiol.*, **29**, 125–7.
Bridie, A. L. and Bos, J., 1971. Biological degradation of mineral oil in sea water. *J. Inst. Petrol. (London)*, **57**, 270–7.
Carthy, J. D. and Arthur, D. R., 1968. *The Biological Effects of Oil Pollution in Littoral Communities*, Field Studies Council, Institute of Petroleum, London.
Colwell, R. R., 1977. Bacteria and viruses — indicators of unnatural environmental changes occurring in the nation's estuaries. *Estuarine Pollution Control and Assessment Proc. Conf.*, Pensacola, Fla., February, 1975. Vol. II, pp. 507–18, Office of Water Planning and Standards, Environmental Protection Agency, Washington, D. C. U.S. Govt. Printing Office: 1977 200–369 1–3.
Colwell, R. R., Carney, J. F., Kaneko, T., Nelson, J. D., and Walker, J. D., 1974. Microbial activities in the estuarine ecosystem. *First Intersectional Congress of International Association of Microbiological Societies*, Vol. 2, Publ. Science Council of Japan, Tokyo, Japan.2 , 410–20.
Colwell, R. R. and Sayler, G. S., 1977. Microbial degradation of industrial chemicals. In *Water Pollution Microbiology* (Ed. R. Mitchell), John Wiley and Sons, Inc., New York, pp. 161–99.
Colwell, R. R., Sayler, G. S., Nelson, J. D., Jr., and Justice, A., 1976a. Microbial mobilization of mercury in the aquatic environment. In *Environmental Biogeochemistry* (Ed. J. O. Nriagu), Vol. 2, Chapter 29, Ann Arbor Science Publ. Inc., Michigan, pp. 473–87.
Colwell, R. R., Walker, J. D., Conrad, B. F., and Seesman, P. A., 1976b. Microbiological studies of Atlantic Ocean water and sediment from potential off-shore drilling sites. *Develop. Ind. Microbiol.*, **17**, 269–82.
Crosby, D. G. and Moilanen, K. W., 1973. Photo-decomposition of chlorinated biphenyls and dibenzofurans. *Bull. Environ. Contam. Toxicol.*, **10**, 372–7.
Crow, S. A., Hood, M. A., and Meyers, S. P., 1975. Microbial aspects of oil intrusion in southeastern Louisiana. In *Impact of the Use of Microorganisms on the Aquatic Environment* (Eds. A. W. Bourquin, D. G. Ahearn, and S. P. Meyers), U.S. Environmental Protection Agency, Washington, D.C., pp. 221–7.
Crow, S. A., Meyers, S. P., and Ahearn, D. G., 1974. Microbiological aspects of petroleum degradation in the aquatic environment. *La Mer*, **12**, 37–54.
Davies, J. E. and Rownd, R., 1972. Transmissible multiple drug resistance in *Enterobacteriaceae*. *Science*, **176**, 758–68.
Davis, J. B., 1967. *Petroleum Microbiology*, Elsevier Publishing Co., New York.
D'Emidio, J. A. (Moderator), 1972. *Proceedings Navy Oil Spill Conference*, Naval Ordnance Laboratory, White Oak, Maryland.
Der Yang, B. and Humphrey, A. E., 1975. Dynamic and steady state studies of phenol biodegradation in pure and mixed cultures. *Biotechnol. Bioeng.*, **7**, 1211–35.

Duce, R. A., Quinn, J. G., Olney, C. E., Piotrowicz, S. R., Ray, R. J., and Wade, T. L., 1972. Enrichment of heavy metals and organic compounds in the surface microlayer of Narragansett Bay, Rhode Island. *Science*, **176**, 161–3.

Engelhardt, G., Wallnoffer, P. R., and Hutzinger, O., 1975. The microbial metabolism of di-*n*-butyl phthalate and related dialkyl phthalates. *Bull. Environ. Contam. Toxicol.*, **13**, 342–7.

Evans, W. H. and David, E. J., 1974. Biodegradation of mono-, di-, and tri-ethylene glycols in river waters under controlled laboratory conditions. *Water Res.*, **8**, 97–100.

Feary, T. W., Sturtevant, A. B., and Lankford, J., 1972. Antibiotic-resistant coliforms in fresh and salt water. *Arch Environ. Health*, **24**, 215–20.

Floodgate, G. D., 1972. Microbial degradation of oil. *Mar. Pollut. Bull.*, **3**, 41–43.

Friede, J., Guire, P., and Gholson, R. K., 1972. Assessment of biodegradation potential for controlling oil spills on the high seas. *U.S. Coast Guard Report*, No. 4110. 1/3.1.

Gibson, D., Venkatanarayana, T., and Davey, J. F., 1974. Bacterial metabolism of *para*- and *meta*-xylene: oxidation of the aromatic ring. *J. Bacteriol.*, **119**, 930–6.

Goldman, P., 1972. Enzymology of carbon–halogen bonds. In *Degradation of Synthetic Organic Molecules in the Biosphere*, National Academy of Sciences Publication, Washington, D. C., pp. 147–65.

Greer, D. E., Keil, J. E., Stillway, L. W., and Sandifer, S. H., 1974. The effect of DDT and PCB on lipid metabolism in *E. coli* and *B. fragilis*. *Bull. Environ. Contam. Toxicol.*, **12**, 295–300.

Guerry, P. and Colwell, R. R., 1977. Isolation of cryptic plasmid deoxyribonucleic acid from Kanagawa positive strains of *Vibrio parahaemolyticus*. *Infec. Immunity*, **16**, 328–34.

Hartung, R. and Klingler, G. W., 1968. Fisheries and wildlife sedimentation of floating oils. In *Papers of the Michigan Academy of Science, Arts, and Letters* (Ed. R. A. Loomis), Vol. 53 (1967 meeting), Zoology Section, Univ. Mich. Press, Ann Arbor, pp. 23–7.

Heidelberger, C., 1975. Chemical carcinogenesis. *Ann. Rev. Biochem.*, **44**, 79–121.

Hepple, P., 1971. Water pollution by oil. *Seminar Proc.*, Inst. Petrol. Eng. Applied Science, London.

Holcomb, R. W., 1969. Oil in the ecosystem. *Science*, **166**, 204–6.

Holmes, R. W. and Dewitt, F. A., Jr., 1970. *Santa Barbara Oil Symposium*, University of California, Santa Barbara.

Hood, M. A., Bishop, W. S., Jr., Bishop, F. W., Meyers, S. P., and Whelan, T., III, 1975. Microbial indicators of oil-rich salt marsh sediments, *Appl. Microbiol.*, **30**, 982–7.

Horn, M. H., Teal, J. M., and Backus, R. H., 1970. Petroleum lumps on the surface of the sea. *Science*, **163**, 245–6.

Hoult, D. P., 1969. *Oil on the Sea*, Plenum Press, New York.

Jannasch, H. W., 1970. Threshold concentrations of carbon sources limiting bacterial growth in sea water. In *Organic Matter in Natural Waters* (Ed. D. W. Hood), University of Alaska Press, College, pp. 321–8.

Jensen, S. and Rosenberg, R., 1975. Degradability of some chlorinated aliphatic hydrocarbons in sea water and sterilized water. *Water Res.*, **9**, 659–61.

Jobson, A., Cook, F. D., and Westlake, D. W. S., 1972. Microbial utilization of crude oil. *Appl. Microbiol.*, **23**, 1082–9.

Johnston, R., 1970. The decomposition of crude oil residues in sand columns. *J. Mar. Biol. Assn. U.K.*, **50**, 925–37.

Kabler, P. W., Clark, H. F., and Geldreich, E. E., 1964. Sanitary significance of coliforms and fecal coliform organisms in surface water. *U.S. Public Health Rep.*, **79**, 58–60.

Kaiser, K. L. E. and Wong, P. T. S., 1974. Bacterial degradation of polychlorinated biphenyls. I. Identification of some metabolic products from 'Aroclor' 1242. *Bull. Environ. Contam. Toxicol.*, **11**, 291–6.

Kator, H., Oppenheimer, C. H., and Miget, R. J., 1971. Microbial degradation of a Louisiana crude oil in closed flasks and under simulated field conditions. In *API/EPA/USCG Conference on Prevention and Control of Oil Spills*, American Petroleum Institute, Washington, D. C., pp. 287–96.

Keil, J. E. and Sandifer, S. M., 1972. DDT and polychlorinated biphenyl ('Aroclor' 1242). Effects of uptake on *E. coli* growth. *Water Res.*, **6**, 837–41.

Kinney, P. J., Button, D. K., and Schell, D. M., 1969. Kinetics of dissipation and biodegradation of crude oil in Alaska's Cook Inlet. In *API/FWPCA Conference on Prevention and Control of Oil Spills*, American Petroleum Institute, Washington, D.C., pp. 333–40.

Kirsch, E. J. and Etzel, J. E., 1973. Microbial decomposition of pentachlorophenol. *J. Water Pollut. Control Fed.*, **45**, 359–64.

Koditschek, L. K. and Guyre, P., 1974. Antimicrobial-resistant coliforms in New York Bight. *Mar. Pollut. Bull.*, **5**, 71–4.

Lunt, D. and Evans, W. C., 1970. Microbial metabolism of biphenyl. *Biochem. J.*, **118**, pp. 54–5.

Marx, W., 1971. *Oilspill*, Sierra Club, San Francisco.

McCann, J., Choi, E., Yamasaki, E., and Ames, B. N., 1975. Detection of carcinogens as mutagens in the *Salmonella*/microsome test: assay of 300 chemicals. *Proc. U.S. Natl. Acad. Sci.*, **72**, 5135–9.

Miller, E. C. and Miller, J. A., 1971. The mutagenicity of chemical carcinogens: correlations, problems and interpretations. In *Chemical Mutagens: Principles and Methods for Their Detection* (Ed. A. Hollaender), Vol. 1, Plenum Press, New York, 83–119.

Mills, A. L. and Colwell, R. R., 1977. Microbial effects of metal ions in Chesapeake Bay water and sediment. *Bull. Environ. Contam. Toxicol.*, **18**, 99–103.

Moulder, D. S. and Varley, A., 1971. *A Bibliography on Marine and Estuarine Oil Pollution*, Mar. Biol. Assn. U.K., Plymouth, p. 129.

Mulkins-Phillips, G. J. and Stewart, J. E., 1974a. Effect of four dispersants on biodegradation and growth of bacteria on crude oil. *Appl. Microbiol.*, **28**, 547–52.

Mulkins-Phillips, G. J. and Stewart, J. E., 1974b. Effects of environmental parameters on bacterial degradation of Bunker C oil, crude oils and hydrocarbons. *Appl. Microbiol.*, **28**, 915–22.

Nelson, J. D., Jr. and Colwell, R. R., 1975. The ecology of mercury-resistant bacteria in Chesapeake Bay. *Microbial Ecol.*, **2**, 191–218.

Nelson, N. and Van Duuren, B., 1975. *Final Report of NSF Workshop Panel to Select Organic Compounds Hazardous to the Environment*, National Science Foundation, Washington, D.C.

Nelson-Smith, A., 1973. *Oil Pollution and Marine Ecology*, Plenum Press, New York.

Novick, R. F., 1969. Extrachromosomal inheritance in bacteria. *Bacteriol. Rev.*, **33**, 210–63.

Rashid, M. A., 1974. Degradation of Bunker C oil under different coastal environments of Chedabucto Bay, Nova Scotia. *Estuarine and Coastal Marine Science*, **2**, 137–44.

Robertson, B., Arheleger, S., Kinney, P. J., and Button, D. K., 1973. Hydrocarbon biodegradation in Alaskan waters. In *The Microbial Degradation of Oil Pollutants* (Eds. D. G. Ahearn and S. P. Meyers), Louisiana State University Publ. No. LSU-SG-73–01, Baton Rouge, La., pp. 171–84.

Sayler, G. S. and Colwell, R. R., 1976. Partitioning of mercury and polychlorinated biphenyl by oil, water, and suspended sediment. *Environ. Sci. Technol.*, 10, 1142–5.

Sayler, G. S., Shon, M., and Colwell, R. R., 1977a. Growth of an estuarine *Pseudomonas* sp. on polychlorinated biphenyl. *Microbial Ecol.*, 3(3), 241–55.

Sayler, G. S., Nelson, J. D., Jr., and Colwell, R. R., 1975. Role of bacteria in bioaccumulation of mercury in the oyster, *Crassostrea virginica*. *Appl. Microbiol.*, 30, 91–6.

Sayler, G. S., Thomas, R., and Colwell, R. R., 1977b. Polychlorinated biphenyl (PCB)-degrading bacteria and PCB in estuarine and marine environments. *Estuarine and Coastal Marine Science*, in press.

Schwarz, J. R., Walker, J. D., and Colwell, R. R., 1974a. Growth of deep-sea bacteria on hydrocarbons at ambient and *in situ* pressure. *Develop. Ind. Microbiol.*, 15, 239–249.

Schwarz, J. R., Walker, J. D., and Colwell, R. R., 1974b. Deep-sea bacteria: growth and utilization of hydrocarbons at ambient and *in situ* pressure. *Appl. Microbiol.*, 28, 982–6.

Schwarz, J. R., Walker, J. D., and Colwell, R. R., 1975. Deep-sea bacteria: growth and utilization of *n*-hexadecane at *in situ* temperature and pressure. *Can. J. Microbiol.*, 21, 682–7.

Sharpley, J. M., 1966. *Elementary Petroleum Microbiology*, Gulf Publishing Co., Houston, Texas.

Sizemore, R. K. and Colwell, R. R., 1977. Plasmids carried by antiobiotic-resistant marine bacteria. *Antimicrobial Agents and Chemotherapy*, 12, 373–82.

Smith, D. H., 1967. R factors mediate resistance to mercury, nickel and cobalt. *Science*, 156, 1114–6.

Summers, A. O. and Silver, S., 1972. Mercury resistance in a plasmid-bearing strain of *Escherichia coli*. *J. Bacteriol.*, 122, 1228–36.

Voll, M. J., Isbister, J., Isaki, L., McCommas, M., and Colwell, R. R., 1977. Effects of microbial activity on aquatic pollutants. *Ann. N.Y. Acad. Sci.*, 298, 104–10.

Walker, J. D., Calomiris, J. J., Herbert, T. L., and Colwell, R. R., 1975a. Petroleum hydrocarbons: degradation and growth potential for Atlantic Ocean sediment bacteria. *Mar. Biol.*, 34, 1–9.

Walker, J. D. and Colwell, R. R., 1973. Microbial ecology of petroleum utilization in Chesapeake Bay. In *API/EPA/USCG Conference on Prevention and Control of Oil Spills*, American Petroleum Institute, Washington, D.C., pp. 685–90.

Walker, J. D. and Colwell, R. R., 1974a. Microbial degradation of petroleum: the use of mixed hydrocarbon substrates. *Appl. Microbiol.*, 27, 1053–60.

Walker, J. D. and Colwell, R. R., 1974b. Microbial degradation of model petroleum at low temperatures. *Microbial Ecol.*, 1, 63–95.

Walker, J. D. and Colwell, R. R., 1974c. Mercury-resistant bacteria and petroleum degradation. *Appl. Microbiol.*, 27, 285–7.

Walker, J. D. and Colwell, R. R., 1974d. Some effects of petroleum on estuarine and marine microorganisms. *Can. J. Microbiol.*, 21, 305–13.

Walker, J. D. and Colwell, R. R., 1975a. Factors affecting enumeration and isolation of actinomycetes from Chesapeake Bay and Southeastern Atlantic Ocean sediments. *Mar. Biol.*, 30, 193–201.

Walker, J. D. and Colwell, R. R., 1975b. Degradation of hydrocarbons and mixed hydrocarbon substrate by microorganisms from Chesapeake Bay. *Progr. Water Technol.*, 7, F83–F91.

Walker, J. D. and Colwell, R. R., 1976a. Long-chain *n*-alkanes occurring during microbial degradation of petroleum. *Can. J. Microbiol.*, 22, 886–91.

Walker, J. D. and Colwell, R. R., 1976b. Oil, mercury, and bacterial interactions. *Environ. Sci. Technol.*, 10, 1145–7.

Walker, J. D. and Colwell, R. R., 1977. Effect of petroleum hydrocarbons on growth and activity of marine bacteria isolated from open ocean water. *Develop. Ind. Microbiol.*, 18, 655–60.

Walker, J. D., Colwell, R. R., Hamming, M. C., and Ford, H. T., 1974a. Extraction of petroleum hydrocarbons from oil-contaminated sediments. *Bull. Environ. Contam. Toxicol.*, 13, 245–8.

Walker, J. D., Colwell, R. R., Hamming, M. C., and Ford, H. T., 1975b. Petroleum hydrocarbons in Baltimore Harbor of Chesapeake Bay: distribution in sediment cores. *Environ. Pollut.*, 9, 231–8.

Walker, J. D., Colwell, R. R., and Petrakis, L., 1975c. Microbial petroleum degradation: application of computerized mass spectrometry. *Can. J. Microbiol.*, 21, 1760–7.

Walker, J. D., Colwell, R. R., and Petrakis, L., 1975d. Bacterial degradation of motor oil. *J. Water Pollut. Control Fed.*, 47, 2058–66.

Walker, J. D., Colwell, R. R., and Petrakis, L., 1976a. Biodegradation rates of petroleum components. *Can. J. Microbiol.*, 22, 1209–13.

Walker, J. D., Colwell, R. R., and Petrakis, L., 1977a. Degradation of hydrocarbons by pure cultures of microorganisms: degradation of crude oil. *Arch. Microbiol.*, in press.

Walker, J. D., Colwell, R. R., and Petrakis, L., 1977b. Degradation of hydrocarbons by pure cultures of microorganisms: degradation of motor oil. *Arch. Microbiol.*, in press.

Walker, J. D., Petrakis, L., and Colwell, R. R., 1976b. Comparison of the biodegradability of crude and fuel oils. *Can. J. Microbiol.*, 22, 598–602.

Walker, J. D., Seesman, P. A., and Colwell, R. R., 1974b. Effects of petroleum on proteolytic, lipolytic and chitinolytic estuarine bacteria. *Mar. Pollut. Bull.*, 5, 186–8.

Walker, J. D., Seesman, P. A., and Colwell, R. R., 1975e. Effect of South Louisiana crude oil and No. 2 fuel oil on growth of heterotrophic microorganisms including proteolytic, lipolytic, chitinolytic and cellulolytic bacteria. *Environ. Pollut.*, 9, 13–33.

Westlake, D. W. S., Jobson, A., Phillippe, R., and Cook, F. D., 1974. Biodegradability and crude oil composition. *Can. J. Microbiol.*, 20, 915–28.

Wong, P. T. S. and Kaiser, K. L. E., 1975. Bacterial degradation of polychlorinated biphenyls. II. Rate studies. *Bull. Environ. Contam. Toxicol.*, 13, 249–56.

Young, L. Y. and Mitchell, R., 1973. Negative chemotaxis of marine bacteria to toxic chemicals. *Appl. Microbiol.*, 25, 972–5.

Zajic, J. E., Supplisson, B., and Volesky, B., 1974. Bacterial degradation and emulsification of No. 6 fuel oil. *Environ. Sci. Technol.*, 8, 664–8.

ZoBell, C. E., 1969. Microbial modification of crude oil in the sea. In *API/FWPCA Joint Conference on Prevention and Control of Oil Spills*, American Petroleum Institute, Washington, D.C., pp. 317–26.

ZoBell, C. E., 1971. Sources and biodegradation of carcinogenic hydrocarbons. In *API/EPA/USCG Conference on Prevention and Control of Oil Spills*, American Petroleum Institute, Washington, D.C., pp. 441–51.

CHAPTER 14

Physical and Chemical Changes in the Environment with Indirect Biological Effects

R. E. MUNN*

Atmospheric Environment Service, Environment Canada, 4905 Dufferin Street, Downsview, Ontario, Canada, M3H 5T4

14.1 INTRODUCTION 295
14.2 CLIMATIC CHANGE 296
    (i) Introduction 296
    (ii) Impact of climate on man and other biota 296
    (iii) Human activities modifying local climate 297
    (iv) Human activities modifying regional climate 298
    (v) Human activities modifying global climate 298
14.3 STRATOSPHERIC OZONE DEPLETION: BIOLOGICAL EFFECTS 301
14.4 ACID RAINS 301
    (i) Freshwater ecosystems 303
    (ii) Forest ecosystems 304
14.5 THERMAL POLLUTION OF WATER BODIES 305
14.6 THE CONSTRUCTION OF LARGE ENGINEERING WORKS 307
14.7 CONCLUSION 308
14.8 REFERENCES 308

14.1. INTRODUCTION

Previous chapters have dealt with the direct effects of pollutants, usually chemical or radioactive, released to the environment as a result of man's activities. In this chapter, a brief overview is given of some of man's impacts on the physical and chemical environment that can indirectly cause biological effects, including those that have been the subject of speculation.

Man is responsible for physical and chemical changes in the environment which have indirect biological effects. The following five examples have been chosen for consideration in this chapter:

— climate change;
— biological effects of stratospheric ozone depletion;

*Present affiliation: *Institute for Environmental Studies, University of Toronto, Toronto, Ontario, Canada M5S 1A4.*

— acid rains;
— thermal pollution of water bodies; and
— the construction of large engineering works.

The relations between causes and indirect effects such as these are usually difficult to quantify because the phenomena can take place over large distances (global, atmospheric, or stratospheric) and against a background of large natural variations. However, if the chapter alerts the reader to the need for quantitative cause and effect data (to permit assessment of the real magnitude of the hazard) and provides him with some current views in this subject area, it will have served its purpose.

14.2. CLIMATIC CHANGE

(i) Introduction

Weather and climate affect man and other forms of life. Man, on the other hand, affects climate, mainly on the local scale but also regionally and possibly globally.

(ii) Impact of Climate on Man and Other Biota

Biometeorology is the study of the interrelationships between living organisms and weather or climate. The subject is so vast that it cannot be summarized, and the reader is referred to the texts by Landsberg (1969) and Munn (1970) and to the *International Journal of Biometeorology*. Only a few examples will be mentioned by way of illustration.

Meteorological stresses are often severe and irreversible. Figure 14.1 shows a schematic dose–response curve for frost damage to tobacco or oranges. In many industrialized regions, the agricultural losses due to frost are in fact much greater than those due to air pollution.

Not only cold but also heat is of concern. In particular, summer heat-waves cause excess mortality (Bridger and Helfand, 1968; Clarke and Bach, 1971; Clarke, 1972). During such spells, cities are hotter and have less natural ventilation than surrounding rural areas, particularly at night; in addition, low-income families, who are generally without air-conditioning equipment, often live near the centre of cities, which may be as much as 5 to 8°C warmer than the surrounding countryside. Clarke and Bach (1971) presented some data on deaths caused by heat in St. Louis, Mo. in July 1966 as a function of average daily temperature lagged by one day. The number of deaths rose from zero at a temperature of 32°C (assumed to be the threshold) to 73 cases at a temperature of 35°C.

Other meteorologically induced stresses are caused by wind, humidity (too high or too low), and precipitation (too much or too little). Living things often modify their own microclimate, e.g. man builds shelters, small animals burrow, trees grow

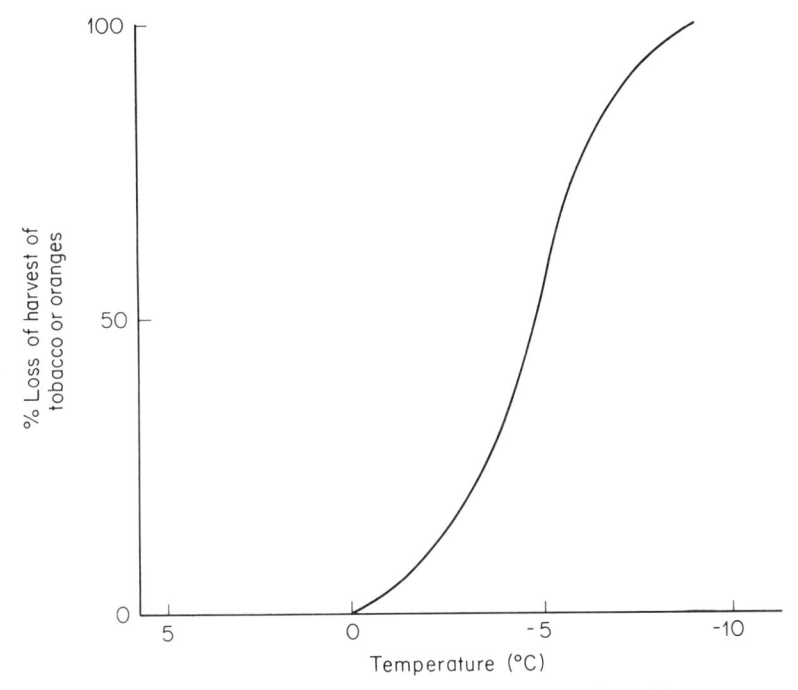

Physical and Chemical Changes in the Environment

Figure 14.1 Schematic dose–response curve for frost damage to tobacco or oranges

closer together (thus reducing wind speeds, providing shade, etc.). Sometimes, however, a rare weather event (hurricane, drought, flood, etc.) is too severe for life to survive. Major shifts in climate (to an ice age in the most extreme case) also cause biological stress but in these situations the biosphere often has time to adapt through various mechanisms including species selection. The rate of change of climate is an important factor in the degree of possible adaptation.

(iii) Human Activities modifying Local Climate

Mankind modifies local climate. The modifications are so well known that it is only necessary to include a list, without explanatory comments: (see, for example, Geiger, 1965).

(a) construction of houses and buildings;

(b) paving of roads and parking lots;

(c) cutting of trees;

(d) ploughing of fields;

(e) erection of snow fences and shelter belts;

(f) frost protection;

(g) drainage of swamps;

(h) irrigation;

(i) water vapour emissions from cooling towers and chimneys (fog and ice fog; deposition of trace salts in adjacent agricultural fields); and

(j) heat emissions.

Climate modifications on the local scale are not in general irreversible.

(iv) Human Activities modifying Regional Climate

Mankind modifies regional climate by changing the earth's surface over a large area, e.g. by cutting forests, ploughing fields, building cities, emitting pollutants, and releasing heat. In some cases, the regional impact is almost irreversible; after a tropical forest has been cut, for example, soil erosion may proceed rapidly and the land may quickly become too impoverished to permit forest regeneration.

Climatic modification may be significant on this scale, although the usual controls are still the dominant ones — latitude, altitude, global weather systems, and proximity to large bodies of open water. There is no evidence that climatic modifications on the regional scale have had any effect on global weather patterns.

(v) Human Activities modifying Global Climate

Impacts on hemispheric and global climate are difficult to detect. Nevertheless, this is a widely discussed topic, partly because of the fact that if one waits for evidence of significant effects, the changes in world climate may already have progressed to the irreversible stage. The implications for world food production and energy requirements are of particular concern.

There are a number of well-established climatic research activities, including those associated with GARP, the ICSU-WMO *G*lobal *A*tmospheric *R*esearch *P*rogramme (Bolin, 1975). One of the GARP objectives is to seek an increased understanding of the physical basis of climate.

Recent speculation about mankind's impact on global climate has focussed on four themes:

(a) the effects of increases in concentrations of CO_2;

(b) the effects of increases in concentrations of suspended particulate matter;

(c) disruption of the stratospheric ozone layer by a number of substances; and

(d) thermal pollution.

The four themes will be discussed briefly in the following paragraphs.

(a) CO_2 Increases

Atmospheric CO_2 concentrations have been rising in recent years at a rate of about 0.6 to 1.0 ppm per year. This increase is due to (a) the combustion of fossil fuels, releasing CO_2 at greater rates than can be taken up by the oceans and land plants, and (b) the destruction of vegetation with consequent reduction in CO_2 fixation.

The present concentration of CO_2 is about 330 ppm. If the rate of increase in the use of fossil fuels were to continue at 4% per year, a CO_2 concentration of about 375 to 400 ppm would be expected by the year 2000. Various scenarios about consumption of fossil fuels during the twenty-first century suggest that within the next 200 years, there could be a fourfold and perhaps even an eightfold increase in the concentrations of atmospheric CO_2 (Bacastow and Keeling, 1973).

Atmospheric CO_2 causes a 'greenhouse' effect, trapping the outgoing long-wave radiation emitted from the surface of the earth, but having no effect on the incoming short-wave radiation. A simulation model of the atmosphere predicts that a doubling in CO_2 concentrations would cause an average surface temperature rise of about 2.5°C (Manabe and Wetherald, 1975), However, the model does not include important mechanisms such as a rise in sea temperature, which would reduce the oceanic uptake rate of CO_2 and tend to increase further the atmospheric concentration. There is therefore a great need for additional investigations on the climatic effects of the predicted increases in CO_2 concentrations.

Added to the concern about CO_2 is the fact that chlorofluorocarbons (from aerosol spray cans, refrigerants, etc.) also cause a greenhouse effect, but in different 'windows' of the long-wave radiation spectrum. These man-made gases are also accumulating in the atmosphere, and so could reinforce the CO_2 effects

(b) Suspended Particulate Matter

Suspended particulate concentrations have been decreasing in cities during recent decades, but have been increasing in the surrounding countryside. However, there is little information on trends in the Third World, where slash burning is important.

Suspended particulates intercept solar radiation, reducing the amount reaching the ground. They also intercept the returning long-wave radiation from the earth. The net result is believed to be a cooling of the earth's surface, although this inference is still somewhat uncertain.

(c) Stratospheric Ozone

A layer of ozone exists in the stratosphere at a height of about 20 km. This has a major effect on the radiation balance of the stratosphere, and thus on the general circulation of the entire atmosphere.

The fragility of the stratosphere has recently been debated, and the following mechanisms that could affect the ozone balance have been proposed:

(a) Natural processes
   1. volcanic emissions
   2. solar flares

(b) Man-made processes
   1. supersonic aircraft emissions
   2. chlorofluoromethanes from aerosol spray cans and refrigerants
   3. nitrous oxide from fertilizers
   4. nuclear bomb debris

The World Meteorological Organization has issued a statement (WMO, 1976) on man-made modification of the ozone layer. The principal conclusions are as follows:

(1) A large fleet of supersonic aircraft flying above 17 km would have a noticeable effect on the ozone layer, and emission standards may have to be accepted internationally. The numbers of currently planned SST flying at and below 17 km do not present a problem.

(2) If present emission rates of chlorofluoromethanes continue, the long-term effect would be about a 10% ozone depletion, with an uncertainty of about a factor of two.

(3) Problems associated with fertilizers warrant attention.

(4) Although a 10% decrease in stratospheric ozone may decrease the upper stratosphere's temperature by as much as 10°C, the climatic consequences cannot be reliably estimated because of the complexity of stratospheric–tropospheric interactions.

(5) An intensified monitoring programme for stratospheric trace substances (including ozone) is needed in order to distinguish between trends caused by man's influence and those arising from natural variability of the stratosphere.

(d) Thermal Pollution

Although the heat arriving on the earth from solar radiation far exceeds man's heat output when averaged over the world, the two quantities can be comparable on a local and sometimes on a regional scale, e.g. within the Boston to Washington megalopolis.

The Great Lakes, which are sources of heat in winter and heat sinks in summer, have a major influence on the movement and intensity of weather systems crossing eastern North America. There has therefore been considerable speculation that anthropogenic thermal releases could cause similar effects. Sawyer (1974) has estimated that the heat output from an area the size of western Europe would have to increase 50-fold before there were climatic changes comparable to natural year-to-year variations on the global scale, and this view is supported by Machta (1975).

14.3. STRATOSPHERIC OZONE DEPLETION: BIOLOGICAL EFFECTS

The stratospheric ozone layer shields the surface of the earth from ultraviolet sunlight, preventing excessive sunburn, skin cancer, and cellular and subcellular damage in people, animals and plants.

A number of mechanisms have been postulated that could affect the stratospheric ozone layer (see section 14.2.v) and quantitative estimates have been made of the resulting depletion. In addition, there are experimental data on skin cancer incidence for various levels of ultraviolet light. Machta (1976) has examined these predictions in terms of the concepts of dose commitment and harm commitment, to provide estimates of the numbers of cases of skin cancers for several scenarios of stratospheric disruption. His results are given in Table 14.1. The estimates are rather uncertain but the methodology is of particular interest because it suggests a way of quantifying the harmful effects of a complex global phenomenon in which the dose–response relations involve a chain of environmental circumstances.

There are, of course, other possible biological effects of a depletion in stratospheric ozone and ultraviolet radiation:

(a) on people (e.g. reduction in the incidence of rickets);

(b) on animals;

(c) on vegetation;

(d) on marine life; and

(e) on terrestrial and aquatic ecosystem behaviour.

Many of the predicted impacts are still speculative but there has been a major surge of research in this field in the last five years. A recent review has been given in a U.S. National Academy of Sciences report (NAS, 1976).

14.4. ACID RAINS

Monthly precipitation samples in Scandinavia have been analysed for pH and for the concentrations of various trace substances since the 1950's. These samples

Table 14.1 Estimated Incidence of and Mortality from, Skin Cancers from Steady-state Release of Selected Chemicals[a]. (Reproduced by permission of MARC Chelsea College from Machta, 1976)

Chemical	Input (g/yr)	Altitude of input (km)	Source	Dose commitment % ozone depletion	Dose commitment % increase u.v.-B	Harm commitment: additional cases per year[b] incidence non-melanoma skin cancers	incidence melanoma skin cancers	mortality melanoma skin cancers
NO	156 × 10¹¹	10.5	400 wide-bodied subsonic aircraft	0.082	0.12	2,000	40	10
NO	1 × 10¹¹	16–18	100 Concorde TU-144 aircraft	0.11–0.57	0.17–0.86	2,500 15,000	50–250	10–50
NO	164 × 10¹¹	19.5	100 large supersonic aircraft	3.27	4.91	75,000	1,500	300
N₂O	2 × 10¹⁴ (N₂)	0	Fertilizers in 2000 A.D. and constant thereafter	<1.8–23	<2.7–35	<40,000 500,000	<900 10,000	<150–2,000
F-11} F-12}	{2 × 10¹¹ 2 × 10¹¹}	0	Aerosols, refrigerants, etc. at 1973 rate	6.5–18	9.8–27	150,000 400,000	30,000	600–1,500

[a]Estimates assume a constant world light-skinned population of 10^9 persons.
[b]Based on current non-melanoma skin cancer incidence of 150/100,000 light-skinned persons per year, melanoma incidence of 4/100,000, melanoma mortality of 1.5/100,000.

show increasing acidity while many of the Scandinavian rivers and lakes have become more acid.

There is strong circumstantial evidence to suggest that the cause of these changes is the increasing consumption of fossil fuels, increasing the emissions of SO_2. After entering the atmosphere, SO_2 is gradually transformed into particulate sulphates and then scavenged by precipitation.

Scandinavia is particularly sensitive to acid rains for two reasons:

(1) orographic precipitation: Rain is often triggered by lifting of air masses over the high ground of Scandinavia, delivering pollution in high concentrations to small target areas.

(2) podsol soils: The soils and lakes of Scandinavia have low buffering capacities and so are sensitive to slight changes in acidity. In areas suffering from salinization, on the other hand, acid rains would be beneficial.

In recent years, the concern about acid rains has spread to the Laurentian Shield of North America and to some podsolized regions in the tropical rain forests. There are in fact increasing numbers of international research and monitoring activities as well as international meetings dealing with these questions. A recent conference on the effects of acid precipitation was held in Telemark, Norway, June 1976, sponsored by the Government of Norway. The main conclusions of that meeting provide a balanced view of the present state of knowledge of the biological effects of acid rains in Scandinavia (Norway, 1976). They are quoted below without change by permission of Dr. Lars Overrein, Director SNSF-project, Oslo, Norway.

(i) Freshwater Ecosystems

Water quality has changed during the last decades in numerous lakes and rivers in southern Scandinavia and eastern North America, pH often falling below 5, with sulphate becoming the most important anion. There is strong evidence that this change is due to acid precipitation. It is associated with the loss of buffering capacity, and the occurrence of additional short-term decreases in pH, related to meltwater from snow or episodic inputs of acid precipitation from polluted air masses.

The acidification of freshwater ecosystems leads to many changes, most of which involve decreases in biological activity and important changes in nutrient cycling. For example, decomposer organisms are less active in acid waters, resulting in increased accumulations of organic matter. When the pH drops below 6, numbers of species in several groups of organisms (phytoplankton and zooplankton, bottom fauna and several other groups of invertebrates), decrease considerably, thus affecting the variety of food for fish and other animals depending on freshwater ecosystems. Shifts have been observed from higher aquatic plants toward mosses, which influence not only the bottom fauna but also nutrient exchange with the sediments.

High acidity (pH < 5.5) seriously affects fish populations, particularly when occurring in waters of low ionic strength. Rapid extinction rates of fish populations inhabiting acidified waters have been observed during the past few decades in southern Scandinavia as well as in parts of eastern North America.

Case studies of several fish populations and experiments clearly indicate that the elimination of fish is often a result of chronic reproductive failure in acid conditions and of damage done to sensitive stages, especially the newly hatched larvae. Such a process is insidious and not readily evident in terms of fishery yield until extinction is imminent. In lakes and streams with soft waters acid stress has also been shown both experimentally and in field studies to cause mortality among adult fish as a result of interference with physiological mechanisms regulating active ion exchange across gill membranes. In this case factors such as size, age, acclimatization history, genetic background and ionic strength of the water interact in complex ways to determine the relative acid tolerance of the fish.

There is strong evidence that the increased acidity of precipitation is now the main cause of these extensive losses of salmonid fish stocks as well as other populations of economic importance both in southern Scandinavia and the northeastern part of the United States and parts of southeastern Canada.

Sensitivity is related to the tolerance of the differing species (reflecting their genetic diversity), to the timing of episodic acid precipitation in relation to the stage in the life cycle, and to the influence of the geological environment, e.g., lakes and rivers on bedrock, overburden and soils highly resistant to chemical weathering. This may also be the case in geological areas of this type elsewhere in the world. Special consideration should be given to the preservation of unique gene pools and habitats. There may be complex interactions with other environmental factors, including some organic compounds. In waters of low pH the adverse effects of heavy metals on fish* and other organisms can be enhanced, whereas at pH values nearer neutrality these substances would have been tolerated.

(ii) Forest Ecosystems

For a long time the effects of sulfur compounds have been observed near sources of emission. Recently, however, interest has also been devoted to regional effects at sites remote from these sources where vegetation is influenced by both 'wet' and 'dry' deposits of sulfur compounds. However,

---

*Editor's note. Landner and Larsson (1976) found that the methylmercury concentration in the fish of Swedish lakes was proportional to the acidity of the water in which they lived.

instead of being exposed to mean annual sulfur dioxide concentrations of up to 200 $\mu g/m^3$, which are associated with the development of obvious foliage blemishes, e.g., chlorotic spots, they are subject to annual average concentrations of possibly 25–30 $\mu g/m^3$ which normally would not be expected to cause blemishes, although other physiological disturbances may occur.

Because tree growth has been shown to be directly related to base saturation – a widely accepted indicator of soil fertility – adverse effects of acid could be expected – base saturation being inversely related to acidity.

In field and/or laboratory experiments, it has been found that acid precipitation:

(a) decreased soil respiration, an indicator of microbial activity;

(b) affected nitrogen mineralization;

(c) increased amounts of minerals leaching from soil;

(d) affected the germination and establishment of conifer seeds and seedlings which were maximal at about pH 5.0;

(e) accelerated cuticular erosion of leaves;

(f) enhanced the leaching of nutrients and organic compounds from leaves;

(g) decreased the activities of associated pathogens, of some beneficial symbionts, also of saprophytes;

(h) produced leaf damage at pH 2.5–3.5.

Although many of these factors might be expected to adversely affect tree growth, it has not yet been possible to demonstrate unambiguously decreased tree growth in the field. However it is possible that acid damage might have been offset by the nutritional benefits gained from nitrogen compounds commonly occurring in acid precipitation. Changes already detected in soil processes may as yet be too small to affect plant growth. Forests are complex. It has been shown that through-fall and stemflow are affected differently by different tree species. Thus the composition of 'precipitation' reaching soil, possibly affecting soil processes and transfer to freshwater systems, could be influenced by the nature of the tree cover.

14.5. THERMAL POLLUTION OF WATER BODIES

Figure 14.2 illustrates the effect on surface water temperature of the warm effluent from a power generating station on the shore of Lake Huron (Kenney, 1973). On this occasion (March 2, 1971), the lake was frozen except in the vicinity

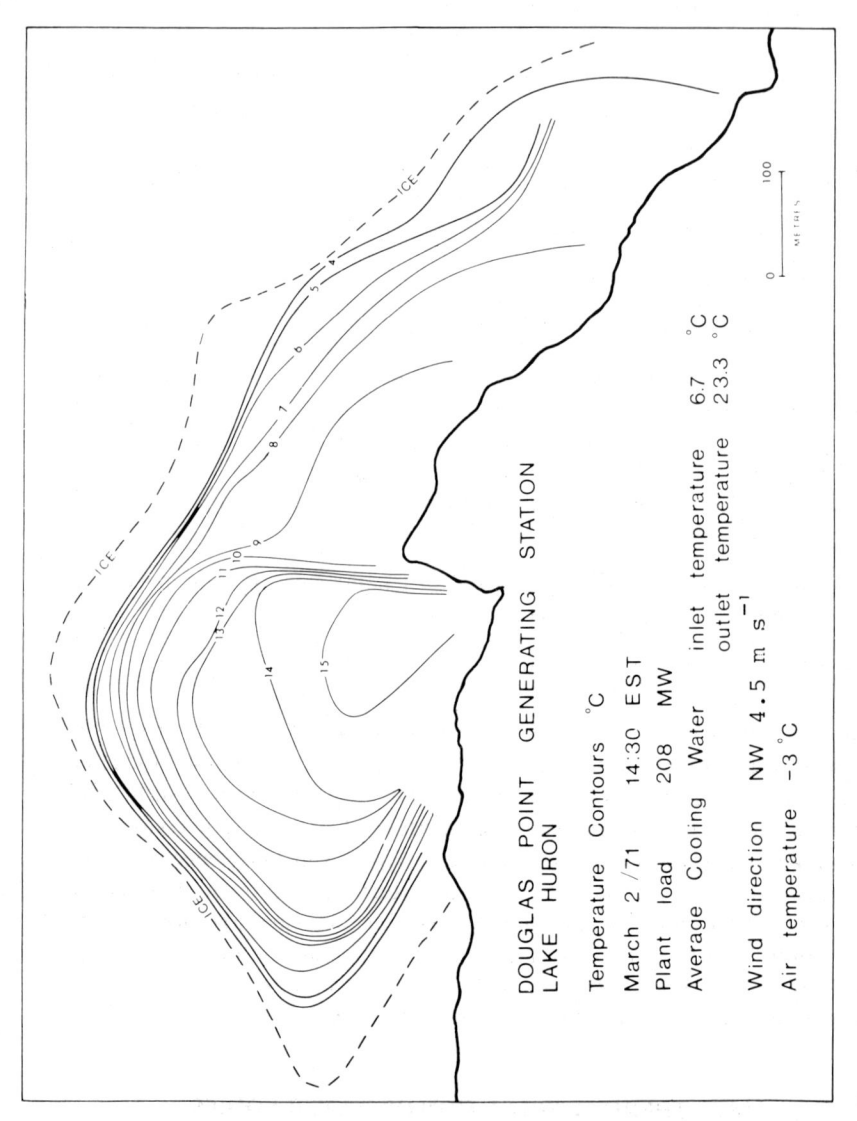

Figure 14.2 Douglas Point generating station, Lake Huron. Temperature contours, °C – March 2, 1971, 14:30 EST. (Reproduced with permission, from Kenney, 1973)

of the station. The dashed line indicates the edge of the ice pack, which had receded as much as one-half kilometre.

Figure 14.2 reveals a very large modification of the physical environment but this is not unusual in lakes where water currents are weak. In rapidly moving rivers and in tidal estuaries, on the other hand, the magnitude is much less, the heat being dissipated more quickly.

The biological effects associated with thermal pollution are mainly local in scale, although a series of effluents on the same watershed could have a wider influence. A survey of the main effects (Dickson, 1975) indicates that they are site-specific, depending on the organisms living in the body of water and on the physical characteristics of the area. In organically enriched waters, chronic exposures to temperatures elevated by a few degrees cause an increase in productivity. There is also a shift towards more heat-tolerant species, but often with a decrease in diversity.

Secondary physical effects may of course modify these general conclusions in particular situations. For example, a thermal plume may have the beneficial effect of preventing the usual winter oxygen depletion from taking place by keeping a lake open.

14.6. THE CONSTRUCTION OF LARGE ENGINEERING WORKS

The construction of large engineering works causes a multitude of biological impacts. Quite often, major effects take place during construction when heavy earth-moving equipment transforms the landscape, and when construction workers roam the area in search of recreational fish and game. On the other hand, some impacts may not become apparent for several decades. For example, a temporary clearing of trees may create a frost pocket, preventing regeneration of the stand.

Environmental impacts occur over three space scales:

(1) locally;
(2) regionally;
(3) continentally.

Sometimes the impact is focussed sharply within a few hectares. On the other hand, the effects may be widely distributed. For example, the diversion of a river to another watershed could have significant biological effects if the 'hardness' of the water were different in the two areas.

The number of biological effects associated with the construction of large engineering works is too large to catalogue. In another SCOPE publication (Munn, 1975), the scientific basis of an environmental impact assessment has been described, as well as the operational procedures to be followed. It should be emphasized that large engineering works should not be constructed until the environmental impacts have been evaluated by biologists and ecologists, as well as by physical scientists.

The underlying principles associated with environmental impact assessments are relevant to all the examples of environmental disruption given earlier in this chapter.

14.7. CONCLUSION

In conclusion, the modifications in the biological environment from climatic change, thermal pollution, and large engineering works are multifold. The changes are not always harmful. For example, the effect of draining a swamp or building a reservoir will be to alter the flora and fauna, a result that may be perceived to be beneficial by many people. Similarly a change in climate would influence the biosphere in favourable as well as unfavourable ways, depending on the point of view.

14.8. REFERENCES

Bacastow, R. and Keeling, C. D., 1973. Changes from A.D. 1700 to 2070 as deduced from a geochemical model. In *Carbon and the Biosphere* (Eds. G. M. Woodwell and E. V. Pecan), Nat. Tech. Inf. Service, U.S.A. pp. 86–133.

Bolin, B. (Ed.), 1975. The physical basis of climate and climate modelling. *GARP Pub.*, No. 16, WMO, Geneva, Switzerland, 265 pp.

Bridger, C. A. and Helfand, L. A., 1968. Mortality from heat during July 1966 in Illinois. *Int. J. Biometeorol.*, 12, 51–70.

Clarke, J. F., 1972. Some effects of the urban structure on heat mortality. *Environ. Res.*, 5, 93–104.

Clarke, J. F. and Bach, W., 1971. Comparison of the comfort conditions in different urban and suburban micro-environments. *Int. J. Biometeorol.*, 15, 41–54.

Dickson, D. R., 1975. *Waste Heat in the Aquatic Environment,*, ISSN 0316–0114, NRCC 14109, National Res. Council, Ottawa, Canada, 39 pp.

Geiger, R., 1965. *The Climate Near the Ground*, Harvard University Press, Cambridge, Mass., 611 pp.

Kenney, B. C., 1973. The physical effects of waste heat input to the Great Lakes. *Scientific Series*, No. 28, Inland Waters Directorate, Canada Centre for Inland Waters, Burlington, Ont., Canada.

Landner, L. and Larsson, P. O., 1976. Biological effects of mercury fall-out into lakes from the atmosphere. *Rep. B115, Swedish Inst. for Water and Air Res.*, IVL, Stockholm, 8 pp.

Landsberg, H. E., 1969. *Weather and Health*, Doubleday and Co. Inc., Garden City, New York, 148 pp.

Machta, L., 1975. Trends in atmospheric properties, *Preprint, WHO Conf. on Env. Sensing and Assessment*, Las Vagas, Nevada, Paper 32–2, 2 pp.

Machta, L., 1976. The ozone depletion problem: an example of harm commitment. *MARC Chelsea College Report*, 459A Fulham Rd., London, U.K. 33 pp.

Manabe, S. and Wetherald, R., 1975. The effects of doubling the CO_2 concentration on the climate of a general circulation model. *J. Atmos. Sci.*, 32, 3–15.

Munn, R. E., 1970. *Biometeorological Methods*, Academic Press, New York, 336 pp.

Munn, R. E. (Ed.), 1975. Environmental impact assessment: principles and procedures. *SCOPE 5*, ICSU-SCOPE, 51 Blvd. de Montmorency, Paris, France, 160 pp.

NAS, 1976. *Halocarbons: Environmental Effects of Chlorofluoromethane Release*, U.S. Nat. Research Council, Washington, D.C., 75 pp.

Norway, 1976. *Report from the International Conference on the Effects of Acid Precipitation*, Telemark, Norway, June, 1976, *Tellus*, 5, 200–1.

Sawyer, J. S., 1974. Can man's waste heat affect the regional climate? *Preprint, IAMAP First Special Assembly*, Melbourne, Australia, Int. Ass. Meteorol. and Atm. Physics.

WMO, 1976. Statement on anthropogenic modification of the ozone layer and some possible geophysical consequences. WMO, Geneva, Switzerland.

SECTION IV
ECOSYSTEM RESPONSE TO POLLUTION

CHAPTER 15

Ecosystem Response to Pollution

P. BOURDEAU

Directorate General for Research Science and Education, Commission of the European Communities, Brussels, Belgium

M. TRESHOW

Department of Biology, University of Utah, Salt Lake City, U.S.A.

15.1	INTRODUCTION	313
15.2	METHODS OF STUDYING ECOSYSTEM RESPONSE	314
15.3	ASSESSMENT OF LEVELS OF POLLUTION	315
	(i) Qualitative	315
	(ii) Quantitative	316
15.4	ASSESSMENT OF EFFECTS ON ECOSYSTEMS	318
	(i) Eutrophication	318
	(ii) Acid rain in lakes	318
	(iii) Oil spills in the marine environment	318
	(iv) Detergents in sewage	319
	(v) Ecosystems with algal and protozoan communities	319
	(vi) Rivers and lakes	321
	(vii) Estuarine ecosystems	321
	(viii) Coral reefs	322
	(ix) Mangrove ecosystems	322
	(x) Terrestrial ecosystems	322
15.5	CONCLUSIONS AND PREDICTIONS	325
	(i) Diversity, stability, maturity of ecosystems	325
	(ii) Effects of pollution on ecosystem characteristics	327
15.6	REFERENCES	328

15.1. INTRODUCTION

The effects of pollutants on plants and animals at the individual and population levels have been discussed in the preceding chapters. Moving one step up the integration scale of natural systems, consideration should be given to the implications of the introduction of chemical and physical agents for ecosystems as a whole. Changes specific to this level of organization find their mechanistic explanations in processes at the levels below, i.e. population and individuals.

Ecosystems are environmental units comprising the communities of living organisms and the abiotic components in a given volume of space (habitat). They are characterized by a defined structure and function, which result from the complex interactions between their components. The organisms may be grouped typically into producers (photosynthetic and chemosynthetic plants), consumers (animals), and decomposers (mostly bacteria and fungi), whereas the abiotic components are the soil and the atmosphere for terrestrial ecosystems, sediment, water, solutes, and suspended matter for aquatic ecosystems. Energy flows unidirectionally through the ecosystem while matter is recycled within its boundaries, although energy may be recycled, in chemical form, to a small extent, and nutrients may leave or enter the system in variable amounts.

The complex interrelations between its numerous components determine the structure of the ecosystem and contribute to its stability. This homeostasis or the ability to reestablish the initial state after disturbance, results from the interplay of various adaptive, feedback or damping mechanisms in the functioning of the system. It does not, however, prevent long-term evolutionary changes. The degree of homeostasis, which seems to depend on certain characteristics of the ecosystem, is obviously of major importance in determining the response to a stress, whether natural or man made. Disturbances may be of many kinds — climatic (extreme heat or cold, drought or flooding), geomorphologic (landslides, erosion, silting), man-related (land clearing for agriculture, grazing, fishing, road building, urbanization, fertilization, pollution, etc.). Only the last will be discussed here, i.e. the effects at the ecosystem level of chemical and physical (ionizing radiation, waste heat) pollutants, although it is probably safe to say that responses of ecosystems to these types of insult do not differ essentially from those to other kinds of disturbances.

Response of the system as a whole to pollution is considered here while the transfer pathways of pollutants in the environment leading to man as a target are discussed elsewhere (Chapters 4 and 5).

15.2 METHODS OF STUDYING ECOSYSTEM RESPONSE

The qualitative and quantitative evaluation of the responses of whole ecosystems to disturbances is fraught with difficulties, not the smallest of which is the insufficient understanding of the mechanisms of operation of 'normal', undisturbed ecosystems. Other problems are related to the estimation of exposure dose, the incomplete knowledge of toxicity of the pollutant at the species level, the relative uniqueness of every ecosystem, the difficulty of experimenting on real-life ecosystems, as well as the logistics and costs involved.

It should also be noted that a pollutant may act directly on the organisms (e.g. a selective herbicide on specific plant taxa) or indirectly through alteration of the physical environment (e.g. by reducing the pH of a water body, as in the case of acid rain).

What is known today about ecosystem response has been revealed either by observation of systems subjected to unintentional exposure or by experimentation with ecosystems deliberately exposed to chemical or physical agents.

To the first approach belong case studies pertaining to major pollution occurrences, e.g. oil spills in the sea, atmospheric emissions around power-generating or industrial plants, release of industrial wastes in rivers or lakes, nuclear explosions, etc.

Experimentation on ecosystem response has been carried out rather exceptionally on life-size systems (e.g. the chronic gamma irradiation of a forest). Most of the data available were obtained by treating small parts of ecosystems (such as plots of agricultural land, or confined volumes of water in lakes) or by establishing model systems, or microcosms (e.g. artificial streams, microorganisms in chemostats) and subjecting them to various stresses.

Although they were originally designed to study basic ecosystem processes, microcosms have received increased attention recently as test systems for investigating the impact of chemical pollution and the screening of potentially hazardous chemicals (Draggan, 1976). A distinction should be made here between artificial food chain models, such as those consisting of an alga, a zooplankton, and a fish which are useful to trace the fate and transport of pollutants in the environment (Cole *et al.*, 1976), and real microcosms which should have achieved a certain balance between producers, consumers, decomposers and their physical-chemical environment. In such a system including algae, grazers, and bacteria, Taub (1976) showed that low levels of pesticides and heavy metals drastically altered trophic relationships.

The major problem with microcosm studies is that of extrapolating their results to natural ecosystems. There are also technological difficulties in the establishment and balancing of such microcosms.

A purely theoretical approach has also been used to predict the effect of disturbances on the structure and stability of mathematical models of ecosystems.

In the remainder of the chapter a number of approaches to assess the response of ecosystems to pollution will be reviewed before attempting the generalizations given at the end of the chapter.

15.3. ASSESSMENT OF LEVELS OF POLLUTION

(i) Qualitative

It may be possible to detect a polluted environment by visual observation; although not quantitative the results may nevertheless form a valid assessment. Three examples will be given.

The existence of air pollution will be evident from the presence in the air of solid particles (soot or dust) or of photochemical smog. Alternatively, some of the effects may be visible, e.g. damage to vegetation.

Oil spills give rise to visible pollution. It has been estimated that about 2.5 million tons of oil per year escape at sea, or are discharged, from ships or from underwater wells. This threatens birds and fisheries and is especially harmful to larval forms of marine organisms. Even if not lethal, the oil may lower resistance of organisms to disease and other stresses. Senses of smell and taste may be impaired thus interfering with normal functioning. Drifting to shore, oil is deposited along the shoreline where the shoreline ecosystems are then subjected to a new environmental factor with which they have not evolved. Furthermore, economic, aesthetic, and recreational damage may be considerable in coastal resort areas (Devanney, 1974). Detergents used to clear up oil pollution have proved to be more damaging than oil itself, as clearly shown in the detailed study of the Torrey Canyon accident (Smith, 1968).

Another form of visible pollution is the accumulation of sediments on the bottoms of lakes, streams, or estuaries. One example is cellulose fibres discharged from pulpmills. Another is dead algae from algal blooms, which decay and settle to the bottom of the stream or lake. The deposition of these and other sediments, especially from silt in run-off, modify the habitat, often making it unsuitable for the natural fauna. The rocky beds of many streams are rendered unsuitable for organisms to cling to, and breeding sites of aquatic insects are often buried. The sludge eliminates the niche for mayflies, gaddisflies, and stone flies, important aquatic food sources. They are often replaced by sludgeworms and mosquito larvae.

(ii) Quantitative

An obvious measure of the state of the environment can be obtained by measuring the amount of pollutant in environmental media (air, water, soil) or in some biological material. Indications of the biota to be analysed may come from signs of adverse effects. Such analyses are in the beginning exploratory as opposed to the routine analyses conducted in a monitoring programme. When a polluted ecosystem or a source of pollutant has been identified it may become necessary to keep the situation under continued surveillance by instituting regular analyses in a monitoring programme. The monitoring should be related as closely as possible to the source, to the route of transport to the critical receptors and to the doses received by these receptors. These principles are elaborated in a SCOPE Report (1977). Four examples of the monitoring that may be carried out for environmental assessment follow.

Integrative measurements of airborne metals were made by enclosing clean and dry samples (1.5 g) of an epiphytic moss *Hypnum* in nylon mesh bags and exposing these at sites of interest (Goodman and Roberts, 1971). These showed elevated levels of metals in the air downwind from smelters in Wales. Studies to correlate the analysed levels on the moss with emission rates and doses to receptors, are continuing (Swansea, 1975).

Oxides of sulphur are major contaminants released into the atmosphere from power plants burning fossil fuels and from smelters. Sulphur dioxide is converted to sulphuric acid and sulphates. Most of this settles to the ground but before doing so may be transported great distances.

In the area around a nickel smelter at Sudbury, Ontario, Gorham and Gordon

(1963) have found a sharp rise in sulphate concentrations in lakes within 6.4 to 8 kilometres of the source. Recent evidence shows that some of this sulphur drifts for hundreds of kilometres and is deposited on ecosystems far from its origin (Odum, 1975). Such is the case with atmospheric sulphur released in the United Kindom and Central Europe some of which deposited in Scandinavian countries.

When deposited on the earth, largely in precipitation, the sulphurous or sulphuric acid enters a number of pathways potentially altering the normal sulphur and acid balance of ecosystems. The effect of this 'acid rain' is most evident so far in aquatic ecosystems. The lakes in many parts of southern Norway and Sweden are on igneous bedrock with a very low buffering capacity. Therefore, they are especially sensitive to additions of acids. The strong acids are principally sulphuric and nitric. An increase of 0.5 ppm of SO_4, as occasionally demonstrated in the precipitation, would be sufficient to increase the hydrogen ion concentration an order of magnitude, from pH 5.6 to 4.6. The increasing acidity of lakes in the south and west of Sweden and in southern Norway is consistent with the atmospheric fall-out of acids (Braekke, 1976).

Four hundred of some 3,000 lakes in the west coast region of Sweden were investigated from 1970 to 1972 (Almer, 1974). Half of these lakes had pH values below 6.0. In the autumn and spring, 36 and 22 per cent of the lakes, respectively, had pH values below 5.0. These are the periods of spawning and hatching when the greatest stress to the fish population occurs. It is in the autumn that frequent rains add to the acidity and in the spring that the melting snow releases the accumulated pollution of a winter's deposition. Comparisons of early pH measurements with recent data indicate a 30- to 60-fold increase in acidity since 1941 in 21 lakes and streams in east-coast Norway. Data from 14 lakes in southwestern Sweden show a similar rate of acidification. The increase is attributed to inputs of acid precipitation. The yearly mean pH in precipitation has been 4.3 to 4.4. Since the 1930's the pH has dropped as much as 1.8 pH units. The sulphate and nitrate contents have increased proportionately indicating a low biological activity. Zinc, copper, and lead in the precipitation have also increased.

The release of radioactive iodine from nuclear installations provides a good example of environmental contamination and its monitoring. The food chain to the receptor (air – grass – cow – milk – child) is sufficiently well quantified (see Chapter 5) that monitoring of release rates will permit calculation of the dose to the thyroids of the affected children. Monitoring of the levels in milk gives data leading to more certain estimates.

The fourth example of monitoring for environmental quality is concerned with the pollution of an aquatic ecosystem with mercury (see pages 109–110). The most relevant measurements of food-chain contamination are the methylmercury content of fish muscle. Other measurements for the assessment of past environmental contamination are of the methylmercury content of birds' (especially fish-eating birds) feathers.

15.4. ASSESSMENT OF EFFECTS ON ECOSYSTEMS

Effects at the ecosystem level may be detected by visual semiquantitative observations or by more quantitative measurements. A few examples are briefly described hereafter.

(i) Eutrophication

Chemicals may adversely affect aquatic ecosystems which they can reach by direct aqueous discharges or by way of ground-water after discharge to the atmosphere and deposition on land. Similarly, nutrients applied as fertilizer or arising as farm sewage may enter aquatic ecosystems by run-off. Incompletely treated urban sewage may make a similar contribution. It was recognized early in the nineteenth century that the addition of excess nutrients to lakes accelerated aging (eutrophication). The excessive algal blooms observed in many European lakes were becoming common with the development of sewage systems in the larger cities that piped sewage directly into lakes and streams plus the normal nitrogen-rich run-off from nearby farms (Rolich, 1969).

More intensive farming plus denser populations along the shore began to contribute more nitrogen in the run-off than the lake water could assimilate without alteration of its aquatic ecosystem. As the algal blooms flourished, the dissolved oxygen content of the water diminished, and the populations of sensitive fish species were reduced. Similar situations have occurred in lakes around the world whenever the nutrient balance has been disturbed.

(ii) Acid Rain in Lakes

The effects of such rain on the lake ecosystems are considerable. Most diatoms and green algae species disappear below pH 5.8; among the zooplankton, most daphnians disappear below pH 6.0. Fish reproduction and populations are affected below pH 5.5. Minnow, arctic char, and trout have disappeared from acidified waters. Shifts have been observed from higher aquatic plants towards mosses, which will also influence nutrient exchange with the sediments. In lakes around Sudbury, Ontario, which may also be affected by direct deposition, the numbers of aquatic plant species were inversely proportional to the concentration of dissolved sulphates.

(iii) Oil Spills in the Marine Environment

The study of the impact of oil on microbial populations has generally concentrated on biodegradation phenomena. As stated by Colwell and Walker (1977) there is not enough information about the effects on the autotrophic microbes involved in the main biogeochemical cycles, nor on viruses, fungi, and yeasts. Reasons for concern are the resistance to biodegradation of the more toxic aromatic components of oil, the heavy oxygen demand of oil biodegradation, and the ability of oil to concentrate heavy metals and chlorinated hydrocarbons.

(iv) Detergents in Sewage

A unique example of interaction from land to sea to air to trees was shown to be taking place in an area near Sidney, Australia (personal observation, Treshow). Norfolk Island pines (*Araucaria excelsa* R. Br.) grown in the area for many years to enrich the beauty of the popular nearby beaches were slowly dying. None of the usual fungus or insect pests was involved; virus diseases seemed unlikely, and air pollution in the area had not reached toxic levels. The species was not out of place, as it had evolved along the neighbouring shores dotting the South Pacific. But near Sidney, a new component had been added to the ecosystem: sewage. Sewage wastes were not released immediately to sea but pumped 300 yards out from land in a large pipe. Here, apparently, detergents in the sewage floated to the surface to be whipped into the air and drift on the winds, often onshore. The detergent spray then settled on the foliage and entered the leaf in amounts sufficient to disrupt the normal waxy layer of the cells that protected the trees from salt injury. With the protective coating gone, the salt spray could penetrate and damage the plant cells to an extent leading to the slow death of the trees.

A more refined assessment of the condition of an ecosystem and the amount of deviation from 'baseline' conditions requires a quantification of the kind of visual observations described above.

(v) Ecosystems with Algal and Protozoan Communities

Ecosystems in which the living components are mostly algae (for effects of pollution on algae, see Chapters 11 and 12) and protozoa exhibit a fairly rapid response to pollution stress. From observations and experiments it has been shown that the imposition of such a stress results generally in a reduction of the number of species present, an increase in the range of numbers of individuals per species, and a shift in dominance favouring some species over others. In fact the change in species diversity may be used as an indicator of pollution. The overall response is usually a reduction in the complexity of the living communities. The total number of species is reduced but there are also greater differences in the numbers of individuals per species which in undisturbed systems often exhibit a log-normal distribution (see Figures 15.1 and 15.2). This is due to the fact that as pollution stress increases, species are eliminated or strongly reduced in numbers while other, more tolerant, species become more abundant under conditions of reduced competition. Of course as pollution increases further, more and more species are affected and diminish in numbers and eventually disappear.

There is thus a pattern of dose-response of the community showing a zone of 'no

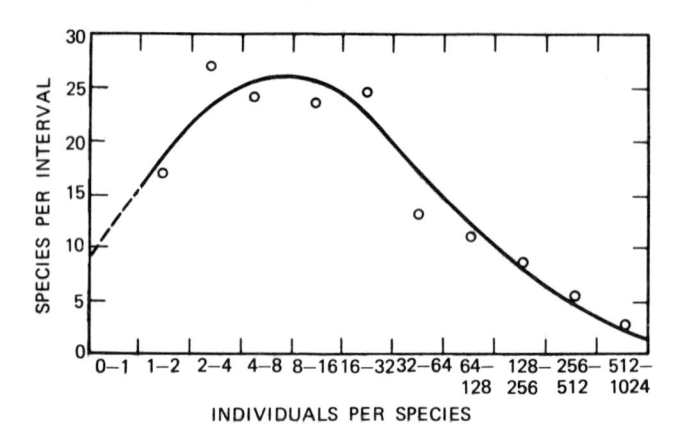

Figure 15.1 Diatom population in a stream not adversely affected by pollution (Ridley Creek, Pennsylvania, U.S.A., November 1951). (Reproduced by permission of John Wiley & Sons, Inc., from Cairns and Lanza, 1972)

detectable' effect, a range of graded response, and a zone of constant response as the pollutant concentration keeps increasing (Cairns and Lanza, 1972). The graded response to increasing pollution stress exhibited by such communities of algae may be used in practice to assess the degree of pollution in streams.

For practical purposes pollution may be said to be any environmental change which alters the species diversity more than a fixed percentage, say 20%, from the empirically determined level for a particular biotope.

As examples of the response of these communities to various types of pollution, one may mention:

(a) for non-toxic organic and inorganic substances: the well-known process of eutrophication (see above) accompanied by increased biomasses, caused by increased levels of nutrients (N,P,C), shift toward blue–green algae and other less oxygen-demanding groups of organisms. In flowing waters, the replacement of the communities by others more tolerant to oxygen depletion or more adapted to mineralizing organic matter results in an increase of the heterotrophic part of the biocenosis and leads to the self-purification process which exemplifies the homeostatic character of such systems.

(b) for thermal pollution: successive replacement of diatom communities by green algae and by blue–green algae, reduction of species diversity of protozoa after a thermal shock, followed by recovery in a few days or hours.

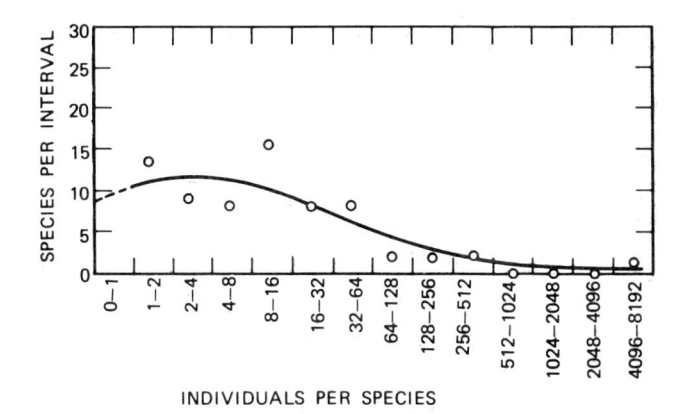

Figure 15.2 Diatom population in a polluted stream (Lititz Creek, Pennsylvania, November 1951). (Reproduced by permission of John Wiley & Sons, Inc., from Cairns and Lanza, 1972)

(c) for toxic chemicals and radioisotopes: large bioconcentration in the unicellular organisms with high surface-to-volume ratios and thus a greater effect on individuals and populations than in larger, more complex organisms. In general, species diversity is reduced, as well as numbers of individuals, although some tolerant species may proliferate; recovery can take place only if the toxin is diluted, or altered to less noxious compounds.

(vi) Rivers and Lakes

To assess the impact of effluents from a chemical factory on a lake (Beak *et al.*, 1959) and on a river (Cooke *et al.*, 1971) the number of freshwater molluscs per unit area of bottom mud were counted. Counts were made at 3-month intervals beginning before manufacturing operations, at various distances from the outfall and upstream of the factory on the river. The results at any one sampling point were plotted on industrial quality control charts to detect significant variations caused by the industrial effluents.

(vii) Estuarine Ecosystems

Even though much information is available on the impact of organic micropollutants on species living in estuaries (Lincer *et al.*, 1976), effects at the community and ecosystem levels have received much less attention.

Here also, loss in species diversity and shifts in dominant taxa have been demonstrated, even at very low pollution levels. Thus estuarine communities dominated by arthropods are replaced, when exposed to 1 to 10 ppb of PCB, by communities poorer in phyla, species and individuals, dominated by tunicates and other chordates. Analogous shifts occur in mixed algal cultures. For instance, when grown together under controlled conditions, a marine diatom (*Thalassiosira pseudonata*) and a green alga (*Dunaliella tertiolecta*) experienced a change in dominance from the former to the latter if exposed to low concentrations (ppb range) of DDT and PCB, which have no apparent effect on single-species cultures (Mosser et al., 1972).

(viii) Coral Reefs

This very specialized ecosystem in which the main primary producers are intimately associated with the consumers has proved to be sensitive to pollution: in areas polluted with sewage, coral may be overcome and replaced by a community dominated by an alga (*Dictyosphaeria cavernosa*), and the coral may be killed if thermal pollution results in an increase of not more than 4°C above ambient temperature (Johannes and Betzer, 1975).

(ix) Mangrove Ecosystems

The responses of mangrove ecosystems to environmental stress are far from uniform: this ecosystem is quite resistant to oxygen depletion or increased turbidity in the water such as might result from sewage outfall (and in fact has been considered as a potential natural sewage treatment system) but it is very vulnerable to oil pollution (Odum and Johannes, 1975). It is also extremely sensitive to defoliants as has been shown in Vietnam. Mangrove response to defoliant has been a reduction in species diversity, loss of productivity, and a reduced inventory of mineral nutrients (Westing, 1971).

(x) Terrestrial Ecosystems

(a) Pesticides

Considering particularly the impact of pesticides in terrestrial ecosystems, Pimentel and Goodman (1974) outline as follows the course of events which has been observed in many instances: the number of species is reduced, which may lead to instability and subsequent population outbreaks of some species. This occurs because the normal check-balance structure of the system has been disrupted, as when a predator or parasite is strongly reduced or eliminated. If, thereafter, the chemical disappears, some species in the lower part of the food chain (e.g. herbivores) may increase to outbreak proportions, hence a new disruption of the system. It is also clear that organisms at the top of food chains are very susceptible to the loss of a species or to wide fluctuations of species populations lower in the food chain.

Pesticides must be used in agriculture because the very much simplified ecosystems created to produce food are especially subject to perturbation. When plant species over a given area are reduced to one, as is the aim in most agricultural and many forestry situations, any adverse stress which the crop cannot tolerate will inevitably cause some disruption. This has always been true where some disease or insect pest has become prevalent and ruined a crop. Since the advent of agriculture, farmers have attempted to manage their land for maximum productivity. So when insects threaten to reduce yields, the natural expediency is to try to reduce the numbers of insects. During the past century, chemical controls of insects have become increasingly effective, but the chemicals often have had effects reaching beyond the agricultural ecosystem. Pesticides, notably insecticides, kill more than the target species. Thus in an agricultural situation, desirable insects such as bees, as well as insects that are parasitic or predatious on the pest species are killed. In this way, the natural biological control, or population balance, of insects is disrupted (Rudd, 1964).

The agricultural crops themselves, like any ecosystem, are also subject to toxic doses of the chemicals used to control pests. The concentrations may be immediately toxic, or they may accumulate in the soil over a period of years. This subject has been thoroughly treated by Rudd (1964) and others, so will be discussed here only briefly to show the interplay of chemicals among different communities in the ecosystem.

The above discussion of pesticide use in farming is also relevant to forest management, where the greatest damage comes from native insects. Their numbers are kept in balance by natural factors. But the fluctuations of their populations are great around a 'norm' and their density may become far greater than forest managers would like. To limit losses, pesticides are then applied. Rather than control the pest, however, they may further disrupt the system. Pesticides sprayed over forests to control spruce budworm or other pests may reduce the pest populations, but they sometimes kill their parasites and predators. The pest species often build back much more rapidly than their predators so that the long-term balance of the system is upset and the damage may become increasingly severe.

It is important to recognize that half or more of the pesticide applied may never reach the target species or even the forest canopy. A significant fraction is picked up by air currents, circulated through the lower troposphere and ultimately deposited on the ground in precipitation, often in areas remote from that where the application was made. If the chemical is persistent and stable (and fat-soluble) as is DDT, it may build up in food chains many miles distant from where it was released and accumulate to toxic levels in top carnivores. For a discussion of global contamination with DDT, see Woodwell *et al.* (1971).

(b) Acid Rain in Terrestrial Ecosystems

Acid precipitation may increase soil acidity (Schofield, 1975) resulting in the likelihood that ions such as aluminium, manganese, or iron, might become more mobile, be leached form the soil system and thereby accumulate in toxic concentrations. The presence of these ions and possibly even the lower pH alone, could alter the soil microfauna and flora present, as well as injure higher plant species in both terrestrial and aquatic systems. Aluminium and iron might have secondary effects in interfering with phosphorus transport in germinating seedlings.

The release of other minerals might also be accelerated. Nitrogen, for example, might become more available, enhancing plant growth in the short term, but becoming deficient in the long run as it becomes bound in perennial vegetation.

The acid rain absorbed into the bark of trees has the further potential of changing the reaction of the bark thus placing a stress on the lichen species growing on this bark (Staxang, 1969; Johnsen and Söchting, 1973). Since certain lichens play a role in absorbing nitrogen from the atmosphere and making it available to the plants, the nitrogen balance of the ecosystem could be adversely altered. Other biogeochemical cycles might also be affected, such as the build-up of sulphur in some ecosystems as it is deposited from industrial processes elsewhere.

It should be stressed, however, that it has not been possible to demonstrate these, or many other effects sometimes postulated, in the field. If they are already occurring, either the effects are balancing each other out, or they are not of sufficient magnitude to measure.

(c) Metals

Heavy metals provide yet another example of the transfer of chemicals from one system to another. When ores, coal or other materials are burned, the unwanted or waste materials are generally released into the atmosphere. Particulate materials containing such heavy ions as lead, zinc, arsenic, cadmium, and copper, tend to settle out of the air within a few hundred yards of the source where the concentrations may build up in the soil to increasingly toxic amounts. The chemicals may then leach into the ground-water to a slight extent, but mostly they remain in the soil where they influence the composition of the plant community — some plant species or ecotypes being more tolerant of the toxic ions than others.

Thus for instance, *Agrostis tenuis* in western Europe, which is quite tolerant to zinc and cadmium as well as to acidity tends to replace the pre-existing vegetation around zinc smelters (Denaeyer-De Smet and Duvigneaud, 1974).

A further threat to the ecosystem arises when the metal ions are taken up by the plants. A build-up in the leaves could be harmful to animals feeding upon them, including humans. This possibility has been suggested for vegetable crops grown near smelters where the chemicals have been accumulating for perhaps a hundred years. Fortunately, the uptake of these ions by the roots of plants is very limited but foliar deposition must always be considered.

Population modifications also influence species diversity as the numbers of sensitive individuals are reduced and the tolerant individuals thrive. An extreme example of this occurred in the area immediately surrounding an aluminium reduction plant near Zwollen in Czechoslovakia (personal observation, Treshow). Within a few years of operation, the native conifers and oaks were killed along with much of the understory, and the plant populations closest to the industry were reduced almost solely to a single species, *Conium macrophylla.*

(d) Ionizing Radiation

Studing the response of a pine—oak forest to gamma radiation, Woodwell (1967) found that diversity expressed as the number of species per unit area was reduced in a continual gradient up to 1,000 roentgens per day. A dose of 160 roentgens per day reduced the diversity to 50 per cent of that of the unirradiated community. The reduced diversity was accompanied by a reduced cover, but the more tolerant species, particularly *Carex* spp. soon expanded over the open land to become dominant. This illustrates the potential importance of less common species in maintaining the stability of the overall community following periods of stress. The productivity along the gradient as measured by total dry weight of plants changed initially very little since the space and energy resources available became used by the more tolerant species. Thus diversity was initially a far more sensitive index of impact than productivity or biomass.

After a few months of exposure to the gamma source, several well-defined vegetation zones were established as a function of the exposure dose. Trees were eliminated first, then the tall shrubs, the low shrubs, the herbs, and finally the lichens and mosses. Parallel changes were observed with respect to diversity, primary production, total respiration, and nutrient inventory. In this case as well as in others involving different stresses (toxic chemicals, fire, etc.) the more hardy species, whether animal or vegetable, were the 'generalists'.

Other terrestrial ecosystems (tropical forest, grassland, etc.) have been acutely or chronically irradiated. One of them is a Mediterranean forest with subtypes dominated respectively by *Quercus pubescens* and *Quercus ilex*. There also, chronic exposure to gamma radiation increased leaching of mineral and organic compounds from litter and soil (Saas *et al.*, 1975).

15.5 CONCLUSIONS AND PREDICTIONS

In the preceding section, case studies of the response of various types of ecosystem to pollution have been described briefly. In attempting to draw general conclusions on this subject, one must refer to current concepts in basic ecological theory.

(i) Diversity, Stability, Maturity of Ecosystems

As mentioned in the Introduction, ecosystems are characterized by structure and function. Function relates to energy and material flows and involves such concepts as productivity (primary and secondary) and the biogeochemical cycles of chemical elements. Structure may be described in terms of trophic levels and ecological 'pyramids' of population and biomass. It also involves such concepts as diversity, stability, and maturity which are less precisely defined.

Diversity is often determined by the number of species per unit area or volume. A variety of indices have been proposed to express it, such as the reciprocal Simpson Index used by E. Odum (1975), which is equal to $1 - \Sigma(P_i)^2$, where P_i is the probability for the occurrence of each species in terms of the ratio of its importance to the total of importance values. Importance is usually based on numbers of individuals but may also be calculated in terms of biomass.

Diversity may also be expressed in terms other than species e.g. life-forms for plants (phanerophytes, geophytes), or subspecific units such as human population groups included in a consideration of an ecosystem, as advocated by Jacobs (1975).

Diversity contains in fact two notions, (a) that of richness or number of species (or other groups), and (b) that of evenness of distribution or the relative abundances of the individuals within each species (or group). It never concerns whole ecosystems but only certain components thereof and is assumed to be a reflection of multiple, dynamic functions in the system (Jacobs, 1975).

Stability is the tendency of a system to remain near an 'equilibrium' condition or to return to it after a disturbance. This concept may imply the ideas of constancy (lack of change), persistence (referring to survival time), inertia (ability to resist external perturbations), elasticity (speed with which a system returns to its former state following a perturbation), amplitude (amount of perturbation from which recovery is possible). Cyclical stability is the property of the system to cycle or oscillate around a central point (e.g. predator—prey systems), whereas trajectory stability refers to the property of moving towards a final end-point despite differences in starting points, such as in plant succession towards a climax (Orians, 1975). A listing of environmental factors and species characteristics which increase the various kinds of stability is given by the same author.

Even though stability (as expressed by constancy) is often found to be positively correlated with diversity, the assumption of any general relationship between these two attributes is not warranted. Much depends on the adaptive characteristics of the organisms present, which are determined in turn by their environment, the past experiences with perturbations, and the continual evolution of species.

The concept of practical stability has been introduced by Harte and Levy (1975) to account for the fact that disturbed ecosystems do not return to precisely their previous states. They may be considered as practically stable if the initial perturbation (which is finite) is not sufficient to push the system beyond tolerable limits. Practical stability lends itself to mathematical treatment by the Liapunov Direct Method (Lasalle and Lefshetz, 1961). Thus these authors have shown theoretically that increasing the number of trophic levels has no effect on stability — except if the biomass pyramid is inverted — and that increasing the number of species at each trophic level leads to a decreasing domain of stability. They also discovered that damage to the decomposers or to the organic or inorganic nutrient pools is a potentially stronger cause of instability than disturbances of the predator–prey components of the system.

Maturity applies to the successional stage of an undisturbed ecosystem approaching a more or less stable climactic stage (Jacobs, 1975). It is associated with an increase in internal effective links, making the system less dependent on inputs from outside. A stabilizing factor in ecological succession is the increasing importance of species which extend their connection in space and time ('K-strategists'). Another feature of succession is a progressive decrease in the ratio of primary productivity to biomass, thus a slowing down of turnover, at least in relatively closed systems, receiving, e.g. only radiation from the outside (Margalef, 1975).

Diversity increases generally with succession, at least in the early stages. Afterwards it may still increase, due to increases in richness or evenness but it may also decrease as observed for instance when single-species, very stable phytocoenoses become established.

(ii) Effects of Pollution on Ecosystem Characteristics

The estimation of doses has been discussed in Chapter 5, and the relations between doses and effects in Chapter 6. When these are used in conjunction with models described in Chapter 4, some predictions of the effects of releasing a pollutant into the environment may be possible. Unfortunately most of the available information is about individual species. Because of a lack of knowledge about fundamental ecosystem ecology and its responses to chemical and physical agents the predictions possible in the present state of knowledge are severely limited. Although it is impossible to make any sweeping generalizations, some tentative statements are presented here.

(a) Acute, accidental exposures to a pollutant often bring succession back to a previous stage, generally, but not always, of lower diversity, i.e. through a decrease in richness of species, or in evenness. Similar effects may be induced by other types of drastic disturbance such as fire.

The resistance of ecosystems to these disturbances is not, however, related to their complexity. In fact, quite simple systems, especially if they have evoked in a stressing environment, may be more able to withstand an acute insult. On the other hand, very complex systems survive strong disturbances of the kind they may encounter under 'natural' conditions.

(b) Chronic exposures to chemicals, or ionizing radiation, or warm effluents, bring about gradual modifications of ecosystem structure to adapt it to the new prevailing conditions. Species, or ecotypes, are replaced by others that can survive and compete under the new set of environmental factors, thereby altering the physiognomy of the biocoenosium (e.g. suppression of the tree stratum in irradiated areas). Trophic functions may shift in importance (e.g. increased heterotrophy in organically polluted rivers, with sometimes a reduction in self-purification resulting from heavy metal poisoning of heterotrophic organisms). Primary productivity of terrestrial systems may be reduced but standing crop will be reduced even more, with a subsequent reduction in the turnover of nutrients. Increased leakage of nutrients may occur as a consequence of the decrease in standing crop.

By and large, one may agree with Woodwell's statement (1970) that pollution in general, although there are exceptions, brings about a simplification of the ecosystem structure and shifts the ratio of gross production to total respiration.

There is a trend from highly specialized species to 'generalists', a reduction in diversity and increase in monotony, a depletion of nutrients in terrestrial system coupled with an excessive input of these in aquatic systems.

The large-scale (i.e. of an order of magnitude similar to that of natural phenomena: e.g. heat release in a city in relation to solar radiation balance, dispersion of metals by industrial and power-generation activities as compared with the natural sedimentary cycle) introduction of pollutants into the environment has also the consequences of bringing down the barriers between natural ecosystems, enhancing exchange of energy and materials and, in a way, making them subsystems of a larger ecosystem in which man must be included.

In view of this, and considering the many uncertainties which still prevail, one should try to minimize the interferences of human ecosystems with 'natural' systems, upon which the balance of the biosphere depends, through the use of 'clean' technologies, recycling, economizing materials. Until we know better, every species should be saved, and one should keep in mind the aphorism of the great conservationist Aldo Leopold: 'The first rule of intelligent tampering is to save all the pieces'.

15.6 REFERENCES

Almer, B., 1974. Effects of acidification on Swedish lakes. *Ambio*, 3, 30–6.

Beak, T. W., De Courval, C., and Cooke, N. E., 1959. Pollution monitoring and prevention by use of bivariate control charts. *Sewage Ind. Wastes*, 31, 1383–94.

Braekke, F. H. (Ed.), 1976. Impact of acid precipitation on forest and fresh-water ecosystems in Norway. *Summary Report on SNSF-project*, Aas, Norway.

Cairns, J. Jr. and Lanza, G. R., 1972. Pollution controlled changes in algal and protozoan communities. In *Water Pollution Microbiology* (Ed. R. Mitchell), Wiley–Interscience, John Wiley and Sons, New York, London, pp. 245–72.

Cole, L. K., Metcalf, R. L., and Sanborn, J. R., 1976. Environmental fate of insecticides in terrestrial model ecosystems. *Int. J. Environ. Stud.*, 10, 7–14.

Colwell, R. R. and Walker, J. D., 1977. Impact of petroleum hydrocarbons on marine microorganisms. *GESAMP Report*, in press.

Cooke, N. E., Cooper, R. M., and Beak, T. W., 1971. Biological monitoring of the effluent from a large chemical works in a river which has sources of pollution upstream. Presented at the *5th International Water Pollution Research Conference*, July-August 1970. (Ed. S. H. Jenkins), Pergamon Press Ltd., Oxford.

Denaeyer-De Smet, S. and Duvigneaud, P., 1974. Accumulation of toxic heavy metals in various terrestrial ecosystems polluted by depositions of industrial origin (in French). *Bulletin Société Royale Botanique de Belgique*, **107**, 147–56.

Devanney, J. W. III, 1974. Key issues in offshore oil. *Technol. Rev.*, **76**, 20–6.

Draggan, S., 1976. The role of microcosms in ecological research. *Int. J. Environ. Stud.*, **10**, 1–2.

Goodman, G. T. and Roberts, T. M., 1971. Plants and soils as indicators of metals in the air. *Nature*, **231**, 287–92.

Gorham, E. and Gordon, A. G., 1963. Some effects of smelter pollution upon aquatic vegetation near Sudbury, Ontario. *Can. J. Bot.*, **41**, 371–8.

Harte, J. and Levy, D., 1975. On the vulnerability of ecosystems disturbed by man. In *Unifying Concepts in Ecology* (Eds. W. H. van Dobben and R. H. Lowe-McConnell), Dr. W. Junk, The Hague, pp. 208–23.

Jacobs, J., 1975. Diversity, stability and maturity in ecosystems influenced by human activities. In *Unifying Concepts in Ecology* (Eds. W. H. van Dobben and R. H. Lowe-McConnell), Dr. W. Junk, The Hague, pp. 187–207.

Johannes, R. E. and Betzer, S. B., 1975. Marine communities respond differently to pollution in the tropics than at higher latitudes. In *Tropical Marine Pollution* (Eds. E. J. Wood and R. E. Johannes), Elsevier, Amsterdam, Oxford, New York, pp. 1–12.

Johnsen, I. and Söchting, U., 1973. Influence of air pollution on the epiphytic lichen vegetation and bark properties of deciduous trees in the Copenhagen area. *Oikos*, **24**, 344–51.

Lasalle, J. and Lefshetz, S., 1961. *Stability by Liapunov's Direct Method*, Academic Press, New York.

Lincer, J. L., Haynes, M. E., and Kelin, M. L., 1976. The ecological impact of synthetic organic compounds on estuarine ecosystems. *Ecological Research Series*, EPA-600/3.76.0.75. September 1976.

Margalef, R., 1975. Diversity, stability and maturity in natural ecosystems. In *Unifying Concepts in Ecology* (Eds. W. H. van Dobben and R. H. Lowe-McConnell), Dr. W. Junk, The Hague, pp. 151–60.

Mosser, J. L., Fisher, N. S., and Wurster, C. F., 1972. Polychlorinated biphenyls and DDT alter species composition in mixed cultures of algae. *Science*, **176**, 533–5.

Odum, E. P., 1975. Diversity as a function of energy flow. In *Unifying Concepts in Ecology* (Eds. W. H. van Dobben and R. H. Lowe-McConnell), Dr. W. Junk, The Hague, pp. 11–4.

Odum, W. E. and Johannes, R. E., 1975. The response of mangroves to man-induced environmental stress. In *Tropical Marine Pollution* (Eds. E. J. Ferguson Wood and R. E. Johannes), Elsevier, Amsterdam, Oxford, New York, pp. 52–62.

Orians, G. H., 1975. Diversity, stability and maturity in natural ecosystems. In *Unifying Concepts in Ecology* (Eds. W. H. van Dobben and R. H. Lowe-McConnell), Dr. W. Junk, the Hague, pp. 139–50.

Pimentel, D. and Goodman, N., 1974. Environmental impact of pesticides. In *Survival in Toxic Environments* (Eds. M. A. Q. Khan and J. B. Bederka), Academic Press, New York, pp. 25–52.

Rolich, G. A., 1969. *Eutrophication: Causes, Consequences, Corrections*, Nat. Acad. Sci., Washington, D.C., 661 pp.

Rudd, R. L., 1964. *Pesticides and the Living Landscape*, Univ. Wisconsin Press, 319 pp.

Saas, A., Bovard, P., and Grauby, A., 1975. The effect of chronic gamma irradiation

on decay of oak (*Quercus pubescens* Willd) and dogwood (*Cornus mas* L.) leaves and subjacent litter. *Radiat. Bot.*, **15**, 141–51.
Schofield, C. L., 1975. Acid precipitation: Our understanding of the ecological effects. In *Proc. of a Conference on Emerging Environmental Problems: Acid Precipitation*, EEP-1, Cornell Univ., New York. pp. 76–81.
SCOPE, 1977. *Environmental Issues* (Eds. M. W. Holdgate and G. F. White), *SCOPE Report 10*, John Wiley and Sons, London, pp. 130–44.
Smith, J. E. (Ed.), 1968. *'Torrey Canyon' Pollution and Marine Life*, University Press, Cambridge, 196 pp.
Staxang, B., 1969. Acidification of bark of some deciduous trees. *Oikos*, **20**, 224–30.
Swansea, 1975. *Report of a Collaborative Study on Certain Elements in Air, Soil, Plants, Animals and Humans in the Swansea–Neath–Port Talbot Area Together with a Report on a Moss Bag Study of Atmospheric Pollution Across South Wales*. Y Swyddfa Gymreig, Welsh Office.
Taub, F. B., 1976. Demonstration of pollution effects in aquatic microcosms. *Int. J. Environ. Stud.*, **10**, 23–33.
Westing, A. H., 1971. Ecological effects of military defoliation on the forests of South Vietnam. *BioScience*, **21**, 893–8.
Woodwell, G. M., 1967. Radiation and the patterns of nature. *Science*, **156**, 461–70.
Woodwell, G. M., 1970. Effects of pollution on the structure and physiology of ecosystems. *Science*, **168**, 429–33.
Woodwell, G. M., Craig, P. P., and Johnson, H. A., 1971. DDT in the biosphere: where does it go? *Science*, **174**, 1101–7.

SECTION V
CONCLUSIONS

CHAPTER 16

Conclusions

G. C. BUTLER

Division of Biological Sciences, National Research Council of Canada, Ottawa, Canada, K1A 0R6

In the Introduction the new science of ecotoxicology was defined. To paraphrase that definition, ecotoxicology is a study of the effects of released pollutants on the environment and on the biota that inhabit it. According to ome, the human species is the most important of the biota and thus the effects of pollutants on human health are central to ecotoxicology; the Introduction explains why a chapter was not devoted specially to this subject. There are two additional reasons for close connection between human activities and ecotoxicology: (a) human beings not only alter the environment but also produce and release pollutants on human health are central to ecotoxicology. The introduction explains or indirectly, the physical, economic, or aesthetic wellbeing of mankind.

The preceding chapters have described the behaviour of different pollutants, a variety of effects caused by them and mathematical procedures for relating those effects to their causes. From a reading of the chapters it becomes evident that the subject of ecotoxicology may be divided into the following five parts:

(1) Chemical and physical form of the pollutant(s) released to the environment and the medium into which they are released.

(2) Transformation of the pollutant(s) by abiotic and biotic processes during its transport from the point of release to the receptor.

(3) Quantitative metabolism of the pollutant in transporting and receptor organisms; this permits calculations of accumulation in food chains and doses to receptors.

(4) The effects of these doses on individuals, populations, and communities of receptors.

(5) The results of these effects on the welfare of the ecosystem.

The views of the writers on the five elements of the subject are consolidated below:

The physical and chemical forms of released pollutants will determine their toxicity and environmental behaviour, especially their transformation and transport. The medium into which they are released will determine which of the candidate receptors are most affected and the most probable routes of intake. It is the role of monitoring to provide information on this subject.

Many chemicals, especially non-persistent pollutants, undergo chemical changes in the environment as a result of abiotic and biotic agents. A series of metabolic alterations may result from several organisms acting in succession. As a result of such changes, pollutants may be converted to more toxic compounds (e.g. the conversion of mercury and fluoride ions to organic compounds) or to compounds more easily accumulated in food chains. These changes should be known before models for environmental transport are constructed. 'Universal' models are not possible since an individual one is usually required for each pollutant in each ecosystem. The models need to comprise the elements of time and space. Models should be tested for 'robustness' and the results should be given as estimates of uncertainty.

Much more quantitative information is needed about the pharmaco-kinetics of pollutants in living organisms. Compartmental analysis leading to time-dependent formulations of pollutant levels is a useful approach.

In studying the relations between doses and responses in single species more attention should be paid to:

— sublethal effects, especially rapid and economical screening tests for mutagens and carcinogens;

— integrative tests such as those for effects on reproduction and growth;

— the long-term effects of low levels of pollutants;

— the variation of lethal and sublethal effects on populations from different geographical areas;

— the interaction of pollutants with environmental variables such as temperature, salinity, and oxygen tension; and

— the variation in the effects of a pollutant acting alone or in concert with others.

One of the most frequently identified needs for improvement is in concepts and methods for using the results of tests on single species to predict the effects on communities existing in nature. A prerequisite is a greater knowledge of community dynamics in undisturbed systems. It has been suggested that first steps should be to establish 'standard communities' under controlled conditions or to study microcosms. There is, however, no confidence that the results of such tests would permit predictions about the effects on whole ecosystems.

With enough knowledge of ecosystem dynamics, models might be constructed, normal variations estimated, and criteria for significant perturbations produced. In this subject lie the greatest possibilities for important developments in the new science of ecotoxicology.

Index

Abbott's correction, 134
Abiotic degradation, 11–12
  models for, 33
Abortion due to pesticides, 164
Absorption
  of radionuclides by man, 93
  of toxicants
    dependence on particle size, 175–176
    dependence on solubility, 175
    in digestive tract, 176
    by plants, 224–225
    in respiratory tract, 176
Accumulation of toxicants
  by aquatic plants, 244–246
  compartmental model for, 170–175
  in soil, 224–225
Acetanilide as allergen, 164
Acetylcholinesterase
  inhibition by toxicants, 153, 176–177
  isozymes of, 177–178
Acidification
  effects
    on ecosystems, 303–304, 323–324
    on forests, 304–305
    on lakes, 317
    on organisms, 318
    on toxicity of soil, 225, 230
    on waters, 303
  by rain
    causes of, 301–303
    water pollution by, 317
Adaptation to pollutants
  by algae, 249–251
  by insects, 139–140
Adaptive response, 179–180
Additive effects
  on algae, 264
  on aquatic animals, 210, 213–214
  on aquatic plants, 247–248
  of toxicants, 164–165, 207–210
Aerosols, atmospheric washout, 15

Air pollution, 315, 316
Aldrin
  behaviour in model ecosystem, 60–61
  conversion products in plants, 53–54
  degradation by plants, 51–52
  distribution among tissues by sex, 43
  epoxy group, effect on metabolism, 63
  metabolism by animals, 48–49
  metabolism by microorganisms, 44–46
  oxidation of, 18
  photoisomerization of, 23
  photomineralization of, 32–33
  reactions with oxygen, 20–21
  stereochemistry, influence on metabolism, 63–64
  transport in soil, 14
Algae
  accumulation of toxicants by, 244–245
  blue–green, 260–262
  copper, toxicity of, 263, 265–266
  DDT, uptake of, 245
  growth rate of, 259–260
  inhibition of photosynthesis, 260
  mercury, toxicity of, 262–263
  organohalogens, uptake of, 245
  photosynthesis, inhibition of, 260
  testing with, 240–241
Alkanes biodegradation of, 278, 283–284
Allergens effects, 164
Aluminium effects on ecosystems, 323–325
Ames test, 284–285
Amines as carcinogens, 163
Anaemia as allergic mechanism, 164
Antagonistic effects
  on algae, 264
  on aquatic animals, 208–209, 211–212
  on aquatic plants, 247–248

Aphlatoxins
  as carcinogens, 163
  as mutagens, 161
Arctic, degradation of petroleum in, 280–281
Aromatics in oil, 278–279
Arsenic
  accumulation in plants, 225
  inhibition of carbohydrate metabolism by, 157
  mutagenic effects of, 161
Asphaltenes in oil, 278–279
Assessment of ecotoxicological effects, 3–4
ATPase, effect of mercury on activity, 192
Avoidance studies, 195–196
Azide
  inhibition by
    of enzymes, 156
    of respiration, 157

Background level of stimulus, 134–135
Bacteria
  marine
    cellulolytic, 282, 288
    chitinolytic, 282
    heterotrophic, 282, 288
    lipolytic, 282
    proteolytic, 282
    removal of oil by, 278–279
Baseline studies, 314, 319
Benzene hexachloride (BHC) isomers, stereochemistry, influence on metabolism, 63–64
Binomial probability, 145
Bioconcentration
  in food chains, 173–174
  of methylmercury, 109
  of toxicants in organisms, 57
  of xenobiotics, 57
Biodegradation
  of alkanes, 283–284
  of hydrophobic contaminants, 276
  of oil, 278–284
    effects of temperature, 280–281
    rate of, 281–282
  of PCB's, 285–286
  of toxicants, 275
Biogeochemical cycles, role of bacteria and fungi in, 275–276
Biometeorology, 296–297

Blood
  analysis of, 194–195
  effects of toxicants on, 194–195
Bone marrow, effect of allergens on, 164
Buturon conversion products in plants, 52

Cadmium, see also Metals, heavy
  effects
    on aquatic plants, 247
    on membranes, 152
    on zooplankton, 268
  inhibition by
    of CO_2 fixation, 159
    of reproduction, 163
  pollution by, 324
Caesium, half-life of retention, 98
Cancer
  effects of ozone layer on skin, 301
  effects of toxicants on, 213
  induction by chemical agents, 161–163
Carbohydrates, inhibition of metabolism by toxicants, 156–157
Carbon dioxide levels in atmosphere, 299
Carbon disulphide, inhibition of microsomal enzyme systems by, 158
Carbon monoxide, inhibition by
  of enzymes, 156
  of microsomal enzyme systems, 158
  of respiration, 157
Carbon tetrachloride
  effect on membranes, 152
  inhibition by
    of microsomal enzyme systems, 158
    of protein synthesis, 158
Carcinogenic polynuclear aromatic hydrocarbons, 283–284
Carcinogens
  activation of, 162–163
  detection of, 284–285
  effects on fish, 202–203
  examples of, 162–163
  identification of, 284
  testing for, 203–204
  types of, 162
Carrying capacity of zooplankton, 268
Cellulolytic bacteria, 282, 288
Chemical form, importance of, 75–76

Chemotactic responses of motile marine bacteria, 283
Chi-square test, 129–133
Chitinolytic bacteria, effect of fuel oil on, 282
Chloralkali plants, sedimentation of mercury from, 16
Chloramphenicol as allergen, 164
Chlordane
  conversion products, 44
  evaporation from soil, 14
  metabolism by microorganisms, 44–46
  sedimentation of, 15
Chlordene
  chlorination of, 26
  dimerization of, 26
  reactions with oxygen, 20–21
Chlorine
  effects on blood of aquatic animals, 194–195
  influence on toxicant behaviour, 61–63
Chloroalkylene-9
  distribution between tissues, 40–41
  excretion by rats, 62–63
  metabolism by biomass, 55
Chlorofluorocarbons in atmosphere, 299
Chromium, see also Metals, heavy
  as a carcinogen, 163
Chromosomes
  effects of heavy metals on, 161
  effects of mutagens on, 160
  effects of toxicants on, 201
Climate
  effect of man on, 297–301
  effect on man, 296–297
  effect on petroleum biodegradation, 280–281
  global modification, 298–301
  local modification, 297–298
  regional modification, 298
Cluster analysis to determine toxic effects, 234
Cobalt retention by man, 98
Cocarcinogens, 213
Communities
  definition, xxii
  influence of toxicants on, 241
  research on, 269–270
  structure in zooplankton, 269
Compartments
  advantages of use, 75, 171–175
  chronic exposure model, 171–173
  description of, 78–79
  determination of variables for, 79–80, 171
  in ecosystems, 75
  interactions between, 80–85
  number of, 78
  use in modelling, 170–175
Competing risks, 137
Concentration, effective, 189
Conditioning in fish, 196
Confidence limits
  calculation of, 133
  of estimated responses, 128
Conformation, stereochemical, influence on toxicant behaviour, 63–64
Conjugation of toxicants, 39
Contaminants, environmental, 276–278
  chemical, in aquatic environments, 277
Conversion
  of chemicals by plants, 51–53
  products formed, 38–39
Copper
  chemical states in water, 243–244
  effects
    on algae, 263
    on aquatic plants, 243
    on fish, 195
    on zooplankton, 251
  inhibition
    of carbohydrate metabolism by, 157
    of growth by, 265–266
  mutagenic effects of, 161
Coral reefs, pollution of, 322
Correlations between plants and pollutants, 249
Coughing and toxicity, 193–194
Criteria
  definition, xxii
  of toxicity, 189
Critical organ
  definition, 99
  importance of, 176
  retention in, 99
Cyanide
  effects on fish, 199
  inhibition by

of enzymes, 154, 156
of respiration, 157
Cyanophytes, growth, 260–262
Cycasin
as carcinogen, 163
as mutagen, 161
Cyclodienes
dechlorination, 25
isomerization, 23
photoisomers, 32

DDE
effects
on aquatic plants, 247–248
on eggshell thickness, 156
on falcon eggs, 180–181
on zooplankton, 268–269
irradiation of, 26–27, 28–31
photomineralization of, 32–33
DDT
analysis in ponds, 57–60
content in biota, 57
dispersion of, 12
effects
on algae, 266
on aquatic plants, 241, 247–248
on eggshell thickness, 179
on estuarine ecosystems, 322
on falcon eggs, 180–181
on fish, 196
on reproduction, 164
on zooplankton, 268–269
inhibition
of growth by, 159
of lipid metabolism by, 157–158
irradiation of, 26–27, 28–31
metabolism by zooplankton, 269
photomineralization of, 32–33
toxicity, dependence on state, 179
transport by air, 323
uptake
by algae, 245
equation for, 103
ingestion by man, 93–94
Degradation
of pollutants
abiotic, 11–12
by aquatic microorganisms, 275–278
of crude oil, 278–284
Deposition velocities, determination of, 74

Detection of sublethal effects, 179
Detergents
and oil pollution, 316
in sewage, 319
Detoxification
interaction between toxicants, 212
of mercury, 244
of toxicants, 38, 234–235
by plants, 38, 234–235
by soil, 38–39
processes for, 39
Diagnosis of toxic effects on plants, 233–234
Diatoms
effect of copper on, 265–266
effect of pollutants on, 247–248
effect of pollution on population, 319–321
2,2'-Dichlorobiphenyl
absorption of, 40, 176
distribution between tissues, 40
excretion by rats, 62–63
hydroxy derivatives, 43–44
metabolism by biomass, 55
reactions with oxygen, 20–21
2,4-Dichlorophenoxyacetic acid (2,4-D), inhibition of growth by, 159
Dieldrin
absorption
by organisms, 40
rate, 171
amounts in blood, liver, and fat, 177
conversion products in animals, 43
effects
on aquatic plants, 242
on falcons and pigeons, 183
on reproduction, 164
evaporation from soil, 14
inhibition by
of growth, 159
of lipid metabolism, 157–158
irradiation of, 27–28
metabolism
by animals, 48
metabolites formed, 45
by microorganisms, 44–46
photoisomerization of, 23
photomineralization of, 32–33
stereochemistry, influence on metabolism, 63–64
transport in soil, 14
Di-2-ethylhexylphthalate in a model ecosystem, 60–61

Diethylsulphate, mutagenic effects, 160
Diffusion
  coefficient, 12–13
  of pollutants, 175–176
Dispersants, effects on biodegradation of oil, 281
Dispersion of pollutants, 12
Distribution
  of organisms in aquatic communities, 250
  of species, effects of toxicants on, 250–251
DNA
  damage by toxicants, 201–202
  reactions with chemical agents, 151, 160–161, 163
Dominance in plankton, 265–266
Dose
  calculation, 92–93
    from excretion equations, 100
    from integral of concentration, 105–107
  collective, 106–107
  commitment, 104–107
    calculation, 104–105
    collective, 106
    definition, 102
    estimation of, 104
  definition, 92
  effective, 105, 233
  estimation of, 92
  level, 116
    allocation of subjects over, 119
    plant toxicity and, 226–227
    selection of, 118–119
  threshold, 105
Dose–effect relation, 117
Dose–response relation, 116–117, 126
  analysis of, 136
  determination, 118–119
    using models, 122–126
  estimation, 122, 126 135
  for quantal responses, 120–121
  time-to-response, 136–140, 203–204
Dusts, accumulation in plants, 224
Dynamics of plant communities, 233–234

Ecosystems
  acidification of, 303–305
  aquatic, 187–188
  characteristics of, 325
  definition, xxii
  diversity, 325–327
  effects of pollution on, 319–321, 327–328
  estuarine, 321–322
  maturity, 327
  stability, 326–327
Ecotoxicology, definition, xx
Effective concentration, 189
Effective dose level, 116, 128–129
Eggs, resistance to toxicity, 200
Eggshells
  causes of thinning, 181–182
  DDT, 181
  insecticides, 164
  effects of thinning, 182–184
Embryos
  effects of toxicants on, 199–200
  examples of toxicants affecting, 200–201
  resistance and sensitivity to toxicants, 200
Endrin
  distribution between tissues, 40–41
  metabolism by microorganisms, 44–46
  stereochemistry, influence on metabolism, 63–64
Enzymes
  effects of inhibition on, 154
  inhibition by a chemical agent, 154–156
Epoxides
  behaviour as toxicants, 63
  carcinogenic action, 162
Equilibrium of ecosystems, 326
Equivalent dose rule, 141
Estimation of ecotoxicological effects, 4
Ethionine, inhibition of protein synthesis by, 158
Ethyl esters as mutagens, 161
O-Ethyl-O-para-nitrophenyl phenylphosphoethionate (EPN), toxic interactions, 165
Eutrophication, 318, 320
  aquatic ecosystems, 188
  effects of pollutants on, 260
  effects of toxicants on, 251
Eve's gastrointestinal model, 93–94
Excretion equations
  determination of, 99–100
  examples of, 97–98

Exposure
  acute, 327–328
  chronic, 328
  definition, xxii, 92
  effects of, 117–118
  model for chronic, 171–173

Falcons, effects of DDT on, 181–184
FAO toxicity tests, 240
Fat reserves, effects of insecticides on, 177–179
Fish
  effects of acidification on, 303–304
  sensitivity to acid, 304
Fluoride
  accumulation in plants, 225
  effects on plants, 226
Fluorocitrate, inhibition of carbohydrate metabolism by, 157
Frost, damage to tobacco or oranges by, 296–297
Fungi, degradation of crude oil by, 281

Gills, effects of toxicants on, 191–192
Global Atmospheric Research Programme (GARP), 298
Glycine, effect of diet on metabolism of, 180
Greenhouse effect, 299
Growth
  effects of pollutants on, 198
  effects of toxins on rate, 240
  inhibition by toxicants, 159
  rate for algae, 259–260

Half-life of pollutants in organisms, 171
Heart, effects of toxicants on, 194
Heptachlor
  conversion products in plants, 47
  epoxy group, effect on metabolism, 63
  evaporation from soil, 14
  hydrolysis, 17
  metabolism by microorganisms, 44–46
Heterotrophic microorganisms
  effects of fuel oil on, 282
  effects of hydrocarbons on, 288
Hexachlorobenzene (HCB)
  absorption, 40
  distribution between animal tissues, 41–43

Hill reaction
  in algae, 266
  inhibition by toxicants, 260
Homeostasis of ecosystems, 314
Hydrocarbons
  biodegradation of, 278, 282–284
  effects on bacteria, 282–283
  oxidation by bacteria, 283
  oxygenation of polycyclic aromatic, 22
Hydrogen sulphide, inhibition of enzymes, 156
Hydrolysis of toxicants, 39

Impaction of pollutants, 175
Imugan distribution among tissues, 40–41
Independent action of toxicants, 208–210, 264
Indicators, bacterial, of environmental pollution, 288
Industry, effects on ecosystems, 307
Infrared photography to determine toxic effects, 234
Inhibition
  of carbohydrate metabolism, 156–157
  of complexes, 155–156
  of enzymes, 154–156
  of growth, 159
  of lipid metabolism, 157–158
  of microsomal enzyme systems, 158
  of protein synthesis, 158
  of reproduction, 163–164
  of respiration, 157
  of thiol groups, 154–155
Insecticides
  adsorption to soil, 14
  effects
    on fat reserves, 177–179
    on fish reproduction, 163–164
    on membranes, 152–153
    on nerve impulses, 183–184
    on populations, 180–181
    on zooplankton, 267–269
  enhanced toxicity, 165
  evaporation from soil, 14
  inhibition by
    of acetylcholinesterase, 176–177
    of microsomal enzyme systems, 158

irradiation, 26–33
isomerization, 23
Intake of toxicants, definition, xxii, 92
Interactions
   additive, 164–165, 207–210, 264
   between acid and other toxicants, 305
   between plant species, 247
   between pollutants, 247–248, 264–265
   between toxicants, 119, 164–165, 205–212
   between toxicants and organisms, 116
   chemical, 165, 205–206, 242
   displacement, 165
   mechanisms, 206
   physiological, 205–206
   sites of, 212
   types, 165
   uptake of toxicants and, 212
Intercalation of toxic mutagens, 160
Inverse sampling rule, 118
Iodide
   absorption by man, 93
   calculation of dose, 108
   pollution by, 317
   retention equation, 99
Iron
   chemical states in water, 264
   displacement from complexes by other metals, 155
   effects of
      on algae, 263–264
      on ecosystems, 323–324
Irradiation, effects on plant communities, 230–231
Irving–Williams series, 155
Isodrin
   conversion products in plants, 46–47
   epoxy group, effect on metabolism, 63
   stereochemistry, influence on metabolism, 63–64

Labilization of cell membranes, 152
Leaching of pollutants, 14
Lead
   accumulation in plants, 224
   as allergen, 164
   effects on nervous system, 153
   enzymatic transformation, 38
   interactions, 165
   retention by man, 98
Lead, triethyl, inhibition of respiration by, 157
Least squares method, 127–128
Lethal dose level, 129
   as measure of toxicity, 179
Liapunov direct method, 327
Lichens, effects of toxicity on, 229–230
Lindane, metabolism in insects, 47–48
Lipids, inhibition of metabolism by toxicants, 157–158
Lipolytic bacteria, effect of fuel oil on, 282
Locomotion, effects of toxicants on, 195–197
Lung, model for intake of toxicants, 94–95, 176
Lysosomes, effects of toxicants on, 152

Macrophytes, effects of toxicants on, 252
Malathion, toxic interactions, 165
Manganese, mutagenic effects of, 161
Mangrove ecosystems, pollution of, 322
Membranes
   effect of toxicants on, 152–153
   permeability, 152
Mercury
   absorption
      inorganic *vs* organic, 77–78
      model for, 77–78
      of vapour, 175
   accumulation in aquatic plants, 246
   chemical forms, 80
   detoxification of, 244
   effects
      on algae, 262–263
      on aquatic animals, 192
      on membranes, 153
      on potassium concentration in algae, 263
   enzymatic transformations, 38
   inhibition by, of carbohydrate metabolism, 157
   in oil, effect on degradation, 286
   sedimentation of, 16
   toxicity
      dependence on state, 179
      to fish, 80
   transport

Index

coefficients for, 82–83
model for, 80–81, 86
uptake by aquatic plants, 244–245
water pollution by, 80, 317
Mercury, dimethyl, effects on algae, 263
Mercury, methyl
  calculation of dose, 109–110
  effects on membranes, 153
  excretion, 100
  ingestion by man, 94
  pollution by, 317
  rate of uptake in fish, 95–96
  retention
    by fish, 98
    by plants, 98
Metabolism, effects of toxicants on, 156–159
Metals, heavy
  absorption by man, 93
  body content, 100
  carcinogenic action, 163
  catalysis of oxidation, 17–18
  chelation and toxicity, 263–264
  displacement from complexes, 155–156
  effects
    on algal growth, 263
    on aquatic animals, 192
    on aquatic plants, 247
    on ecosystems, 324–325
    on fish growth, 198
    on plant communities, 247
    on zooplankton, 268–269
  ingestion by man, 93–94
  inhibition by
    of microsomal enzyme systems, 158
    of reproduction, 163
    of thiol groups, 154–155
  leaching, 324
  in oil pollutants, 283
  pollution, 324–325
    air, 316
    water, 243–244
  uptake
    by aquatic plants, 244–245
    by plants, 96
Methylation of metals, 287
Microbial populations, 278
Microcosms, use of, 315
Microsomal enzyme systems, effects of chemical agents on, 158

Microsomes, liver, 284
Mineralization
  of pollutants, 11–12
    definition, xxii
  role of microorganisms in, 275–276
Models
  definition, 76
  examples, 77–78
    abiotic degradation, 33
    carbon dioxide in atmosphere, 299
    chrysanthemums and rotenone toxicity, 129–130
    lung model in reference man, 94–95, 102–103, 176
    mercury transport, 80–81
    mice and skin tumours, 138
    pollutant transport, 12
    Rainbow trout and hepatoma, 131–133
  types, 76, 123–124
    chronic exposure, 171–173
    compartmental, 170–175
    Eve's gastrointestinal, 93–94
    log-logistic, 126–127
    log-normal time-to-occurrence, 137
    logistic curve, 123
    multi-hit, 123–124
    multi-stage, 124
    sine curve, 123–124
    single-hit, 123–124
    steady state, 171–173
    time-to-occurrence, 136–140
  use
    difficulties in, 79–80
    dose–tolerance distribution, 122–123
    estimation of parameters, 138
    examples of, 85
    factors in, 71–72
    in the laboratory, 60–61
    multifactorial studies, 119–120
    value of, 72
  validity of, 87, 123, 124–125
    process of validation, 87
Mustard gas, mutagenic effects of, 160
Mutagens, correlation with carcinogenicity, 284–285
Mutation
  induction by chemicals, 160–161
  testing for, 201–202

toxicants responsible for, 201
types of, 201

Nervous system, effects of toxicants on, 153
Newton—Raphson method, 146
Nickel
  effects on aquatic plants, 242—243
  uptake by aquatic plants, 245
Nitrate, effect on degradation, 280
  of oil in sea water, 282
Nitrogen
  and nutrient balance, 318, 324
  uptake by plants, 39
Nitrogen dioxide
  in photochemical smog, 18—19
  role in photoreactions, 32
Nitrogen oxide, reaction with ozone, 18—19
Nitrosamines as carcinogens, 163

Ocean, biodegradation of oil in, 281—282
Oil
  biodegradation of, 278—284
  carcinogenicity, 284—285
  effects on bacteria, 282—283
  effects on marine growth, 266
  effects on microorganisms, 278, 282—283
  oxidation of, 283
  pollution by, 315—316
Oil pollution
  in oceans, 278
  role of microorganisms in biodegradation, 278
Olefins
  irradiation, 30
  reactions with oxygen, 21
Organic compounds, sources of, in sea, 242
Organochlorines, effects on reproduction, 164
Organohalogens, uptake
  by algae, 245
  by aquatic plants, 244—245
Osmotic regulation, adaptation to toxicity, 191—192
Oxidants, effects of
  on agricultural systems, 232
  on plant communities, 228—229

Oxidation of toxicants, 38—39
Oxygen
  effects of toxicants on supply, 193
  requirement in bacterial oil degradation, 283
  role in photoreactions, 32
  states, 32
  transport in aquatic animals, 194
Ozone
  depletion of layer, 301
    effects of depletion, 301
  effects
    on agricultural systems, 232
    on growth, 159
    on plant communities, 228—229
    on plants, 226
  effects of man on, 300
  in photochemical smog, 18—19
  role in photoreactions, 32
  reactions
    with olefins, 18—19
    with nitrogen oxide, 18—19
  stratospheric, 300

Paraffinic oil, 279, 282
Pentachlorophenol
  irradiation, 28—31
  photomineralization, 32—33
Perception by fish, 197
  effects of toxicants on, 197
Peroxyacyl nitrates, effects on plants, 231—232
Pesticides
  dechlorination of, 30
  degradation microbiologically, 43—45
  effects
    on agricultural systems, 233
    on ecosystems, 322—323
    on the peregrine falcon, 181—184
  hydrolysis, 16 17
  inhibition of growth by, 159
  irradiation, 28—31
  leaching in soil, 14
  oxidation, 17—18
  transport by air, 323
  use of
    in farming, 323
    in forestry, 323
Petroleum components, persistence in environment, 278

Phaeodactylum tricornutum
   nickel-binding capacity, 242–243
   zinc uptake and binding, 245
Phosphate effects of
   on binding of copper and iron, 263
   on binding of nickel, 242–243
Phosphorus, effect on petroleum degradation, 280, 282
Photodegradation, formation of chlorinated dibenzofurans by, 276
Photodieldrin
   effects on zooplankton, 268–269
   irradiation of, 27–28
   photodegradation products, 28
Photosynthesis
   effects of temperature on, 240
   influence of toxicants on rate, 240
   inhibition in algae, 259–260, 266
   measurement of rate, 240
Pine, effects of toxicants on communities, 228–229
Plankton
   communities
     effects of pollutants on, 260–262
     relations between, 258
     studies of, 252–253
   effects of toxicants on, 251–252
   types, 257–258
Plants
   accumulation of toxicants by, 245–246
   effects of heavy metals on, 247
   effects of toxicants, 249–251
     on growth, 240
     on physiology, 225–226
   metabolism of, 46–47
   uptake of toxicants by, 244–245
Plasmids, 287–288
Plume, thermal, 307
Podsol soils, and acid rains, 303
Pollutants
   abiotic degradation, 11–12
   amounts released, 4–5
   behaviour
     in the environment, 7
     in water, 242–244
   chemical state
     importance of, 75–76
     in sea water, 242–244
   definition, xix
   dispersion, 12–13
   effects
     on algae, 259–260
     on aquatic plants, 242–244
     on ecosystems, acute exposure 327
     on ecosystems, chronic exposure, 327–328
     on sea water, 242
     on zooplankton, 267–269
   leaching, 14
   metabolism by zooplankton, 269
   mineralization, 11–12
   sources, 4–6, 257
   toxicity, properties affecting, 7–8
   transport
     deposition in ground, 74
     in test systems, 241–242
     in water, 74
   types of, 257
Pollution
   causes
     artificially induced, 314–315
     oil, 315–316
     sewage, 319
     unintentional, 314–315
   definition, xix, 320
   effects
     on ecosystems, 319–324, 327–328
     on waters, 305–307
   study of, 314–315, 316
   types
     air, 315, 316
     thermal, 300–301, 305–307, 320
     water, 315–317
Polychlorinated biphenyls (PCB's)
   abiotic degradation, 27
   chlorine content, effect on behaviour, 61–63
   degradation by animals, 63
   degradation by microorganisms, 285–286
   distribution among tissues, 40–41
   effects
     on algae, 266
     on aquatic plants, 247–248
     on eggshell thickness, 156
     on estuarine ecosystems, 321–322
     on zooplankton, 267–268
   excretion
     by monkeys, 62–63
     by rats, 62
   metabolism by biomass, 53–55
   uptake calculation, 103

Population
  definition, xxii
  dynamics
    effects of pollutants on, 180–181
    in zooplankton, 267
Potentiation of compounds to toxicity, 211–212
Precipitation, acid rains, 301–303
Predation
  effects of pollution on, 267–268
  in zooplankton, 267
Predictions
  errors in, 86
    estimation of, 86–87
  of toxicity, 6–7, 189, 327–328
    models for, 8
    on plants, 235
  value of, 86
Pressure, effect on microbial degradation of hydrocarbons, 280
Prill's acid as a conversion product, 47
Promutagens in the Ames test, 284
Proteins, inhibition of synthesis, 158
Proteolytic bacteria, effect of fuel oil on, 282
Psychrophilic microorganisms, 280
Pumps, ion, effects of toxicants on, in cells, 152–153

Radiation, gamma, effects on forests, 325
Radioactive particles, atmospheric washout of, 15
Rate, of pollutant transfer, 170–171
Rate constants
  definition, 82–83
  determination, 79–80, 171
  use in compartmental models, 170–171
Reduction of toxicants, 39
Reference Man, 95
Reproduction, inhibition by toxicants, 163–164, 198–199
Research priorities, 4
Resistance
  to antibiotics, 287–288
  in later generations, 139–140
Respiration
  adaptation to toxicity, 192–194
  effects of temperature on, 240
  inhibition by toxicants, 157, 193–194
  monitoring of, 193–194

Response
  by communities, 319–321
  estimation
    confidence limits, 128–129
    of probability, 127–128
    of rate, 127–128
  for frost damage, 296–297
  probability, 121–122
  spontaneous, 133–135
  variability, 122
Retention equations
  determination, 96–98
  from excretion equations, 97–98
  problems in, 99
  examples, 98
Risk, definition, xxii
Rotenone
  maximum likelihood estimates for toxicity, 147–148
  toxicity to chrysanthemum aphis, 129–130

Salt, effects on plants, 227
SCOPE, history of, xix
Sediment
  marine, degradation of crude oil in, 282
  methylation of mercury in, 284
Selection
  of dose levels, 118–119
  of test species, 117
  of type of exposure, 117–118
Sensitivity
  analysis, 87
  coefficient, 87
Sensitization to toxicity, 211–212
Sewage
  coliform bacteria in, 287
  effect of dumping in ocean, 287–288
  pollution by, 319
Shannon–Weaver Index, 267
Silicosis, 152
Silver, mutagenic effects, 161
Simpson Index, 326
Sodium–potassium pump
  effect of mercury on, 192
  effect of toxicants on, 152–153
Soil, and acid rains, 303
Spontaneous response
  effect on validity of models, 135
  rate, 134
    determination of, 134–135

Stabilization of cell membranes, 152
Standard water, tests for toxicity using, 252–253
Steady state, model for, 171–173
Stereochemical conformation, influence on toxicant behaviour, 63–64
Stochastic effects, definition, 107
Stratosphere, condition of, 300
Strontium
    dispersion, 12
    rate of uptake in plants, 96
    retention by man, 98
Sublethal effects, 188–191
    on aquatic animals, 189–199
        osmoregulation, 191–192
        respiration, 192–194
    detection, 179
    on populations, 180
Sulphur
    accumulation in plants, 224–225
    air pollution by, 316–317
    effect on soil acidity, 225
Sulphur dioxide
    and acid rains, 303
    detoxification, 234–235
    dose
        calculation of dose–effect relation, 232–233
        threshold, 232
    effects
        on agricultural systems, 232–233
        on clover and forage, 235
        on forest ecosystems, 304–305
        on lichen, 230–231
        on plant communities, 228
        on plants, 225, 226
    inhibition of growth by, 159
Surface area rule, 140–141
Suspended particulate matter, 299
Swimming
    determination of rate, 196
    effects of toxicants on, 196–197
Synapses, effects of toxicants at, 153
Synergistic effects
    on algae, 264
    on aquatic animals, 208–209, 211–212
    on aquatic plants, 247–248
    in cancer, 213
    permissive, 212
Synergists, action of, 165

Target, definition, xxii
Technetium, excretion, 100
Teratogens, examples, 200–201
Tests
    of bioconcentration, 56
    of biodegradation, 55–56
    of toxicity, 240–241
        on aquatic animals, 189–191
        on aquatic plants, 249–250
        of carcinogens, 203–204
        on plants, 226–227
        selection of dose levels, 118–119
        selection of exposure type, 117–118
        selection of species, 117, 189–190
        significance of, 56
Thiol groups
    effect of toxicants on, 152
    inhibition of, 154–155
D-Threose-2,4-diphosphate, inhibition of growth by, 159
Threshold
    definition, 105
    levels, 125–126, 189
Thyroid, role in growth, 159
Tin, triethyl
    effects on nervous system, 153
    inhibition of respiration by, 157
Tolerance
    definition, 120
    determination, 122–123, 124
    distribution, 120–121
    in later generations, 138–139
    to toxicants (aquatic animals), 210
Toxicants
    adaptation of organisms to, 139–140
    behaviour of, 188
    critical maximum dose of, 251
    effects
        diagnosis of, 233–234
        differences between species, 244
        factors modifying, 119–120
        on fish communities, 188–189
        on fish ecosystems, 187–188
        on growth, 240
        on macrophyte communities, 252
        on offspring and later generations, 139–140
        on organisms, 140–141
        on plankton communities, 251–252

on plant communities, 241
on plant distribution, 251
on plants (aquatic), 241
prediction of, 235
testing for, 226
interactions between 119–120
long-term effects, 249–251
uptake by aquatic plants, 244–245
Toxicity
adaptation to, 189–191
dependence on age, 226–227
dependence on chemical form, 75–76
difficulty in evaluating, 189–191, 314
effects
on agricultural systems, 231–233
on aquatic ecosystems, 187–188
on communities, 227–231
on growth, 227
effects of dose level on, 226
extrapolation from laboratory to ecosystems, 140–141
interpretation of test results, 249–250
methods of testing, 240, 252–253
sublethal, 188–189
testing for, 189–191
Transfer
coefficients for mercury transfer, 82–83, 246
of pollutants
from atmosphere to ground, 74
from ground to atmosphere, 74
Transport
of pollutants, 12–13
aerial, 73–74
aquatic, 73–74
prediction of rate, 73–74
types of, 73
Tritium
absorption by man, 93
retention by man, 98
Trophic levels
advantage of concept, 75
effects of toxicants on, 250–251
Tumours
and contaminants, 202–203
and viruses, 203
induction by chemical agents, 162–163
initiation of, 162
promotion of, 162
Turnover rates for elements in natural ecosystems, 251–252

Ultraviolet irradiation, effect on degradation, 16
Understory plants effects of toxicants on, 229
UNSCEAR, xix, 104, 106
Uptake of toxicants
definition, xxii, 92
dynamics, 265
interaction, 212
rates
in algae, 266
in fish, 95–96
routes
in aquatic plants, 244–245
in fish, 95–96
in plants, 51, 96, 224–225
in zooplankton, 268
types
chronic, 102
declining, 102–103
several in a limited period, 103–104
single isolated, 101

Validity of models, 123, 124–125
case of spontaneous response, 135
process of validation, 87
Vapour pressure, 15
Variability
between models, 124–125
of dose response, 117, 122
Variables, determination of, in compartments, 79–80
Variance
analysis of, 133
in estimation of risk, 141
in log dose, 129
in log-logistic model, 127–128
Variations, significant, 321
Ventilation of aquatic animals, 192–194
Volatilization of toxicants from plants, 53

Wastes, chemical, 187–188
Water
copper species in solution, 243

dissolved organic compounds,
242–244
heavy metals in solution, 243–244
pollution of, 315–317
standards for quality, 213–214
Wildlife, environmental data, 57–60

Xenobiotic chemicals, definition, xxii

Yeasts, degradation of crude oil by, 281

Zinc
displacement by other metals,
155–156
effects on fish, 195–196
inhibition of CO_2 fixation by, 159
pollution by, 324
uptake by aquatic plants, 245
Zooplankton
effects of copper on, 251
effects of pollutants on population,
267–269
metabolism of pollutants by, 269
types, 267